In Praise of *Customizable Embedded Processors*

This book represents a significant contribution to understanding and using configurable embedded processors. The reader will find all design aspects described in detail by the experts in the field, and thus this book will serve as the "standard reference" on this topic.

This contribution is very up to date and complete, as it covers modeling, analysis and design of both the hardware and software components of customizable processors. A unique feature is the gathering, under the same cover, several topics that are different in nature but intimately related.

Giovanni De Micheli, Professor and Director of the Integrated Systems Centre at Ecole Polytechnique Fédérale de Lausanne (EPFL), Lausanne, Switzerland, and President of the Scientific Committee of CSEM (Centre Suisse d'Electronique et de Microtechnique SA), Neuchâtel, Switzerland.

Customizable Embedded Processors provides an objective, state-of-the-art treatment of emerging tools for the design of complex systems-on-a-chip. The contents of the book are provided by well-known experts in the field who have decades of real-world experience. Understanding the pitfalls, limitations, and advantages of SoC design alternatives is a must in today's competitive environment where we cannot afford to re-invent the wheel every few years. I highly recommend the book for anyone who practices advanced architecture development for energy-efficient, low-cost SoC's.

Don Shaver, Senior Member IEEE, Texas Instruments Fellow

The concept of configurable processors has emerged from the lab and entered the mainstream for high-volume electronic systems. Customizable Embedded Processors is the best overview to date of this rapidly evolving field. It pulls together the work of the leading researchers and practitioners from around the world into a single consistent exploration of the benefits, challenges and techniques for creating efficient application-specific processors. It explores both the widely proven benefits and the theoretical limits to customization. Configurable processors have become an essential building block for modern system-on-chip design, and this book is an essential tool for understanding the leading-edge of configurable processor research.

Chris Rowen, CEO, Tensilica, Inc.

It is clear that application-specific instruction-set processors (ASIPs) will be a key building block of the emerging multi-processor system-on-chip platforms of the next decade. For many critical functions, the competing requirements of flexibility and efficiency will favor their use over general-purpose processors or hardwired logic. As a longtime advocate for ASIPs, I am delighted to discover this book which covers the key aspects of their design and use from a variety of perspectives: from leading edge industry R&D labs, through startups, to academic research leaders. It covers the technical challenges in ASIP use, but also many of the important non-technical issues. Most of the leading innovative ASIP projects are covered here, so this book may well become the definitive reference for this increasingly important domain.

Pierre Paulin, Director, SoC Platform Automation, STMicroelectronics Inc.

Standard Processors have been the fuel for the computer revolution over the last 50 years. A new technology called Custom Embedded Processors is fast becoming the key enabling technology of the next 50 years. In this book the reader will learn the basic theory, and practical examples of how this powerful new approach has been put to use. It's a must read far anyone hoping to build the next big thing in the 21st century.

Alan Naumann, President and CEO, CoWare, Inc.

CUSTOMIZABLE EMBEDDED PROCESSORS

The Morgan Kaufmann Series in Systems on Silicon
Series Editor: Wayne Wolf, Princeton University

The rapid growth of silicon technology and the demands of applications are increasingly forcing electronics designers to take a systems-oriented approach to design. This has lead to new challenges in design methodology, design automation, manufacture and test. The main challenges are to enhance designer productivity and to achieve correctness on the first pass. *The Morgan Kaufmann Series in Systems on Silicon* presents high-quality, peer-reviewed books authored by leading experts in the field who are uniquely qualified to address these issues.

The Designer's Guide to VHDL, Second Edition
Peter J. Ashenden

The System Designer's Guide to VHDL-AMS
Peter J. Ashenden, Gregory D. Peterson and Darrell A Teegarden

Readings in Hardware/Software Co-Design
Edited by Giovanni De Micheli, Rolf Ernst and Wayne Wolf

Modeling Embedded Systems and SoCs
Axel Jantsch

ASIC and FPGA Verification: A Guide to Component Modeling
Richard Munden

Multiprocessor Systems-on-Chips
Edited by Ahmed Amine Jerraya and Wayne Wolf

Comprehensive Functional Verification
Bruce Wile, John Goss and Wolfgang Roesner

Customizable Embedded Processors: Design Technologies and Applications
Edited by Paolo Ienne and Rainer Leupers

Networks on Chips: Technology and Tools
Edited by Giovanni De Micheli and Luca Benini

Designing SOCs with Configured Cores: Unleashing the Tensilica Diamond Cores
Steve Leibson

VLSI Test Principles and Architectures: Design for Testability
Edited by Laung-Terng Wang, Cheng-Wen Wu, and Xiaoqing Wen

Contact Information

Charles B. Glaser
Senior Acquisitions Editor
Elsevier
(Morgan Kaufmann; Academic Press; Newnes)
(781) 313–4732
c.glaser@elsevier.com
http://www.books.elsevier.com

Wayne Wolf
Professor
Electrical Engineering, Princeton University
(609) 258 1424
wolf@princeton.edu
http://www.ee.princeton.edu/~wolf/

CUSTOMIZABLE EMBEDDED PROCESSORS
DESIGN TECHNOLOGIES AND APPLICATIONS

Paolo Ienne
Ecole Polytechnique Fédérale de Lausanne (EPFL)

Rainer Leupers
RWTH Aachen University

AMSTERDAM • BOSTON • HEIDELBERG • LONDON
NEWYORK • OXFORD • PARIS • SAN DIEGO
SAN FRANCISCO • SINGAPORE • SYDNEY • TOKYO
Morgan Kaufmann is an imprint of Elsevier

Publishing Director   Denise Penrose
Acquisitions Editor   Chuck Glaser
Publishing Services Manager   George Morrison
Project Manager   Brandy Lilly
Assistant Editor   Michele Cronin
Composition   diacriTech
Copyeditor   Multiscience Press, Inc.
Proofreader   Multiscience Press, Inc.
Indexer   Multiscience Press, Inc.
Interior printer   Sheridan Books Group
Cover printer   Phoenix Color

Morgan Kaufmann Publishers is an imprint of Elsevier.
500 Sansome Street, Suite 400, San Francisco, CA 94111

This book is printed on acid-free paper.

Library of Congress Cataloging-in-Publication Data

ISBN 13: 978-0-12-369526-0
ISBN 10: 0-12-369526-0

For information on all Morgan Kaufmann publications,
visit our Web site at www.mkp.com or www.books.elsevier.com

Printed in the United States of America
06   07   08   09   10   5   4   3   2   1

Working together to grow
libraries in developing countries

www.elsevier.com | www.bookaid.org | www.sabre.org

ELSEVIER   BOOK AID International   Sabre Foundation

Cover image created by Dr. Ute Müller, RWTH Aachen University

To Lorenz – Rainer

To Malgari and Giorgio – Paolo

CONTENTS

3   Customizing Processors: Lofty Ambitions, Stark Realities
Joseph A. Fisher, Paolo Faraboschi, and Cliff Young   **39**

Part II: Aspects of Processor Customization

4   Architecture Description Languages
Prabhat Mishra and Nikil Dutt   **59**

5   C Compiler Retargeting
Rainer Leupers   **77**

6 Automated Processor Configuration and Instruction Extension

David Goodwin, Steve Leibson, and Grant Martin   **117**

7 Automatic Instruction-Set Extensions

Laura Pozzi and Paolo Ienne   **145**

10 Datapath Synthesis
Philip Brisk and Majid Sarrafzadeh   **233**

11 Instruction Matching and Modeling
Sri Parameswaran, Jörg Henkel, and Newton Cheung   **257**

12 Processor Verification
Daniel Große, Robert Siegmund, and Rolf Drechsler — 281

13 Sub-RISC Processors
Andrew Mihal, Scott Weber, and Kurt Keutzer — 303

Part III: Case Studies

14 Application Specific Instruction Set Processor for UMTS-FDD Cell Search
Kimmo Puusaari, Timo Yli-Pietilä, and Kim Rounioja   **339**

LIST OF CONTRIBUTORS

Gerd Ascheid
RWTH Aachen University

Gerd Ascheid is a full professor at the Electrical Engineering and Information Technology department of RWTH Aachen University, heading the Institute for Integrated Signal Processing Systems. He was co-founder of CADIS GmbH, which commercialized COSSAP. Prior to joining the university in 2003 he was Senior Director for Professional Services in Wireless and Broadband Communications at Synopsys, Inc., Mountain View, California. His research interest is in wireless communications, MPSoC architectures and design tools.

Göran Bilski
Xilinx, Inc.

Göran Bilski is a principal engineer at Xilinx where he has architect and designed the soft processor MicroBlaze. He is also involved in the overall direction of the embedded processing at Xilinx. His prior works includes system engineering for space applications, processor architecture design, ASIC and FPGA developments. He has 20 years of industry experience of embedded processing and holds several US patents in this area. He earned his M.S. in Computer Science from Chalmers Technical University, Sweden.

Philip Brisk
University of California, Los Angeles

Philip Brisk is currently a Ph.D. candidate in Computer Science at the University of California, Los Angeles. He earned his Bachelor's and Master's Degrees in Computer Science from the University of California, Los Angeles, in 2002 and 2003 respectively, while working in the Embedded and Reconfigurable Systems Laboratory. His research interests include compilers and synthesis for embedded systems.

Newton Cheung
VaST Systems Technology Corporation

Newton Cheung received his B.E. degree in Computer Science and Electrical Engineering from the University of Queensland in 2002, and the Ph.D. degree in Computer Engineering from the University of New South Wales in 2005. He has been working at VaST Systems Technology Corporation as an engineer since 2005. His research interests include design automation methodology, customizable/extensible processor, and high-speed simulation technology.

Rolf Drechsler
University of Bremen

Rolf Drechsler received his diploma and Dr. phil. nat. degree in computer science from the J.W. Goethe-University in Frankfurt am Main, Germany, in 1992 and 1995, respectively. He was with the University of Freiburg, Germany from 1995 to 2000. He joined the Corporate Technology Department of Siemens AG, Munich in 2000, where he worked as a Senior Engineer in the formal verification group. Since October 2001 he has been with the University of Bremen, Germany, where he is now a full professor for computer architecture.

Nikil Dutt
University of California, Irvine

Nikil Dutt is a Professor of CS and EECS and with the Center for Embedded Computer Systems at UC Irvine. He received best paper awards at the following conferences: CHDL89, CHDL91, VLSI Design 2003, CODES+ISSS 2003, and ASPDAC-2006. He is currently the Editor-in-Chief of ACM TODAES and an Associate Editor of ACM TECS. His research interests are in embedded systems design automation, computer architecture, optimizing compilers, system specification techniques, and distributed embedded systems.

Paolo Faraboschi
HP Labs

Paolo Faraboschi is a Principal Research Scientist at HP Labs, since 1994. Paolo's interests are at the intersection of computer architecture, applications and tools. He currently works on system-level simulation and analysis in the Advanced Architecture group. In the past, he was the principal architect of the Lx/ST200 family of embedded VLIW cores. Paolo received an M.S. (1989) and Ph.D. (1993) in Electrical Engineering and Computer Science from the University of Genoa, Italy.

Joseph A. Fisher
HP Labs

Joseph A. Fisher is a Hewlett-Packard Senior Fellow at HP Labs, where he has worked since 1990. Josh studied at the Courant Institute of NYU (Ph.D. in 1979), where he devised the Trace Scheduling compiler algorithm and coined the term Instruction-level Parallelism. At Yale University, he created and named VLIW Architectures and invented many of the fundamental technologies of ILP. In 1984, he co-founded Multiflow Computer. Josh won an NSF Presidential Young Investigator Award in 1984, and in 2003 received the ACM/IEEE Eckert-Mauchly Award.

David Goodwin
Tensilica, Inc.

David Goodwin is a Chief Engineer at Tensilica working on system-level design methodologies and tools, and ASIP technology and tools. Previously, David worked in the Architecture and Compiler Group at Digital. David holds a Ph.D. in Computer Science from the University of California, Davis, and a B.S. in Electrical Engineering from Virginia Tech.

Daniel Große
University of Bremen

Daniel Große received the Diploma degree in computer science from the Albert-Ludwigs-University, Freiburg, Germany in 2002. He is currently pursuing the Ph.D. degree at the University of Bremen, Bremen, Germany. He has been with the University of Bremen since September 2002. His research interests include verification, micro processors, and high-level languages like SystemC.

Jörg Henkel
University of Karlsruhe

Jörg Henkel received the Ph.D. degree ("Summa cum laude") from Braunschweig University. Afterwards, he joined the NEC Labs in Princeton, NJ. In between, he had an appointment as a visiting professor at the University of Notre Dame, IN. Dr. Henkel is currently with Karlsruhe University (TH), where he is directing the Chair for Embedded Systems CES. He is also the Chair of the IEEE Computer Society, Germany Section, and a Senior Member of the IEEE.

Kurt Keutzer
University of California, Berkeley

Kurt Keutzer is Professor of Electrical Engineering and Computer Science at the University of California at Berkeley. Prior to this Kurt was Chief Technical Officer and Senior Vice-President of Research at Synopsys, Inc. and a Member of Technical Staff at AT&T Bell Laboratories. His current research interests focus on the design and programming of programmable multiprocessor platforms. He is a Fellow of the IEEE.

Steve Leibson
Tensilica, Inc.

Steve Leibson is the Technology Evangelist for Tensilica, Inc. He formerly served as Editor in Chief of the Microprocessor Report, EDN magazine, and Embedded Developers Journal. He holds a BSEE from Case Western Reserve University and worked as a design engineer and engineering manager for leading-edge system-design companies including Hewlett-Packard and Cadnetix before becoming a journalist. Leibson is an IEEE Senior Member.

Grant Martin
Tensilica, Inc.

Grant Martin is a Chief Scientist at Tensilica, Inc. in Santa Clara, California. Before that, Grant worked for Burroughs in Scotland for 6 years; Nortel/BNR in Canada for 10 years; and Cadence Design Systems for 9 years, eventually becoming a Cadence Fellow in their Labs. He received his Bachelor's and Master's degrees in Mathematics (Combinatorics and Optimisation) from the University of Waterloo, Canada, in 1977 and 1978. He is interested in IP-based design, configurable processors, system-level design and platform-based design.

Heinrich Meyr
RWTH Aachen University

Heinrich Meyr is presently a Professor of electrical engineering at RWTH Aachen University, Germany, and Chief Scientific Officer of CoWare Inc., San Jose. He pursues a dual career as researcher and entrepreneur with over 30 years of professional experience. He is a Fellow IEEE and has received several best paper and professional awards. His present research activities include cross-disciplinary analysis and design of complex signal processing systems for communication applications.

Andrew Mihal
University of California, Berkeley

Andrew Mihal is a Ph.D. Candidate in Electrical Engineering and Computer Science at the University of California at Berkeley. His research focuses on the deployment of concurrent applications on multiprocessor platforms composed of fast and lightweight programmable elements.

Andrew received his B.S. in Electrical and Computer Engineering from Carnegie Mellon University in 1999 with University Honors.

Prabhat Mishra
University of Florida

Prabhat Mishra is an Assistant Professor in the Department of Computer and Information Science and Engineering at University of Florida, USA. Mishra has a BE from Jadavpur University, M.Tech. from the Indian Institute of Technology, Kharagpur and Ph.D. from the University of California, Irvine—all in Computer Science. He has received the EDAA Outstanding Dissertation Award in 2004. His research interests include functional verification and design automation of embedded systems.

Sundararajarao Mohan
Xilinx, Inc.

Sundararajarao Mohan is a Principal Engineer at Xilinx, San Jose, USA. He has been responsible for setting the technical direction for embedded processing solutions including both hardware and software, for the last few years. His prior work includes FPGA architecture, CAD software, and circuit design. He holds numerous patents and has published extensively. He earned his B.Tech. from the Indian Institute of Technology, Madras, and his M.S. and Ph.D. from the University of Michigan, Ann Arbor, USA.

Sri Parameswaran
University of New South Wales

Sri Parameswaran is a faculty member at the University of New South Wales in Australia. His research interests are in System Level Synthesis, Low power systems, High Level Systems and Network on Chips. He has served on the Program Committees for International Conferences, such as Design and Test in Europe, the International Conference on Computer Aided Design, the International Conference on Hardware/Software Codesign and System Synthesis, and the International Conference on Compilers, Architectures and Synthesis for Embedded Systems.

Laura Pozzi
University of Lugano (USI)

Laura Pozzi (M.S. 1996, Ph.D. 2000) is an assistant professor at the University of Lugano (USI), Switzerland. Previously she was a post-doctoral researcher at EPFL, Lausanne, Switzerland, an R&D engineer with STMicroelectronics in San Jose, California, and an Industrial Visitor at University of California, Berkeley. She received a best paper award at the Design Automation Conference in 2003. Her research interests include automatic processor-customisation, high performance compilers, reconfigurable computing.

Kimmo Puusaari
Nokia, Technology Platforms

Kimmo Puusaari received his M.Sc. degree in Electrical Engineering with distinction from University of Oulu in 2005. He joined Nokia full-time in spring 2004, and has since been researching new design methodologies and technologies for wireless baseband modems. His special focus is on different commercial and academic ASIP design methodologies.

Kim Rounioja
Nokia, Technology Platforms

Kim Rounioja received his M.Sc. degree in Electrical Engineering from University of Oulu in 2001. He joined Nokia in spring 1999, and has since been working in various wireless baseband modem research and development projects. His special focus is on hardware and software architectures for baseband signal processing.

Majid Sarrafzadeh
University of California, Los Angeles

Majid Sarrafzadeh received his B.S., M.S., and Ph.D. in 1982, 1984, and 1987 respectively from the University of Illinois at Urbana-Champaign in Electrical and Computer Engineering. He joined Northwestern University as an Assistant Professor in 1987. In 2000, he moved to the University of California, Los Angeles. He has authored approximately 250 papers, and is an Associate Editor of several ACM and IEEE journals. He is also the Co-founder of Hier Design, Inc.

Daniel Schmidt
University of Kaiserslautern

Daniel Schmidt studied computer science at the Technical University of Kaiserslautern and received his diploma in 2003. He now is a fellow researcher with Prof. Norbert Wehn at the department of Microelectronic System Design and is working towards his Ph.D. at the same university. His research interests include low-power techniques for embedded systems and embedded software.

Robert Siegmund
AMD Saxony LLC & Co. KG

Robert Siegmund received his diploma and Dr.-Ing. degrees in Electrical Engineering from Chemnitz University of Technology, Germany, in

1998 and 2005, respectively. In 2003 he joined the Dresden Design Center of AMD Saxony LLC & Co. KG in Germany. As a member of the design verification team he is working on SoC verification and SystemC/SystemVerilog based verification methodology.

David Stewart
CriticalBlue, Ltd.

David Stewart co-founded CriticalBlue in 2002 following over 20 years in the EDA and semiconductor industries. This includes 10 years at Cadence Design Systems where he was as a founder and the Business Development Director of the System-on-Chip (SoC) Design facility at the Alba Campus in Scotland. He also worked at startups Redwood Design Automation and Simutech. Before Cadence he was a chip designer, with spells at LSI Logic, NEC Electronics and National Semiconductor. He has a first class honours degree in Electronics from the University of Strathclyde.

Richard Taylor
CriticalBlue, Ltd.

Richard Taylor co-founded CriticalBlue in 2002 to exploit his innovations in compilation and automatic VLIW architecture customization and he is the Chief Technical Officer. He has worked for several years in the areas of compiler optimization technology and engineering management for embedded software and custom ASIC development. He is the author of several patents in the areas of compilation, microarchitecture and memory systems and has a master's degree in Software Engineering from Imperial College, London.

Nigel Topham
University of Edinburgh

Nigel Topham is chief architect for ARC International and Professor of Computer Systems at the University of Edinburgh. He has led a number of processor design teams, including the ARC600. He is the founding director of the Institute for Computing Systems Architecture at Edinburgh University, where he researches into new design methodologies for architectures and micro-architectures and teaches computer architecture. He holds B.Sc. and Ph.D. degrees in Computer Science from the University of Manchester.

Scott Weber
University of California, Berkeley

Scott Weber received his Ph.D. in Electrical Engineering and Computer Science from the University of California at Berkeley in 2005.

His dissertation research focused on the design and simulation of sub-RISC programmable processing elements. Scott received a B.S. in Electrical and Computer Engineering and a B.S. in Computer Science from Carnegie Mellon University in 1998.

Norbert Wehn
University of Kaiserslautern

Norbert Wehn is professor at the Technical University of Kaiserslautern and holds the chair for Microelectronic System Design in the department of Electrical Engineering and Information Technology. He has published more than 150 papers in international conferences and journals in various areas of CAD and VLSI design. He received the Design Contest Award at DATE 2004 and 2005, respectively. In 2006 he was awarded IEEE "Golden Core Member" and received the "Outstanding Contribution Award" from the IEEE Computer Society.

Ralph Wittig
Xilinx, Inc.

Ralph Wittig is a Director in the Embedded Processing Division at Xilinx, San Jose, USA. He has been responsible for defining the vision and managing the development of embedded processing solutions for the last few years. His prior work includes FPGA architecture and software. He holds numerous patents in these areas. He earned his Bachelors and Masters from the University of Toronto, Canada. His Masters thesis on integrating processors with FPGAs is widely recognized.

Timo Yli-Pietilä
Nokia Research Center

Timo Yli-Pietilä received his M.Sc. degree in Digital and Computer Technology in 1986 and Lic. Tech. in 1995. He joined Nokia in autumn 1995, and has since been working in wireless baseband modem research and development projects. His special focus is on reconfigurable computing platforms.

Cliff Young
D. E. Shaw Research, LLC

Cliff Young works for D. E. Shaw Research, LLC, a member of the D. E. Shaw group of companies, on projects involving special-purpose, high-performance computers for computational biochemistry. Before his current position, he was a Member of Technical Staff at Bell Laboratories in Murray Hill, New Jersey. He received A.B., S.M., and Ph.D. degrees in computer science from Harvard University in 1989, 1995, and 1998, respectively.

About the Editors

Paolo Ienne is a Professor at the Ecole Polytechnique Fédérale de Lausanne (EPFL), Lausanne, Switzerland, where he heads the Processor Architecture Laboratory (LAP). He received the Dottore in Ingegneria Elettronica degree from Politecnico di Milano, Milan, Italy, in 1991, and the Ph.D. degree from the EPFL in 1996. From 1990 to 1991, he was an undergraduate researcher with Brunel University, Uxbridge, U.K. From 1992 to 1996, he was a Research Assistant at the Microcomputing Laboratory (LAMI) and at the MANTRA Center for Neuro-Mimetic Systems of the EPFL. In December 1996, he joined the Semiconductors Group of Siemens AG, Munich, Germany (which later became Infineon Technologies AG). After working on datapath generation tools, he became Head of the embedded memory unit in the Design Libraries division. In 2000, he joined EPFL. His research interests include various aspects of computer and processor architecture, reconfigurable computing, on-chip networks and multiprocessor systems-on-chip, and computer arithmetic. Dr. Ienne was a recipient of the DAC 2003 Best Paper Award. He is or has been a member of the program committees of several international workshops and conferences, including Design Automation and Test in Europe (DATE), the International Conference on Computer Aided Design (ICCAD), the International Symposium on High-Performance Computer Architecture (HPCA), the ACM International Conference on Supercomputing (ICS), the IEEE International Symposium on Asynchronous Circuits and Systems (ASYNC), and the Workshop on Application-Specific Processors (WASP). He is a cofounder of Mimosys, a company providing tools for the automatic customization of embedded processors.

Rainer Leupers is a Professor for Software for Systems on Silicon at RWTH Aachen University, Germany. He received the Diploma and Ph.D. degrees in Computer Science with honors from the University of Dortmund, Germany in 1992 and 1997. From 1997 to 2001 he was the chief engineer at the Embedded Systems group at the University of Dortmund. During 1999–2001 he was also a project manager at ICD, where he headed industrial compiler and processor simulator tool projects. Dr. Leupers joined RWTH in 2002. His research and teaching activities comprise software development tools, processor architectures, and electronic design automation for embedded systems, with emphasis on compilers for application specific processors. He authored several books and numerous technical papers on design tools for embedded processors, and he served in the program committees of leading EDA and compiler conferences, including the Design Automation Conference (DAC), Design Automation and Test in Europe (DATE), and the International Conference on Computer Aided Design (ICCAD). Dr. Leupers received several scientific awards, including Best Paper Awards at DATE 2000 and DAC 2002. He has been a cofounder of LISATek, an EDA tool provider for embedded processor design, acquired by CoWare Inc. in 2003.

PART I

OPPORTUNITIES AND CHALLENGES

CHAPTER 1

FROM PRÊT-À-PORTER TO TAILOR-MADE

Paolo Ienne and Rainer Leupers

By now, processors are everywhere. Really everywhere. Not only as microcontrollers in our washing machines and microwave ovens. Not only to read the mechanical displacement of our computer mice or to handle the user interfaces of our VHS video recorders. They have been there for decades, but now they are also truly in control of the combustion process in the engine of our cars. They also recreate and postprocess the film frames recorded on our DVDs and displayed by our digital TVs, recompute the audio signals throughout our Hi-Fi equipment, capture the images in our cameras and camcorders, send our voice on the radio waves through our cell phones, and encode and decode several megabits per second of our data so that they can be reliably transmitted on old telephone lines. They are even inside the components of other major processor-based equipment: they are key to encoding and decoding data to and from the magnetic media of the hard disks inside our iPods. They are in the WiFi module, which is in turn one of many components of our electronic agenda. And yes, almost incidentally by now, processors also happen to be where they have always been: processors are still the CPUs of our computers.

Independently from this shift in applications, processor architecture has evolved dramatically in the last couple of decades: from microprogrammed finite state machines, processors have transformed into single rigid pipelines; then, they became parallel pipelines so that various instructions could be issued at once; next, to exploit the ever-increasing pipelines, instructions started to get reordered dynamically; and, more recently, instructions from multiple threads of executions have been mixed into the pipelines of a single processor, executed at once. This incredibly exciting journey brought processors to a respectable maturity, growing from the 2,400 transistors of the Intel 4004 to nearly a billion transistors, in only 30 to 40 years of life. But now something completely different is changing in the lives of these devices: on the whole, the great majority of the high-performance processors produced today address relatively narrow classes of applications but must be extremely low in cost and energy efficient. Processors are traditionally rather poor at either of these metrics individually, let alone at both simultaneously. Processors, or more specifically the methodologies used to match and fit them to their application domains, need to change radically if they are to proficiently enter this new phase of their life.

This book is a first attempt to explore comprehensively one of the most fundamental trends that slowly emerged in the last decade: to treat processors not as rigid fixed entities, which designers include as they are in their products, but rather to build sound methodologies to tailor-fit processors to the very needs of such products. This new era has already started, and it is a good time to overview the complexity of the task and understand what is solved, what the permanent challenges and pitfalls are, and where there are still opportunities for researchers to explore. In this chapter, we will briefly overview some of the basic considerations leading to the idea of *customizable processors*. After this brief introduction, we will pass the pen on to numerous distinguished colleagues, all of whom are among the pioneers and/or the world experts of the different aspects of the technology in object.

1.1   THE CALL FOR FLEXIBILITY

The complexity of a large share of the integrated circuits manufactured today is impressive: devices with hundreds of millions of transistors are not uncommon at the time of this writing. Unsurprisingly, the nonrecurrent engineering costs of such high-end application-specific integrated circuits is approaching a hundred million U.S. dollars—a cost hardly bearable by many products individually.

It is mainly the need to increase the flexibility and the opportunities for modular reuse that is pushing industry to use more and more software-programmable solutions for practically every class of devices and applications. *Flexibility* implies a reduction of design risk (if applications change with time) and a potential cost share across a broader class of products based on the same hardware design. *Reuse* has a number of advantages at various levels: it helps in coping with complexity by assembling chips largely made of predefined hardware components that are software-adapted for the very application at hand. For essentially the same reason, reuse reduces development time and maximizes the chances to seize markets opportunities. And, finally, it helps amortize development costs by having part of a design survive a single generation of products.

Generally speaking, with product life cycles in the consumer electronic market often close to one year, development needs to be sped up considerably while complexity increases too. This implies that designers' efficiency must grow dramatically, and increasing the share of software is the most promising direction in a majority of cases.

1.2   COOL CHIPS FOR SHALLOW POCKETS

Modern applications require significant computing power, be it for advanced media processing, complex data coding, or other uses. Speed comes from parallelism, and parallelism can be exploited in multiple forms: instruction-level parallelism, multiple threads, homogeneous and heterogeneous multiprocessors, and so forth. All of them have appeared in commercial products, but those most common in general-purpose processors are essentially "brute-force" approaches with a high cost in silicon area and a low energy efficiency, making them prohibitive in most battery-operated applications. It has been estimated that across four generations of Alpha processors, if technology improvements are factored out, while performance has barely doubled, silicon real estate has increased by a factor of 80 and energy consumption by a factor of 12 [1]. Clearly, the classic way to high performance in processors is not viable for most embedded applications.

Typically, designers of application-specific integrated circuits try to exploit parallelism more explicitly. One common way is to design hardware accelerators or co-processors that address the computational kernels; processors are left in place mostly for flexibility and control of the various accelerators—a sort of flexible glue logic between the accelerators. The other way to improve the efficiency of processors is to use architectures that are ideally fit to the application. This is qualitatively like the choice between a digital signal processor and a microcontroller but can go down to a much finer resolution, such as the availability of vector operations, dedicated register files, multiple pipelines, special functional units, and so forth. But licensing and maintaining an architecture is a high cost for a company and a long-term investment in term of personnel qualification to use it proficiently. If the multiplicity of pre-designed architecture could help in achieving the desired performance efficiently, the solution would be impractical if more than a handful of cores are required—traditionally, even large semiconductor companies are restricted to little more than a couple of architectures, and then, often, business issues dominate the choice among them.

1.3   A MILLION PROCESSORS FOR THE PRICE OF ONE?

The question is, therefore, how to obtain the very best processor for an application without having to select every time from among a myriad of completely different architectures that all together are too expensive to maintain. An answer—the answer that this book addresses—consists of developing design methodologies that, ideally, make it possible to maintain a very large family of processors with a wide range of features at a cost comparable to that of maintaining a single processor. Whether this is a realistic expectation is a question that will reappear in the following chapters, either explicitly, in Chapter 3, or implicitly, throughout Part II, through the analyses of the complexity of the various design customization tasks.

The basic idea revolves around the possibility of introducing some flexibility in the two key tools of a processor tool chain: simulator and compiler. A retargetable compiler that can produce good quality code for a wide range of architectures described in some programmatic form is the keystone of this strategy. If the compiler is the keystone, at least in terms of difficulty of implementation, all other elements of the tool chain must be retargetable too and must all depend on a unique description of the target processor. The shape and boundaries of the architectural space covered by the tool chain differentiate the several approaches attempted in the recent years. Roughly, one can identify three trends:

▪ *Parameterizable processors* are families of processors belonging to a single family and sharing a single architectural skeleton, but in which some of the characteristics can be turned on or off (presence of multipliers, of floating point units, of memory units, and so forth) and others can be scaled (main datapath width, number and type of execution pipelines, number of registers, and so forth).

▪ *Extensible processors* are processors with some support for application-specific extensions. The support comes both in terms of hardware interfaces and conventions and in terms of adaptability of the tool chain. The extensions possible are often in the form of additional instructions and corresponding functional pipelines but can also include application-specific register files or memory interfaces.

▪ *Custom processor development tools* are frameworks to support architects in the effort to design from scratch (or, more likely, from simple and/or classic templates) a completely custom processor with its complete tool chain (compiler, simulator, and so forth). Ideally, from a single description in a rich architectural description language, all tools and the synthesizable description of the desired core can be generated.

One should note that these approaches are not mutually exclusive: A parameterizable processor may also be extensible—and in fact most are. A template processor in a processor development framework can be easily parameterized and is naturally extended. We group all these approaches under the name of *customizable processors*. Often, we will also refer to them as *application-specific instruction processors*, or *ASIPs* in short, without any major semantic difference.

1.4 PROCESSORS COMING OF AGE

Customizable processors have been described as the next natural step in the evolution of the microprocessor business [2]: a step in the life of a new technology, where top performance alone is no longer sufficient to guarantee market success. Other factors become fundamental, such as time to market, convenience, and ease of customization. A number of companies thrive in this new space, and many are represented in this book. In the next chapters we will try to look from a neutral standpoint at various aspects of the technology needed to develop, support, and use such processors.

One of the earliest industrial attempts to customize processors was the *custom-fit processors* project in the early 1990s at HP Labs. More than a decade later, a few original customizable processors have been on the market for some years. Also, most widespread licensable embedded-processor cores have been retrofitted with proper interfaces for instruction-set extensions, and the tools have been adapted accordingly. Licences for design-automation tools that can help in the efficient development of application-specific processors can be purchased from leading electronic system-level design tool providers. Field programmable gate array (FPGA) vendors, owners of an ideal technology for customizable processors, have introduced lightweight, highly customizable processors. Despite this evident commercial maturity, we believe that, on one side, there are significant challenges to successfully deploying customizable processors; on the other, there are abundant opportunities for research in new methodologies to support users in the specialization process. With this in mind, we have assembled the many contributions in this book.

1.5 THIS BOOK

Part I of the book deals with the motivation and the challenges of customizable processors. The next chapter, by Ascheid, Meyr, and Aachen looks at the upcoming challenges in one of the most lively domains of electronic innovation: wireless systems. The authors identify ASIPs as the key answer to the challenges and make a case for customizable processors—not without a word of caution against excessive attempts at automation. Chapter 3, by some of the pioneers of customizable processors—Fisher, Faraboschi, and Young—takes a more critical stand on different grounds: rather than technical challenges, they look at some logistical and sociotechnological issues to warn about the difficulties of using proficiently customizable processors in a productive environment.

Part II, the scientific core of the book, looks systematically at the various aspects of the technologies involved in designing and maintaining application-specific processors. Chapter 4, by Mishra and Dutt, deals with what everything else will exploit: languages that succinctly capture the features of an architecture so that the complete tool chain can be customized automatically from such a description. Among the various tools available, we have already mentioned compilers as the keystone of a tool chain: Chapter 5, by Leupers, discusses how the fundamental challenges have been solved in this critical domain. Once the basic elements of a tool chain have been taken care of, Chapter 6, by Goodwin, Leibson, and Martin, focuses on how users can be helped in the selection of the fittest architecture of a parametrizable processor, out of millions of possible configurations. The authors exemplify the solution with experiments on one of the best known commercial customizable architectures. Chapter 7, by Pozzi and Ienne, continues along the same train of thoughts but focuses on the instruction-set extensibility issue: how can we automatically infer the best instruction-set extensions from user code? Complementarily, Chapter 8, by the chief architect of a widespread commercial customizable processor, Topham, argues that current techniques for automatic instruction-set extension cannot help but miss some significant opportunities—an interesting viewpoint, especially for researchers looking for significant problems to tackle. Taylor and Stewart, in Chapter 9, discuss the alternate way to achieve performance efficiently: adding co-processors. Interestingly, they address the automatic design of co-processors, which are themselves instances of rich ASIPs; thus, they discuss their take on classic challenges in parallelism extraction, processor synthesis, and code mapping. Many of the approaches discussed in these chapters require the design of datapaths for various nonstandard functions; Chapter 10, by Brisk and Sarrafzadeh, illustrates the solution to several challenges of efficiently merging and implementing several such datapaths. Chapter 11, by Parameswaran, Henkel, and Cheung, looks into two specific aspects of the process of customization: instruction matching and efficient modelling of custom instructions. Verification is often the Achilles' heel of new design methodologies, and Große, Siegmund, and Drechsler discuss in Chapter 12 the challenges of verification with respect to customizable processors in particular. Part II concludes with Chapter 13, by Mihal, Weber, and Keutzer, which argues in favor of a different *grain* of custom processor design and illustrates a methodology to conceive systems out of tiny sub-RISC processors based on appropriate high-level programming paradigms.

Once the most important pieces of technology to build a custom processor have been studied in detail, in Part III we conclude by presenting some concrete examples of the applications of previously addressed concepts. Chapter 14, by Puusaari, Yli-Pietilä, and Rounioja, shows some real-world issues in designing an ASIP for cell search in a third-generation cellular-phone application. Chapter 15, by Wehn and Schmidt, tackles a computing-intensive problem of utmost practical importance: the implementation of channel encoding for wireless systems using Turbo encoding and decoding. The authors discuss several solutions to meet the real-world constraints of third-generation wireless standards. One of us—Leupers, in Chapter 16—puts to work techniques similar to those of Chapter 8 and shows how one can achieve practical results on a real and broadly used extensible architecture. The book concludes with Chapter 17, by Bilski, Mohan, and Wittig, which looks into the world of FPGAs and describes the opportunities and challenges of designing one of the two best known lightweight customizable processors commercially available for FPGAs.

1.6 TRAVEL BROADENS THE MIND

This book is a journey through a multitude of technical problems and successful solutions, with the occasional open issue for academics to tackle. We hope you will enjoy reading through the following chapters at least as much as we had the pleasure in assembling them for you. Have a good journey!

CHAPTER 2

OPPORTUNITIES FOR APPLICATION-SPECIFIC PROCESSORS: THE CASE OF WIRELESS COMMUNICATIONS

Gerd Ascheid and Heinrich Meyr

We presently observe a paradigm change in designing complex systems-on-chip (SoCs) such as occurs roughly every 12 years due to the exponentially increasing number of transistors on a chip. This paradigm change, as all the previous ones, is characterized by a move to a higher level of abstraction. Instead of thinking in register-transfer level (RTL) blocks and wires, one needs to think in computing elements and interconnect. There are technical issues as well as nontechnical issues associated with this paradigm change. Several competing approaches are discussed in the context of this paradigm change.

The discussion of the pros and cons of these competing approaches is controversial. There are technical as well as nontechnical dimensions to be considered. Even for the technical dimensions, no simple metrics exist. A look at the classic book by Hennessy and Patterson [1] is very instructive. Each chapter closes with sections entitled "fallacy" and "pitfall," which is an indication that no simple metric exists to characterize a good processor design. This is in contrast to communication engineering. To compare competing transmission techniques, the communication engineers compare the bit rate that can be transmitted over a bandwidth B for a given SNR. Hence, the metric used is bit per second per Hertz bandwidth. A system A is said to be superior to a system B if this metric is larger.

While one might hope that the future brings this metric for the technical dimensions, the case is hopeless with respect to the nontechnical ones. While designing a SoC, the following issues have to be considered:

- Risk management

- Project management

- Time to market

- Organizational structure of the company

- Investment policy

A design discontinuity has never occurred based on purely rational decisions. Nothing could be farther from reality. Experience has shown that the design discontinuity must be demonstrated with only few engineers. These engineers work on a project with a traditional design methodology and have blown the schedule one time too many. They have two options: the project gets killed or they adopt a new design paradigm.

In our opinion the next design discontinuity will lead to different solutions, depending on the application. We dare to make the following core proposition for wireless communications: *future SoC for wireless communications will be heterogeneous, reconfigurable Multi-Processor System-on-Chip (MPSoC)*.

They will contain computational elements that cover the entire spectrum, from fixed functionality blocks to domain-specific DSPs and general-purpose processors. A key role will be played by ASIPs. ASIPs exploit the full architectural space (memory, interconnect, instruction set, parallelism), so they are optimally matched to a specific task. The heterogeneous computational elements will communicate via a network-on-chip (NoC), as the conventional bus structures do not scale. These MPSoC platforms will be designed by a cross-disciplinary team.

In the following, we will substantiate our proposition. We begin by analyzing the properties of future wireless communication systems and observe that the systems are computationally extremely demanding. Furthermore, they need innovative architectural concepts in order to be energy efficient. We continue by discussing the canonical structure of a digital receiver for wireless communication. In the third part of this chapter, we address the design of ASIPs.

2.1   FUTURE MOBILE COMMUNICATION SYSTEMS

Fig. 2.1 provides an overview of current and future wireless systems.

Today's communication systems demand very high computational performance and energy-efficient signal processing. Future 4G systems will differ from the existing ones both qualitatively and quantitatively. They will have ultrahigh data rates reaching 1 Gbps, which is multiplexed for densely populated areas to allow mobile Internet services. 4G systems will offer new dimensions of applications. They will be multifunctional and cognitive.

Cognitive radio is defined as an intelligent wireless communication system that is aware of its environment and uses the methodology of understanding by building to learn from the environment and adapt to statistical variations in the input stimuli with two primary objectives in mind: highly reliable communications whenever and wherever needed, and efficient utilization of the radio spectrum, using multiple

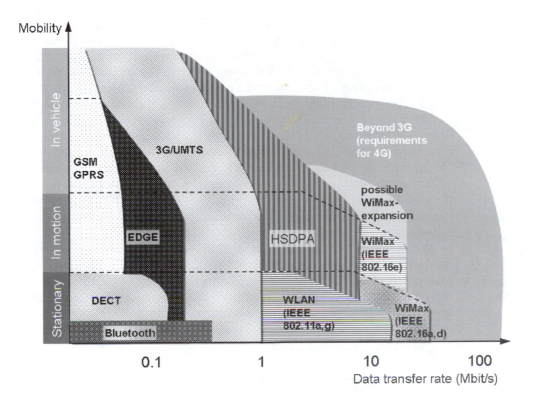

Mobility

In vehicle

In motion

Stationary

GSM
GPRS

3G/UMTS

EDGE

DECT

Bluetooth

Beyond 3G
(requirements
for 4G)

possible
WiMax-
expansion

HSDPA

WiMax
(IEEE
802.16e)

WLAN
(IEEE
802.11a,g)

WiMax
(IEEE
802.16a,d)

0.1   1   10   100

Data transfer rate (Mbit/s)

 FIGURE 2.1

Wireless systems today and in the future.

antennas (**MIMO**) and exploiting all forms of diversity—spatial, temporal, frequency, and multiuser diversity.

The fundamental tasks to be addressed in the context of cognitive radio are:

- Radio scene analysis (e.g., channel estimation, load, capacity, requirements, capabilities)

- Resource allocation (e.g., rate power, spectrum, diversity, network elements)

- System configuration (e.g., multiple access, coding, modulation, network structure)

The cognitive approach will lead to a new paradigm in the cross-layer design of the communication networks. For example, on the physical layer, the increasing heterogeneity requires flexible coding, modulation, and multiplexing. Current state-of-the-art transmission systems, which are designed for particular sources and channels, may fail in changing conditions. A new flexibility will be introduced: the traditional static design will be replaced by a cognitive approach with, for example, adaptive self-configuration of coding and modulation, taking the instantaneous statistics of the channel and the sources into account.

All these techniques will be introduced with a single objective in mind: to optimally utilize the available bandwidth. It is obvious that these techniques demand ultracomplex signal processing algorithms. The introduction of these complex algorithms has been made economically possible by the enormous progress in semiconductor technology. The communication engineer has learned to trade physical performance measures for signal processing complexity. On top of that, necessary support of legacy systems and multiple (competing) standards further increases processing complexity.

We will discuss in the next section how these algorithms will be processed on a heterogeneous MPSoC, which is designed to find a compromise between the two conflicting objectives of flexibility and energy efficiency. Any MPSoC comprises two dimensions of equal importance, namely processors and interconnects. In this chapter we focus on the processors.

2.2   HETEROGENEOUS MPSoC FOR DIGITAL RECEIVERS

Digital receivers incorporate a variety of functions; basic receiver and transmitter functions include the signal processing of the physical layer, a protocol stack, and a user interface. A cell phone, however, offers an increasing amount of additional functions, such as localization and multimedia applications. Throughout this chapter, physical layer processing is used to make the case, but arguments and concepts apply for the other digital receiver functions as well.

2.2.1   The Fundamental Tradeoff Between Energy Efficiency and Flexibility

We have discussed in the previous section that the algorithmic complexity of advanced wireless communication systems increases exponentially in order to maximize the channel utilization measured in bps per Hz bandwidth. Until the year 2013, the International Technology Roadmap for Semiconductors (ITRS) estimates an increase of performance requirements by a factor of almost 10,000. Thus, the required computational performance measured in millions of operations per second (MOPS) must increase at least proportionally to the algorithmic complexity, which by far cannot achieved by the standard architectures.

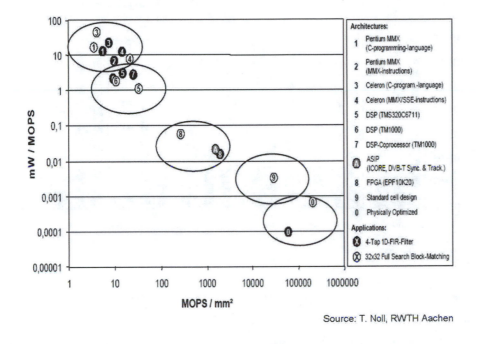

■ FIGURE 2.2

Energy per operation (measured in mW/MOPS) over area efficiency (measured in MOPS/mm^2).

Consequently, there is a need for innovative computer architectures, such as MPSoCs, to achieve the required computational performance at a predicted increase of the clock frequency of a factor of 5 only. However, performance is only one of the drivers for novel architectures. The other is energy efficiency. It is well known that battery energy remains essentially flat. Thus, if the numbers of MOPS must increase exponentially, energy efficiency must grow with at least at the same exponential rate.

We are now confronted with two conflicting goals. On the one hand, one desires to maximize flexibility, which is achieved by a fully programmable processor. However, the energy efficiency of this solution is orders of magnitude too low, as will be demonstrated shortly. Maximum energy efficiency is achieved by a fixed-functionality block with no flexibility at all. A SoC for wireless communications is, therefore, a compromise between these conflicting goals. Conceptually there are two approaches, both of which have been applied to industrial designs. In the first approach, one specifies a maximum energy per task and maximizes flexibility under this constraint. This favors an almost fully

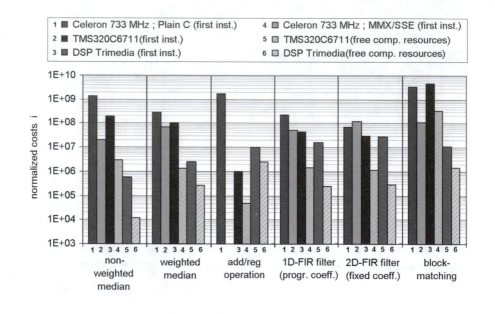

Source: T. Noll, RWTH Aachen

▪ **FIGURE 2.3**

Normalized cost for execution of sample signal processing algorithms on different processors. Source: [2].

programmable solution. In the second approach, one minimizes the hardware flexibility to a constraint set so as to achieve maximum energy efficiency.

To compare architectures, the useful metrics are *area efficiency* and *energy efficiency*. In Fig. 2.2, the reciprocal of energy efficiency measured in mW/MOPS is plotted versus area efficiency measured in MOPS/mm^2 for a number of representative signal processing tasks for various architectures.

The conclusion that can be drawn from this figure is highly interesting. The difference in energy/area efficiency between fixed and programmable functionality blocks is orders of magnitude. An ASIP provides an optimum tradeoff between flexibility and energy efficiency not achievable by standard architectures/extensions.[1]

To assess the potential of complex instructions, as well as the penalty for flexibility expressed in the overhead cost of a processor, we refer to Fig. 2.3, where we have plotted the normalized cost *area* × *time* ×

---

[1] The data point A is included for qualitative demonstration only, as it performs a different task.

energy / per sample for various processors and tasks. The numbers differ for other processors; nevertheless, the figure serves well to make the point.[2]

Let us examine the block-matching algorithm. We see that more than 90 percent of the energy is spent for overhead (flexibility). For other algorithms in this figure, the numbers are even worse. From the same figure we see that the difference in cost between the two DSPs is more than two orders of magnitude. The explanation for this huge difference is simple: the Trimedia has a specialized instruction for the block-matching algorithm, while the TI processor has no such instruction.

2.2.2 How to Exploit the Huge Design Space?

While the design space is huge, the obvious question is: can we exploit this gold mine? Before we attempt to outline a methodology to do so, it is instructive to examine Fig. 2.4.

Fig. 2.4 shows network processors based on their approaches toward parallelism. On this chart, we have also plotted iso-curves of designs that issue 8, 16, and 64 instructions per cycle. Given the relatively narrow range of applications for which these processors are targeted, the diversity is puzzling. We conjecture that the diversity of architectures is not based on a diversity of applications but rather a diversity of backgrounds of the architects. This diagram, and the relatively short lives of these architectures, demonstrate that design of programmable platforms and ASIPs is still very much an art based on the design experience of a few experts. We see the following key gaps in current design methodologies that need to be addressed:

- Incomplete application characterization: Designs tend to be done with the representative computation kernels and without complete application characterization. Inevitably this leads to a mismatch between expected and delivered performance.

- Ad hoc design space definition and exploration: We conjecture that each of the architectures represented in Fig. 2.4 is the result of a relatively narrow definition of the design space to be explored, followed by an ad hoc exploration of the architectures in that space. In other words, the initial definition of the set of architectures to be explored is small and is typically based on the architect's

---

[2] By "first instantiation," it is understood that the processor performs only one task. By "free computational resource," it is understood that the processor has resources left to perform the task. The latter is a measure for the useful energy spent on the task. The difference between "first instantiation" and "free computational resources" is the penalty paid for flexibility [2].

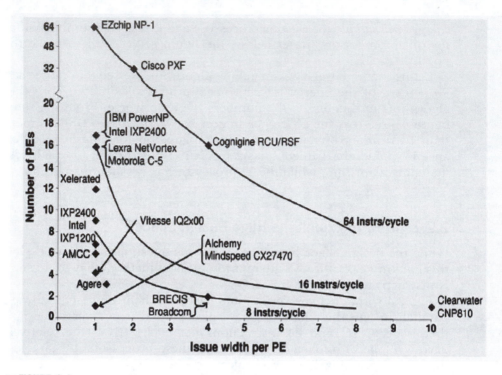

■ **FIGURE 2.4**

Number of processing elements (PE) versus issue width per processing element for different network processors. Source: [3].

prior experience. The subsequent exploration, to determine which configuration among the small set of architectures is optimal, is done in completely informal way and performed with little or no tool support.

This clearly demonstrates the need for a methodology and corresponding tool support. In MESCAL, such a methodology is discussed in detail. The factors of the MESCAL methodology [3] are called elements, not steps, because the elements are not sequential. In fact, an ASIP design is a highly concurrent and iterative endeavor. The elements of the MESCAL approach are

1. Judiciously apply benchmarking

2. Inclusively identify the architectural space

3. Efficiently describe and evaluate the ASIPs

4. Comprehensibly explore the design space

5. Successfully deploy the ASIP

Element 1 is most often underestimated in industrial projects. Traditionally, the architecture has been defined by the hardware architects based on ad hoc guessing. As a consequence, many of these architectures have been grossly inadequate.

A guiding principle in architecture, which is also applicable to computer architecture, is

Principle: Form follows function. (Mies van der Rohe)
From this principle immediately follows

Principle: First focus on the applications and the constituent algorithms, and not the silicon architecture.

Pitfall: Use a homogeneous multiprocessor architecture and relegate the task mapping to software at the later point in time.
Engineers with a hardware background tend to favor this approach. This approach is reminiscent of early DSP times, when the processor hardware was designed first and the compiler was designed afterward. The resulting compiler-architecture mismatch was the reason for the gross inefficiency of early DSP compilers.

Fallacy: Heterogeneous multiprocessors are far too complex to design and a nightmare to program. We have known that for many years.
This is true in the general case. If, however, the task allows the application of the "divide and conquer" principle, it is not true—as we will demonstrate for the case of a wireless communication receiver.

2.2.3 Canonical Receiver Structure

To understand the architecture of a digital receiver, it is necessary to discuss its functionality. Fig. 2.5 shows the general structure of a digital receiver. The incoming signal is processed in the analog domain and subsequently A/D converted. In this section, we restrict the discussion to the digital part of the receiver.

We can distinguish two blocks of the receiver called the *inner* and *outer* receivers. The outer receiver receives exactly one sample per symbol from the inner receiver. Its task is to retrieve the transmitted information by means of channel and source decoding. The inner receiver has the task of providing a "good" channel to the decoder. As can be seen from Fig. 2.5, two parallel signal paths exist in the inner receiver, namely, detection and parameter estimation path. Any digital receiver employs the principle of synchronized detection. This principle states that in the detection path, the channel parameters estimated from the noisy signal in the parameter estimation path are used as if they were the true values.

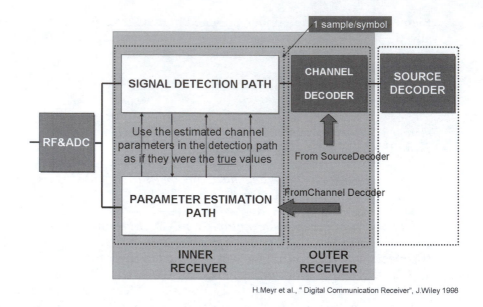

■ **FIGURE 2.5**

Digital receiver structure. Source: [4].

Inner and outer receivers differ in a number of key characteristics:

Property of the outer receiver
The algorithms are completely specified by the standards to guarantee interoperability between the terminals. Only the architecture of the decoders is amenable to optimization.

Property of the inner receiver
The parameter estimation algorithms are not specified in the standards. Hence, both architecture and algorithms are amenable to optimization.

The inner receiver is a key building block of any receiver for the following reasons:

- Error performance (bit error rate) is critically dependent on the quality of the parameter estimation algorithms.

- The overwhelming portion of the receiver signal processing task is dedicated to channel estimation. A smaller (but significant) portion is contributed by the decoders. To maximize energy efficiency, most of the design effort of the digital receiver is spent on the algorithm and architecture design of the inner receiver.

2.2.4   Analyzing and Classifying the Functions of a Digital Receiver

Any definition of a benchmark (element 1 of MESCAL) must be preceded by a detailed analysis of the application. There are several steps involved in this analysis.

Step 1: Identify structural and temporal properties of the task
The signal processing task can be naturally partitioned into decoders, channel estimators, filters, and so forth. These blocks are loosely coupled. This property allows us to structurally map the task onto different computational elements. The single processing is (almost) periodic. This allows us to temporally assign the tasks by an (almost) periodic scheduler.

Example: DVB-S Receiver
Fig. 2.6 is an example of a DVB-S receiver.

Step 2: From function to algorithm
The functions to be performed are classified with respect to the structure and the content of their basic algorithmic kernels. This is necessary to define algorithmic kernels that are common to different standards in a multistandard receiver.

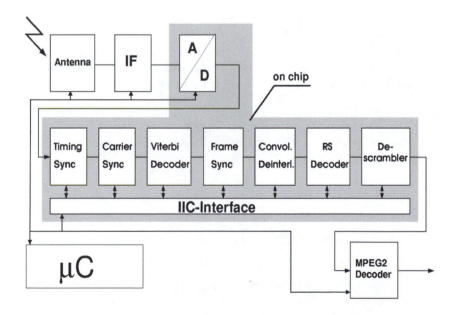

■ **FIGURE 2.6**

DVB-S receiver architecture.

- Butterfly unit
  - Viterbi and MAP decoder
  - MLSE equalizer
- Eigenvalue decomposition (EVD)
  - Delay acquisition (CDMA)
  - MIMO Tx processing
- Matrix-Matrix and Matrix-Vector Multiplication
  - MIMO processing (Rx and Tx)
  - LMMSE channel estimation (OFDM and MIMO)
- CORDIC
  - Frequency offset estimation (e.g., AFC)
  - OFDM post-FFT synchronization (sampling clock, fine frequency)
- FFT and IFFT (spectral processing)
  - OFDM
  - Speech post processing (noise suppression)
  - Image processing (not FFT but DCT)

Example: Cordic Algorithm
The cordic algorithm is an efficient way to multiply complex numbers. It is frequently used for frequency and carrier synchronization where complex signals must be rotated by time-varying angles.

Step 3: Define algorithmic descriptors
The properties of the algorithms must be determined with respect to suitable descriptors.

- Clock rate of processing elements (1/Tc)
- Sampling rate of the signal (1/Ts)
- Algorithm characteristic
  - Complexity (MOPS/sample)
  - Computational characteristic
    * Data flow
      · Data locality
      · Data storage
      · Parallelism
    * Control flow

- Connectivity of algorithms
  - Structural
  - Temporal
- Quantization and word length

Step 4: Classification of the algorithm

Since the signal processing task is (almost) periodic, it is instructive to plot complexity (measured in operations per sample) versus the sampling rate; see Fig. 2.7. The iso-curves in such a graph represent MOPS. A given number of MOPS equals the product of

$$\text{Operations/sec} = (\text{Operations/sample}) \times (\text{samples/sec})$$

This trivial but important relation shows that a given number of MOPS can be consumed in different ways: either by a simple algorithm running at a high sampling rate or, conversely, by a complex algorithm running at a low sampling rate. As a consequence, if a set of algorithms is to be processed, then the total amount of MOPS required is a meaningful measure only if the properties of the algorithms are similar. This is not the case in physical layer processing, as illustrated by a typical example.

Example: Baseband Processing for a 384-Kbps UMTS Receiver

The first observation to be made from Fig. 2.7 is that the algorithms cover a wide area in the complexity-sampling rate plane. Therefore, the total number of operations per second is an entirely inadequate measure. The algorithms requiring the highest performance are in the right lower quadrant. They comprise of correlation, matched filtering (RRC matched filter), interpolation and decimation, and path searcher to select a number of representative algorithms. But even the algorithms with comparable complexity and sampling rate differ considerably with respect to their properties, as will be demonstrated subsequently.

The correlator and the matched filter belong to a class with similar properties. They have a highly regular data path allowing pipelining, and they operate with very short word lengths of the data samples and the filter coefficients without any noticeable degradation in bit error performance of the receiver [4]. As the energy efficiency of the implementation is directly proportional to the word length, this is of crucial importance. A processor implementation using a 16-bit word length would be hopelessly overdesigned. Interpolation and decimation are realized by a filter with time-varying coefficients for which other design criteria must be observed [4]. The turbo-decoder is one of the most difficult blocks to design with entirely different algorithmic properties than the matched filter or the correlator discussed earlier. Of particular

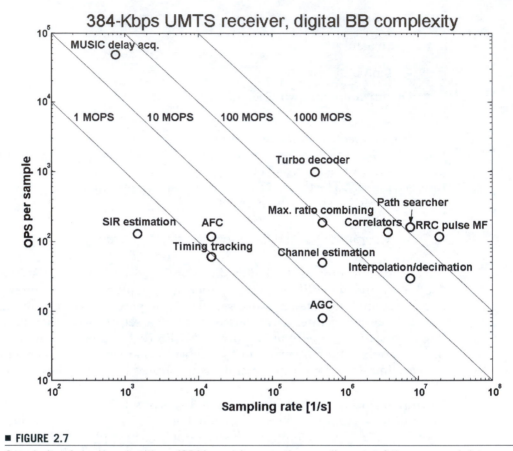

Complexity of receiver algorithms (OPS/sample) versus the sampling rate of the processed data.

importance is the design of the memory architecture. A huge collection of literature on the implementation of this algorithm exists. We refer the reader to [5–8] and to Chapter 17 of this book.

As an example of a complex algorithm running at low sampling rate (left upper corner), we mention the MUSIC algorithm, which is based on the Eigenvalue decomposition.

As a general rule for the classification of the baseband algorithm, we observe from Fig. 2.7 that the sampling rate decreases with increasing "intelligence" of the algorithms. This is explained by the fact that the number crunchers of the right lower corner produce the raw data that is processed in the intelligent units tracking the time variations of the channel. Since the time variations of the channel are only a very small fraction of the data transmission rate, this can be advantageously exploited for implementation.

In summary, the example has demonstrated the heterogeneity of the algorithms. Since the complete task can be separated into subtasks with simple interconnections, it can be implemented as a heterogenous MPSoC controlled by a predefined, almost periodic scheduler.

Example: UWB Receiver

The findings are identical to the one of the previous example. Due to the much higher data rate of 200 Mbps, the highest iso-curve is shifted toward 100 GOPS.

Example 3: Multimedia Signal Processing

An analogous analysis has been published by L. G. Chen in [9].

2.2.5 Exploiting Parallelism

Extensive use of parallelism is the key to achieving the required high throughputs, even with the moderate clock frequencies used typically in mobile terminals. On the top level there is, of course, the task level parallelism by parallel execution of the different applications. This leads to the heterogenous MPSoC, as discussed before. But even the individual subtasks of the physical layer processing can often be executed in parallel when there is a clear separation between the subtasks, as suggested by the block diagram of the DVB-S receiver shown in Fig. 2.6.

For the individual application-specific processor, all known parallelism approaches have to be considered and may be used as appropriate. Besides the parallel execution by pipelining, the most useful types of instruction parallelism are: very long instruction word (VLIW), single instruction multiple data (SIMD), and complex instructions. Within instruction parallelism, VLIW is the most flexible approach, while the use of complex instructions is the most application-specific approach.

Since each of these approaches has specific advantages and disadvantages, a thorough exploitation of the design space is necessary to find the optimum solution. VLIW (i.e., the ability to issue several individual instructions in parallel as a single "long" instruction) inherently introduces more overhead (e.g., in instruction decoding) but allows parallel execution even when the processing is less regular. With SIMD, there is less control overhead, since the *same* operation is performed on multiple data, but for the same reason it is only applicable when the processing has significant regular sections.[3] Finally, grouping several operations into a single, complex instruction is very closely tied to a specific algorithm and, therefore, the least flexible approach. The main advantage is its efficiency, as overhead and data transfer is minimized.

---

[3] The term "regular sections" here denotes sections where a particular sequence of arithmetic operations is applied to multiple data.

As discussed in the previous section, the algorithms in the lower right corner of Fig. 2.7 are number crunchers, where the same operations are applied to large amounts of data (since the data rate is high). For these algorithms, SIMD and complex instructions are very effective. More "intelligent" algorithms, which are found in the upper part of the diagram, are by definition more irregular. Therefore, VLIW typically is more effective than the other two approaches here.

Since the key measures for optimization are area, throughput, and energy consumption, exploitation of parallelism requires a comparison not only on the instruction level but also on the implementation level. For example, a complex instruction reduces the instruction count for the execution of a task but may require an increased clock cycle (i.e., reduce the maximum clock frequency).

2.3   ASIP DESIGN

2.3.1   Processor Design Flow

The design of a processor is a very demanding task, which comprises the design of the instruction set, micro architecture, RTL, and compiler (see Fig. 2.8). It requires powerful tools, such as a simulator, an assembler, a linker, and a compiler. These tools are expensive to develop and costly to maintain.

In traditional processor development, the design phases were processed sequentially as shown next. Key issues of this design flow are the following:

- Handwriting fast simulators is tedious, prone to error and difficult.

- Compilers cannot be considered in the architecture definition cycle.

- Architectural optimization on RTL is time consuming.

- Likely inconsistencies exist between tools and models.

- Verification, software development, and SoC integration occur too late in the design cycle. Performance bottlenecks may be revealed in this phase only.

This makes such a design flow clearly unacceptable today.

We conjectured earlier that future SoCs will be built largely using ASIPs. This implies that the processor designer team will be different, both quantitatively and qualitatively, from today. As the processors become application-specific, the group of designers will become larger, and the processors will be designed by cross-disciplinary teams to capture the expertise of the domain specialist, the computer architect, the compiler designer, and the hardware engineer. In our opinion, it is a

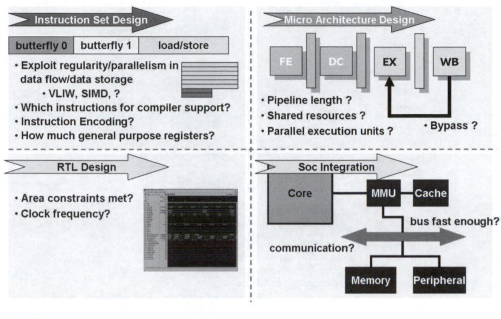

■ FIGURE 2.8

Tasks in processor design.

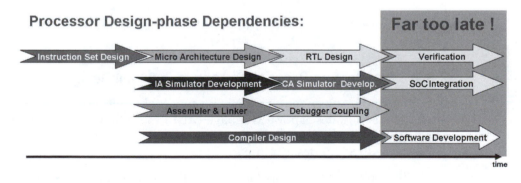

■ FIGURE 2.9

Traditional processor design flow.

gross misconception of reality to believe that ASIPs will be automatically derived from C-code provided by the application programmer.

The paradigm change (think in terms of processors and interconnects) requires a radically different conception of how and by whom the designs will be done in the future. In addition, it requires a tool

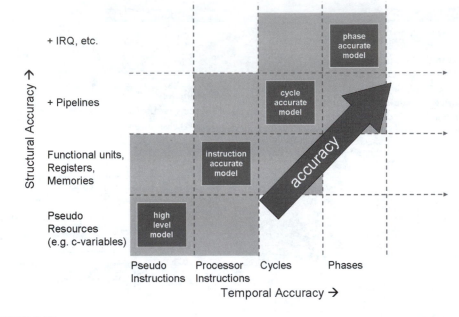

■ **FIGURE 2.10**

Levels of model accuracy.

suite to support the design paradigm, as will be discussed in the next section.

2.3.2   Architecture Description Language Based Design

In our view, the key enabler to move up to a higher level of abstraction is an architecture description language (ADL) that allows the modeling of a processor at various levels of structural and temporal accuracy.

From the LISA2.0 description:[4]

1. The software (SW) tools (assembler, linker, simulator, C compiler) are fully automatically generated.

2. A path leads to automatic hardware (HW) implementation. An intermediate format allows user-directed optimization and subsequent RTL generation.

---

[4] The ADL language LISA2.0, which is the base of the LISATek tool suite of CoWare, was developed at the Institute for Integrated Signal Processing Systems (ISS) of RWTH Aachen University. The LISATek tool suite is described in an overview in the book, where further technical references can be found. For this reason we do not duplicate the technical details of LISATek here.

3. Verification is assisted by automatic test pattern generation and assertion-based verification techniques [10].

4. An executable SW platform for application development is generated.

5. A seamless integration into a system simulation environment is provided, thus enabling virtual prototyping.

This is illustrated in Fig. 2.11.

We like to compare the LISATek tool suite with a workbench. It helps the designer efficiently design a processor and integrate the design into a system-level design environment for virtual prototyping. It does not make a futile attempt to automate the creative part of the design process. Automation will be discussed next.

2.3.3   Too Much Automation Is Bad

A few general comments about automation appear to be in order. We start with the trivial observation: design automation per se is of no value. It must be judged by the economic value it provides.

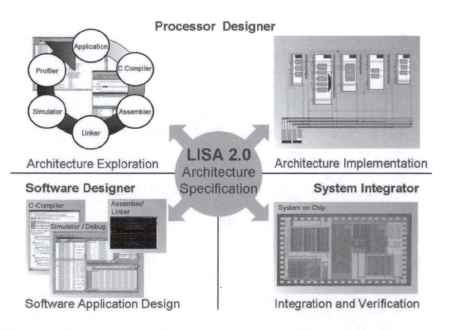

■ **FIGURE 2.11**

LISA 2.0 architecture specification as central design base.

Fallacy: Automate to the highest possible degree.

Having conducted and directed academic design tools research that led to commercially successful products over many years[5], we believe that one of the worst mistakes of the academic community is to believe that fully automated tools are the highest intellectual achievement. The companions of this belief in the EDA industry are highly exaggerated claims of what the tools can do. This greatly hinders the acceptance of a design discontinuity in industry, because it creates disappointed users burned by the bad experience of a failed project.

Principle: You can only automate a task for which a mathematical cost function can be defined.

The existence of a mathematical cost function is necessary to find an optimum. Finding the minimum of the cost function is achieved by using advanced mathematical algorithms. This is understood by the term automation. Most often, finding an optimum with respect to a given cost function requires human interaction. As an example, we mention logic synthesis and the tool design compiler.

Pitfall: Try to automate creativity.

This is impossible today and in the foreseeable future. It is our experience, based on real-world designs, that this misconception is based on a profound lack of practical experience, quite often accompanied by disrespect for the creativity of the design engineer. By hypothetically assuming that a "genie approach" works, it is by no means clear what the economic advantage of this approach would be, compared to an approach where the creative part of a design is left to an engineer supported by a "workbench." The workbench performs the tedious, time-consuming, and error-prone tasks that can be automated.

2.3.4   Processor Design: The LISATek Approach

The LISATek tool suite follows the principles of a workbench. The closed loop shown in Fig. 2.12 is fully automated: from the LISA2.0 description, the SW tools are automatically generated.

In the *exploration phase*, the performance of the architecture under investigation is measured using powerful profiling tools to identify hotspots. Based on the result, the designer modifies the LISA2.0 model of the architecture, and the software tools for the new architecture variant

---

[5] The toolsuite COSSAP was also conceived at ISS of RWTH Aachen University. It was commercialized by the start-up Cadis, which the authors of this chapter co-founded. Cadis was acquired in 1994 by Synopsys. COSSAP has been further developed into the "Co-Centric System Studio" Product. The LISATek tool suite was also conceived at ISS. It was commercialized by the start-up LISATek, which was acquired by CoWare in 2003.

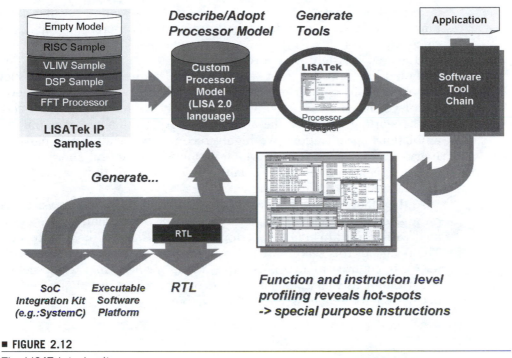

Empty Model
RISC Sample
VLIW Sample
DSP Sample
FFT Processor

LISATek IP Samples

Describe/Adopt Processor Model

Custom Processor Model (LISA 2.0 language)

Generate Tools

LISATek

Processor Designer

Application

Software Tool Chain

Generate...

RTL

SoC Integration Kit (e.g.:SystemC)   **Executable Software Platform**   *RTL*

Function and instruction level profiling reveals hot-spots -> special purpose instructions

■ **FIGURE 2.12**
The LISATek tool suite.

are generated in a few seconds by a simple key stroke. This rapid modeling and retargetable simulation and code generation allow us to iteratively perform a joint optimization of application and architecture. In our opinion, this fully automatic closed loop contributes most to the *design efficiency*. It is important to emphasize that we keep the *expert* in the loop.

A workbench must be *open* to attach tools that assist the expert in his search for an optimum solution. Such optimization tools are currently a hot area of research [11–17] in academia and industry. It is thus of utmost importance to keep the workbench open to include the result of this research. To achieve this goal, we have strictly followed the principle of *orthogonalization of concerns*: any architecture optimization is under the control of the user. Experience over the years with attempts to hide optimization from the user has shown to be a red herring, since it works optimally only in a few isolated cases (which the tool architect had in mind when he crafted the tool) but leads to entirely unsatisfactory results in most other cases.

For example, a tool that analyzes the C-code to identify patterns suitable for the fusion of operations (instruction set synthesis) leading to complex instruction is of the greatest interest. Current state of the art includes the Tensilica XPRES tool, the instruction-set generation developed at the EPFL [18, 19], and the microprofiling technique [20] developed at the ISS. All these tools allow the exploration of a large number of alternatives automatically, thus greatly increasing the *design efficiency*.

However, it would be entirely erroneous to believe that these tools automatically find the "best" architecture. This would imply that such a tool optimally combines all forms of parallelization (instruction level, data level, fused operations), which is clearly not feasible today. C-code as the starting point for the architecture exploration provides only an ambiguous specification. We highly recommend reading the paper of Ingrid Verbauwhede [21], in which various architectural alternatives of an implementation of the Rijndahl encryption algorithm based on the C-code specification are discussed.

Rather, these instruction set synthesis tools are *analysis tools*. Based on a given base architecture, a large number of alternatives are evaluated with respect to area and performance *to customize* the architecture. The results of this search are given in form of Pareto curves (see Chapter 6). In contrast to an analysis tool a true *synthesis tool* would deliver the structure as well as the parameters of the architecture. For a detailed discussion, the reader is referred to Chapter 7.

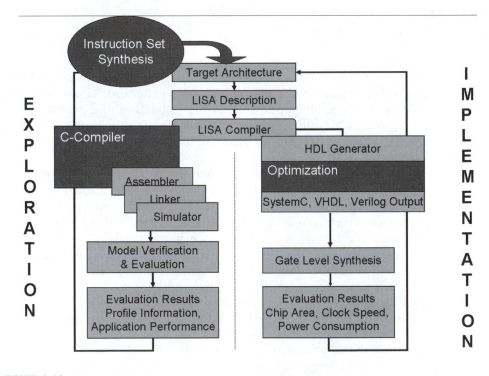

■ FIGURE 2.13

Design flow based on the LISATek tool suite.

Once the architecture has been specified, the LISATek workbench offers a path to hardware implementation. The RTL code is generated fully automatically. As one can see in Fig. 2.13, the implementation path contains a block labelled "optimization." In this block, user-controlled optimizations are performed to obtain an efficient hardware implementation. The optimization algorithms exploit the high-level architectural information available in LISA; see [22–24] for more detail.

2.3.5 Design Competence Rules the World

While it is undisputed that the design of an ASIP is a cross-disciplinary task that requires in-depth knowledge of various engineering disciplines—algorithm design, computer architecture, HW designs—the conclusion to be drawn from this fact differs fundamentally within the community. One group favors an approach that in its extreme implies that the application programmer customizes a processor, which is then automatically synthesized. This ignores the key fact (among others) that architecture and algorithm must be jointly optimized to get the best results.

Example: FFT Implementation
Numerous implementations of the algorithm are available in the literature. The cached FFT algorithm is instructive because of its structure. The structure allows the exploitation of data locality for higher energy efficiency, because a data cache can be employed. On the architecture side, the ASIP designer can select the type and size of the data cache. The selection reflects a tradeoff between the cost of the design (gate count) and its energy efficiency. On the algorithm side, a designer can alter the structure of the algorithm, for instance, by changing the size of the groups and the number of epochs. This influences how efficiently the cache can be utilized for a given implementation. In turn, possible structural alterations are constrained by the cache size. Therefore, a good tradeoff can only be achieved through a joint architecture/algorithm optimization, through a designer who is assisted by a workbench. No mathematical optimization will find this best solution, because there is no mathematical cost function to optimize. For the final architecture, RTL code will be automatically generated and verified and, subsequently, optimized by the hardware designer.

Fallacy: The cross-disciplinary approach is too costly
This conjecture is supported by no practical data. The contrary is true, as evidenced by successful projects in industry. As we have discussed earlier, one of the most common mistakes made in the ASIP design is to spend too little effort in analyzing the problem and benchmarking (element 1 of the MESCAL methodology); see [3]. The design efficiency is only marginally affected by minimizing the time spent on this task, at the expense of a potentially grossly suboptimal solution. The overwhelming improvement of design efficiency results from the workbench.

However, the cross-disciplinary project approach today is still not widespread in industry. Why is this? There are no technical reasons. The answer is that building and managing a cross-disciplinary team is a very demanding task. Traditionally, the different engineering teams involved in the different parts of a system design have had very little interaction. There have been good reasons for this in the past when the tasks involved could be processed sequentially, and thus the various departments involved could be managed separately with clearly defined deliverables. For today's SoC design, however, this approach is no longer feasible.

Budgets and personnel of the departments need to be reallocated. For example, more SW engineers than HW designers are needed. However, any such change is fiercely fought by the departments that lose resources. Today the formation of a cross-disciplinary team takes place only when a project is in trouble.

Finally we need to understand that in a cross-disciplinary project, the whole is more than the sum of the parts. In other words, if there exist strong interactions between the various centers of gravity, decisions are made across disciplinary boundaries. Implementing and managing this interaction is difficult. Why? Human nature plays the key role. Interaction

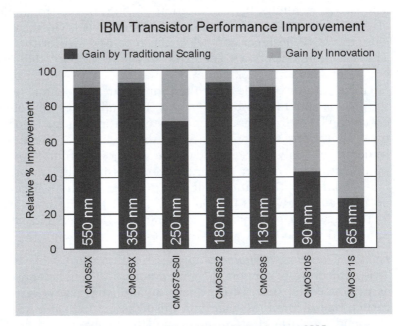

Source: Lisa Su /IBM: MPSoC 05 Conference 2005

■ **FIGURE 2.14**

Gain by scaling versus gain by innovation [25].

between disciplines implies that no party has complete understanding of the subject. We know of no engineer, including ourselves, that gladly acknowledges that he does not fully understands all the details of a problem. It takes considerable time and discipline to implement a culture that supports cross-disciplinary projects, whether in industry or academia.

We conjecture that in the future, mastering a cross-disciplinary task will become a key asset of the successful companies. Cross-discipline goes far beyond processor design. This has been emphasized in a talk by Lisa Su [25] about the lessons learned in the IBM CELL processor project. Su also made the key observation that in the past, performance gain was due to scaling, while today pure scaling contributes an increasingly smaller portion to the performance gain (Fig. 2.14).

Integration over the entire stack, from semiconductor technology to end-user applications, will replace scaling as the major driver of increased system performance. This corroborates the phrase "design competence rules the world."

2.3.6   Application-Specific or Domain-Specific Processors?

In Fig. 2.16 we have summarized the evolution of processor applications in wireless communication. In the past we have found a combination of general-purpose RISC processors, digital signal processors, and fixed wired logic. Today, the processor classes have been augmented by the class of configurable and scalable processors. In the future an increasing number of true ASIPs exploiting all forms of parallelism will coexist with the other classes.

The key question is, what class of architecture is used where? As a general rule, if the application comprises large C-programs such as in multimedia, the benefit of customization and scalability might be small. Since the processor is used by a large number of application programmers with differing priorities and preferences, optimization will be done by highly efficient and advanced compiler techniques, rather than architectural optimization. The application is not narrow enough to warrant highly specialized architectural features. For this class a *domain-specific processor* is probably the most economical solution.

Configurable (customizable) processors will find application when the fairly complex base architecture is justified by the application and, for example, when a customized data path (complex instruction) brings a large performance gain. Advantages of a customized processor are the minimization of the design risk (since the processors is verified) and the minimization of design effort.

The class of *full ASIPs* offers a number of key advantages, which go far beyond instruction-set extensions. The class includes, for example, application-specific memory and interconnect design. Technically, exploiting the full architectural options allows us to find an optimum tradeoff between energy efficiency and flexibility. If the key performance

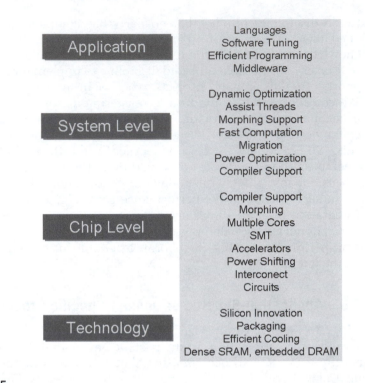

■ FIGURE 2.15

Future improvements in system performance will require an integrated design approach [25].

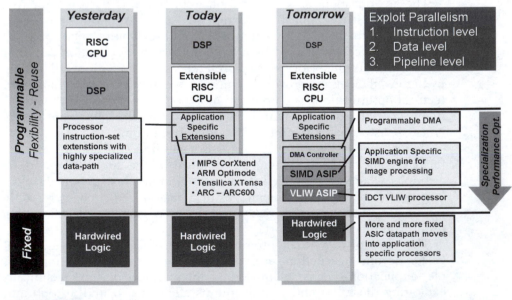

■ FIGURE 2.16

Evolution of processor applications in wireless communications.

metric is energy efficiency, then the flexibility must be minimized ("just enough flexibility as required"). In the case of the *inner receiver* (see Fig. 2.5), most functionality is described by relatively small programs done by domain specialists. For these applications, a highly specialized ASIP with complex instructions is perfectly suited. Quite frequently, programming is done in assembly, since the number of lines of code is small (see for example: [26] and [27]).

Economically, the key advantage of the full ASIP is the fact that no royalty or license fee has to be paid. This can be a decisive factor for extremely cost-sensitive applications with simple functionality for which a streamlined ASIP is sufficient. Also, it is possible to develop a more powerful ASIP leveraging existing software by backward compatibility on the assembler level. For example, in a case study a pipelined microcontroller with 16/32-bit instruction encoding was designed, which was assembler-level compatible with the Motorola 68HC11. The new microcontroller achieved three times the throughput and allowed full reuse of existing assembler code.

Finally, we should also mention that most likely in the future processor processors will have some degree of reconfigurability in the field. This class of processors can be referred as *reconfigurable ASIPs* (rASIPs). In wireless communication application, this post-fabrication flexibility will become essential for software-configurable radios, where both a high degree of flexibility and high performance are desired. Current rASIP design methodology strongly relies on the designer expertise without the support of a sophisticated workbench (like LISATek). Consequently, the design decisions, like amount of reconfigurability, processing elements in the reconfigurable block, or the processor-FPGA interface, are purely based on designer expertise. At this point, it will be interesting to have a workbench to support the designer taking processor design decisions for rASIP design.

CHAPTER 3

CUSTOMIZING PROCESSORS: LOFTY AMBITIONS, STARK REALITIES

Joseph A. Fisher, Paolo Faraboschi, and Cliff Young

In this chapter we describe some of the lessons learned in almost a decade of custom-fit processors research, from the research goals initially developed at HP Labs in the early 1990s through the development of the Lx/ST200 family of customizable VLIW processors.

We believe (and show with examples) that many of the technically challenging issues in customizing architectures and their compilers have either been solved or are well along the way to being solved in the near future. Most of the other chapters of this book are dedicated to describing the details of these technologies, so we refer the readers to them for a deeper understanding of their workings. We also believe that a customizable processor family can be an excellent fit for certain application domains, as was made clear in Chapter 2 about the wireless telecommunication market.

However, it is important to understand that technology is only one half of the picture, and several, largely nontechnical, barriers to customization still exist. In the second half of the chapter, we show where reality sinks in to make processor customization difficult to deploy or economically not viable. We believe that this critical—perhaps controversial—view is necessary to guide future research to focus on what some of the real problems are.

We say that we *customize* or *scale* architectures when we tune them for some particular use. In our usage, these terms have slightly different meanings. *Scaling* is adding, subtracting, or changing the parameters of "standard," general-purpose components of the architecture. This includes, for example, adding more integer addition functional units, removing a floating point unit, changing the latency of a memory unit, or changing the number of ports in a register bank. *Customizing* is adding a particular piece of functionality to accomplish a more specialized and

The content of this chapter is distilled from the authors' book: *Embedded Computing: A VLIW Approach to Architecture, Compilers and Tools*, Morgan Kaufmann, 2004 [1].

narrow task. Adding a "population count" instruction (count the number of 1's in a register) is a classic example of customization. An instruction-set architecture (ISA) that readily admits these changes is called *scalable* or *customizable*. These terms are frequently used interchangeably, and in any case the difference between them is subjective and sometimes hard to determine. In the remainder of the chapter, we often use the term customization as an umbrella for both.

There are a number of strong motivations for scaling or customizing, which roughly correspond to the optimization areas that are most important in the embedded world: performance, cost, and power.

- Hardware performance can improve because the customized hardware fits the application better. It might offer an instruction-level parallelism (ILP) mix that is consistent with the application, it might do certain critical operations faster by implementing them directly in hardware, or it might cycle faster because its critical path is simplified when unnecessary functionality is removed.

- Die size and cost can decrease because functionality that is not important for the application could be removed (e.g., emulated in software, such as what might happen to the floating point unit for an integer-dominant application) or could be implemented with slower, size-optimized circuitry.

- Power can decrease because of the same reasons mentioned in the other two points: smaller hardware, more optimized, with less "redundant" functionality, consumes less power. This is especially true as we move to smaller geometries where static (leakage) power begins to contribute to larger and larger fractions of the total processor power budget.

On the other hand, scaling and customizing have a number of hidden costs, some of which are so severe that they make customization rare in the general-purpose computing world. Code incompatibility arises when the changes are visible in the ISA, and programs must be recompiled to run on a new, customized processor. This disadvantage is unacceptable in many environments, though techniques that solve this problem are beginning to mature. Software toolchain development is expensive, and a new ISA likely requires a new toolchain to support it. User unfamiliarity with the visible aspects of the new ISA can cause the cost of using it to be severe. And, finally, a customized processor might be used in fewer applications, thus lowering potential target volumes and increasing per-part manufacturing costs, although the economics of SoCs can change this equation in favor of customization.

Sometimes these disadvantages are critical and make customization impossible. When they can be overcome, it is often only with great difficulty. Nonetheless, there is a strong temptation to customize and scale in the embedded world, because a processor very often runs only one or a few applications. When that happens, customizing can yield a very substantial gain.

In this chapter, we play devil's advocate and present a critical view of the area of processor customization, based on our long experience in the arena (which led to the design of the Lx/ST200 family of VLIW embedded cores). We begin by reviewing some of the motivations and achievements of the "Custom-Fit Processors" (CFP) project at HP Labs and continue by discussing a set of lessons that we learned during the project that point to the obstacles (not always technical) that have slowed the adoption of customizable processors. Note that we largely refer to socio-technological issues, rather than the challenges of automating the customization, which is well covered in Chapter 8.

3.1 THE "CFP" PROJECT AT HP LABS

In 1994, when HP Labs[1] decided to start the CFP project [2–5], the goals were very ambitious: develop a technology that could generate an individual custom-fit VLIW processor for every embedded product that HP sold, together with the corresponding set of compilers and tools needed to operate it.

The time was ripe for these kinds of processors for three important reasons. First, the cost of silicon had become sufficiently small that one could afford to build a moderately parallel VLIW (4–8 issue) in the area typically allocated for high-end embedded products (printers, digital cameras). Second, the complexity of building and verifying ASICs (and the software to use them) had grown so much that it was becoming by far the most critical component in the time to market of a new product generation. And third, high-performance embedded computing had entered the mass market, where market-driven changes were rapid and unpredictable.

This combination of factors made the advantages of using programmable, configurable, and customized high-performance processors very appealing to everyone who was part of the decision chain (and user chain) of the electronics platform of an embedded product. Managers could reduce time to market and exploit multilevel reuse in future generations. Hardware engineers could apply their expertise to the customization process without suffering the horrors of ASIC hardware and timing verification. Firmware/software writers would benefit from targeting a platform that could last across several product generations (or several products), increasing the level of software reuse.

---

[1] For the sake of brevity, this section sometimes uses "we" to refer to two of the authors of this chapter (Faraboschi and Fisher), who initiated the CFP project at HP Labs.

These messages were so powerful that the CFP research ideas very quickly transformed from an HP Labs research project to a product-oriented development project. The CFP technology became the basis of the HP-STMicroelectronics partnership, which eventually led to the development of the Lx/ST200 family of VLIW embedded cores [2].

From a research standpoint, the technology behind CFP was based on three cornerstones: (1) tools to automatically generate compilers and associated compilation tools, (2) tools to automatically generate processor architectures, and (3) tools to quickly identify the "unique" processor representing the best fit for a given application. The CFP project attacked and made tremendous progress on all three.

Concerning compilers and tools, we believe that the most successful approaches transported the ideas behind the development of the VLIW compilers of the 1980s to the embedded multimedia world. In particular, the CFP project started from HP's descendant of the Multiflow C compiler, heavily modified to fit the target domain. The new "Lx" compiler was capable of generating code for a very wide space of VLIW architecture variants, simply by changing a few configuration parameters in what we called the "machine model tables." This allowed us to quickly assess the performance of literally thousands of different machine configurations on every application that we wanted to optimize. Given that compiler reconfiguration took only seconds, a brute-force search (with very simple pruning) would in many cases find the best configuration in a matter of hours or days. Note that, in addition to the compiler, an important component of this technology was a fast and accurate architecture simulator driven by the same set of information used to retarget the compiler.

Through this iterative configuration-compilation-simulation process (see Fig. 3.1), the CFP project showed that one could automatically find the set of nearly optimal ISA-visible parameters for a *fixed* application. In the case of CFP, ISA-visible parameters include the number and type of functional units and registers, the latency of operations, the VLIW issue width, and the most important parameters of the memory architecture. In other words, this approach could quickly reach a full *architecture* specification for a processor tuned to an application described through its C implementation.

From the architecture specification and using a *microarchitecture template*, it is then possible to quickly generate RTL in Verilog or VHDL. This automatically generated RTL would include the full semantics of the decoding circuitry and of the functional blocks, as well as the basic pipelining information. Even though the CFP project did not perform this step automatically, other projects, such as HPL's PICO [6–8], Tensilica [9–11], and ARC Cores [12], showed that this endeavor was well within reach. Together with the tremendous advances of the high-level synthesis techniques developed in the CAD community, this means that one could automatically construct a synthesized netlist for a customized processor with minimal human intervention.

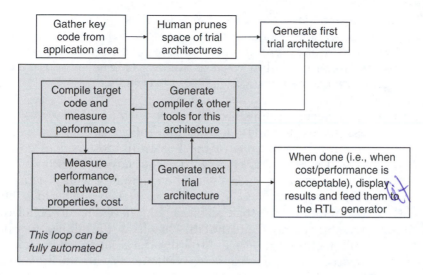

■ **FIGURE 3.1**

Automatic customization. Several research groups advocate an automated approach to customization, based on a loop similar to what the picture illustrates. While many of the techniques have matured (retargetable compilers, architecture space exploration, and so forth), full automation in a realistic environment is a far away goal (derived from [1]).

Note that, so far, we have only discussed *scalability*, which was the primary focus of the CFP project. *Customizability* was the target of several other (academic and industrial) projects, most notably the Tensilica Xtensa architecture [9–11], together with other more DSP-oriented approaches such as ARC Cores [12] and Improv System's "Jazz" [13]. Customizability involves a larger initial manual effort, because today only domain experts can figure out what new operations are truly profitable for an application. The role of the compiler, albeit important, is mostly relegated to mapping language idioms into these special-purpose instructions, the trivial case being the use of explicit *intrinsics*. On the other hand, in the case of scalability, the challenge is to develop a compiler that can find large amounts of ILP in the application, even in the presence of a configuration layer. While companies like Tensilica attacked (and solved) the customizability problem, we at HPL attacked (and solved) the scalability problem.[2]

In summary, with the technology developed in the CFP project and other projects' efforts [14–18] and its early success, we initially thought

---

[2] Subsequently, Tensilica and others have also added some scalability (in the form of ILP and short-SIMD vector parallelism) to their products. Conversely, the Lx/ST200 also permits some degree of manual customizability.

that we had paved the way for a long list of successful customized VLIW processors. Looking back 10 years later, we can see that reality has turned out differently than we expected. In practice, what we have observed can be summarized as follows (we will discuss the reasons later; here we simply state what we observed):

- Even when architectures support customizability and scalability, users are reluctant to fully exploit this capability. For example, the five implementations of the Lx family designed to date (ST210, ST220, ST230, ST231, ST240) are all largely ISA-compatible (with very minor differences) with the original, 4-wide, 1-cluster ST210. This happened despite the fact that the initial architecture and compilation tools had all the hooks necessary to achieve smooth scalability and configurability, such as the ability to extend the architecture to multiple clusters or the ability to add special-purpose instructions. What happened is that users were mostly satisfied with the performance and cost of a single cluster, and this caused the tools to slowly (but inexorably) lose their scalability and customizability features.

- The maintenance effort for a processor toolchain is huge and does not scale nicely. Even for a "captive" developer community, as you might find in a typical high-volume embedded processor, you need a large team to develop and support each instance of the compilation-simulation toolchain. In addition to that, every processor variant requires changes (and verification) in all the supported operating systems (OSs), some of the heavily tuned application kernels, and so on.

- Once you have the fully verified, synthesizable initial RTL of a processor, it still takes a very long time and a very large effort to reach a tuned, releasable core, ready to be seamlessly integrated into an SoC. This implies that the lifetime of a processor (customizable or not) needs to grow accordingly, and to be planned well ahead of time over a product roadmap. As this occurs, the appeal of customization (because of broader applicability and less target predictability) consequently diminishes.

In what we described here, we (deliberately) omitted a few areas, such as the tuning of the applications, the tuning of the microarchitecture, the overall hardware and software verification effort, the ramp-up effort of a new processor, and what it really means to target a *fixed* application. These are the areas that we believe have never really been addressed in the context of customized processors and that constitute the true fundamental barriers. We also believe that the consequences are deep and long-lasting, and that the *one processor per application* view will be neither practical nor economical for a long time to come. This does not mean that there is no space for customized processors, but it implies that we may only be able to successfully exploit *one processor per domain*, assuming that we can efficiently leverage tool and application development. In the rest of this chapter, we discuss each of these areas individually to substantiate our view.

3.2 SEARCHING FOR THE BEST ARCHITECTURE IS NOT A MACHINE-ONLY ENDEAVOR

As we described, if we can start from a *fixed* application, we have shown that we can build an automated procedure that will find the "best" architecture according to a certain cost function. Unfortunately, the task of defining the application is much harder than it seems. This should be easy in an embedded system—after all, it is being built to do one specialized thing over and over again. Sadly, while embedded workloads are easier to characterize than general-purpose ones, the characterization task itself remains complicated. Two different factors make characterization hard.

First, the idea of having in hand, well in advance, the application that will run on the processor is only a dream (see Fig. 3.2). In a typical product lifecycle, several months pass between the selection of a core and the shipment of a product incorporating the selected embedded core. During this period of time, the application developers change the application at will, being motivated by changes in market condition, emerging or changed standards, improved algorithms, or even caprice. Worse yet, in many environments, applications can change long after the product is shipped. This is increasingly common because of the prevalence of Flash-based firmware that enables vendors to update and upgrade the capabilities of devices in the field simply by providing a new version that can be downloaded from their Web site.

Second, just as ISAs can be changed to suit an application, the application can be changed to suit the ISA (see Table 3.1 for a comparison of DCT algorithms). A typical scenario has the application programmer looking at the derived custom ISA and saying "What?! You're willing to give me TWO integer multipliers? If I had known that, I would have used such-and-such algorithm and rewritten that critical kernel." No known automation can begin to capture this process. The important lesson is this: an exploration system must have the interactive services of an application domain expert.

Finally, if we go back to Fig. 3.2, something else that is becoming a necessity is the availability of a reference hardware platform. These have been provided for commercial off-the-shelf (COTS) processors for many generations, but recently even the vendors of embeddable processor cores have begun to provide reference hardware platforms. This

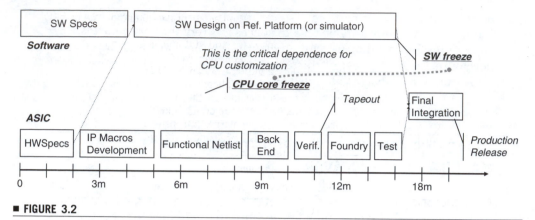

■ FIGURE 3.2

Embedded product lifecycle. The graph shows a typical design lifecycle of a complex embedded product over an approximate 18-month development period. The bottom line shows a very simplified breakdown of the phases of an ASIC design, assuming that the ASIC includes one or more CPU cores together with other IP blocks. The top line shows application software development. The problem of customizing the CPU core comes from the fact that the application is still evolving by the very early time we have to freeze the CPU core design. Waiting for the application to settle would add a dependence between the end of the software development and the start of the ASIC netlist design. This would be a time-to-market killer and is a choice that no engineering manager ever has the luxury of making in the real world.

is because their customers want to reduce the development effort of the application programmers, and they also want to enable concurrent engineering of software and hardware by decoupling them to the largest possible extent. The concept of a reference hardware platform for a customizable processor is a bit of an oxymoron, because of the changing nature of the customization. Software simulation (perhaps accelerated by expensive accelerators) may be the only option, alas much less effective and trusted than a real evaluation board.

3.3   DESIGNING A CPU CORE STILL TAKES A VERY LONG TIME

One of the major achievements of CAD tools in the last decade is that highly complex RTL descriptions can be synthesized to a netlist of gates with a few mouse clicks. A well crafted synthesis recipe can produce a reasonably efficient circuit for an average ASIC design, and it is common practice among skilled ASIC designers to reach timing closure after a few iterations of tuning of the design. However, timing closure is rarely attained quickly when synthesizing a processor,[3] and it is even more

---

[3] Note that the timing problems appear only for aggressive clock cycle targets, but these are really necessary to achieve a competitive CPU design. A VLSI designer being asked the question "can you fit this operation in one cycle?" will usually answer "yes, definitely," without of course committing to the duration of that cycle.

TABLE 3.1 ■ Comparison of DCT Algorithms shows a comparison of the computational complexity of a handful of known DCT algorithms, expressed in number of operations (multiplications and additions) for a 2D-DCT of an 8×8 block (typical of many video and still-imaging compression applications). Even for a well known algorithm such as DCT, customization would produce very different results, depending on the technique we start with. For example, you can see that the ratio between additions and multiplications varies dramatically from 1024 multiplications all the way to a multiplication-free technique [19–21]. The "best" custom processor for one of these would be very different from the others, and we claim that—today—no automated technique could "intelligently" pick the optimal. The problem is that these kinds of choices are not combinatorial, but instead require human intelligence and knowledge of the target application that can only be found in domain experts.

Algorithm	2D (row-column)		2D (direct)	
	Multiplications	Additions	Multiplications	Additions
Theoretical	1024	896	512	682
Ligtemberg and Vetterli	208	464	104	466
Arai, Agui, Nakajaima	208	464	104	466
Arai, Agui, Nakajaima with quant. table	144	464	n/a	n/a
Chen and Fralick	512	512	256	490
Chen and Fralick using DA	0	1024	0	746

difficult when the processor is a core offering significant amounts of ILP. Completely explaining the reasons behind timing problems goes beyond the scope of this chapter, but some of the areas that are not friendly to synthesis include the heavily multiported register file, the messy interconnection network among the computational units, and the delicate balance among the pipeline stages.

In practice, attempting to synthesize a VLIW core (customizable or not) out of a naïve RTL description (such as what could be produced by an automatic generation tool) gets you a noncompetitive design in terms of both area and speed.

The consequence is that, in order to get to a reasonably optimized microarchitecture, a lot of manual work is still necessary, even for a "simple" VLIW that includes only basic architectural features. For example, the design of a balanced pipeline is largely manual (and, unfortunately, technology dependent in many cases), and so is the optimization of the register file array, the placement of the functional units, the architecture of the forwarding (bypass) logic, and so on. What is even worse for customization is that it normally takes two or three design generations for a microarchitecture to "mature" and reach a stable state where pipeline stages are balanced and match the "sweet spot" of the implementation technology.

Finally, it is important to remember that, unfortunately, the design (including tuning) time is only a small fraction of the overall lifecycle of a processor core. Verification alone takes 60% or more of the project time and resources (and even longer for deep submicron designs). One number to keep in mind is the so-called *null change turnaround time*, which is the time it takes for a design to go through the entire timing closure and verification phases without any change in the RTL. It is not uncommon for this number to be on the order of four to six weeks, indicating the massive inertia of a complex design such as a high-performance core.

If we add all of these concerns together, it should become obvious that the *one CPU per product* view is deeply flawed, and that the design of a CPU core must necessarily live much longer than the lifetime of a single product. At least, the microarchitecture "template" and the synthesis recipes need to be planned so that they can be reused for several generations with only minor changes across them. Again, this works against customization and favors the adoption of fewer processor variants, where the variations happen far away from the computational engine.

3.4   DON'T UNDERESTIMATE COMPETITIVE TECHNOLOGIES

Many engineers with computer-science backgrounds approach embedded problems by adding "stuff" to the main system processor, with the goal of migrating most of the computation onto that processor. In this case, the idea of customizing the processor so that the special tasks at hand can be executed much more efficiently has much appeal. However, in many cases, there are other—perhaps less elegant—approaches that might be more efficient, cheaper, and faster to market. In other words, it is common for advocates of the "customized processor" approach to underestimate the competition that comes from other architectures that might be applied to the same problem.

Without claiming to be exhaustive, we cover two competing approaches here: heterogeneous multiprocessing and nonprogrammable accelerators.

Heterogeneous multiprocessing, in this context, means throwing multiple, simpler processors at the problem and having them work independently on completely different aspects of the embedded application, often operating at completely different times. At first glance, this might seem counterintuitive. For example, why would I want to use two processors for a task that I could do with one more powerful processor? The reasons come from legacy software and time-to-market considerations. Cases where a product gets designed from scratch are rare, and often products evolve from previous generations. One common form of evolution integrates formerly separated functions into a single piece of silicon. Another form adds programmability to the original functionality by replacing hardwired logic with a flexible microprocessor. In both cases, the functionality already exists and is well encapsulated in a stand-alone piece of firmware, so it would take longer to reengineer it as part of the application running on the main processor than to add an extra processor dedicated to it. Moving everything to the main CPU involves revisiting the thread model, dealing with real-time considerations (if any), and verifying that no nasty interactions (e.g., races or hazards) exist with the newly added functionality. Some of these considerations are the basis of the success of heterogeneous platforms today. In many cases, even VLIW fanatics like us admit that using multiprocessing is easier than tweaking the application to widen the available instruction-level parallelism. Multiprocessing also helps to separate software components whose coexistence might be problematic with the rest of the code (such as real-time tasks).

Nonprogrammable logic (i.e., "good old" ASIC design, in modern SoC terminology) is also a formidable competitor to customizable processors. A dedicated hardware pipeline executing a single task is usually at least one order of magnitude more efficient, faster, and less power-hungry than an equivalent implementation executed on a programmable processor. This is true, for example, of imaging pipelines and of compression/decompression engines. Of course, hardware pipelines are not flexible, require a long time to design, and have several other disadvantages. However, when they are a good fit for an application, they tend to be unbeatable, and replacing them with a programmable processor can be suicide.

Too often advocates of customizable processors argue against the competition without understanding that in some cases it is a losing battle. As we discuss in the next sections as well, selling flexibility is tough, and only a few domains are ready to make the jump to a fully programmable and clean platform (and pay the upfront price for the transition). An alternate, more effective approach would build customizable processors that coexist with nonprogrammable blocks and other processors, including ensuring tools interoperability with them. Providing an easy and gradual migration path (as opposed to a full replacement) is the only viable way to lower the entry barrier to a new embedded (customizable or not) microprocessor. (Advocates of the programmable approach can think of this pejoratively as a "viral" approach. Get your foot in the door and then gradually suck up more and more functionality.)

3.5 SOFTWARE DEVELOPERS DON'T ALWAYS HELP YOU

One of the foundations behind the idea of customizable processors is that the application at hand is well tuned and optimized, so that automatic analysis tools can find the "best" architecture for it. The implicit underlying hypothesis is that the software development team in charge of the application has spent some time to optimize the code base for performance, by keeping in mind the idiosyncrasies of the architecture target and corresponding compiler.

Unfortunately, the real world couldn't be more different, and it is very easy to overestimate the performance tuning efforts that an average software team will put in the development of an embedded application. The reason is—once again—time to market. When the pressure is on, and requests for new features or bug fixes come in, performance tuning usually gets put aside, is allocated as the last phase of the project, and often never happens because of other delays in deliverables. It is, for example, not uncommon to have large parts of the code compiled with debugging flags, simply because changing the *makefile* would imply reverifying the entire code base, possibly exposing latent application bugs that halt the entire project. Sadly, basic *fear of failure* dominates much of software development, and spending time to optimize the code for an obscure compiler or architecture feature tends to have a very low priority in the mind of an average developer.

Even simple optimizations that can potentially yield major gains in ILP processors (such as the use of C99's *restrict* keyword to help alias analysis or other hints of the same nature) are fiercely resisted and rarely used in practice. The appearance of higher-level languages in the embedded world (such as Java, C++, or Matlab) makes this worse, because it moves the semantics of the application away from the architecture, making it difficult to pinpoint fine-grain events such as where a memory access really occurs.

As a consequence, compiler performance on "generic" code becomes much more important, even in embedded applications, than one could picture in an ideal world. To exploit advanced architecture and/or compiler features, optimized libraries are likely to be a necessary evil.[4] Unfortunately, libraries do not capture the full expressiveness of an architecture, and many applications may not take advantage of them because of small mismatches or the steep learning curve that they require.

Finally, a rarely discussed aspect of legacy code is the allegiance of software developers to a development environment. New toolchains are usually feared and—sadly—often for good reasons. New toolchains tend to have higher defect rates than long-established compilers, and one might argue that a compiler for a customized processor is *always* a new compiler. Thus, developers' resistance to adopt new compiler platforms is understandable, even when the compiler works perfectly, and is a fact that cannot be ignored. As a consequence, the availability of a

---

[4] In general, we don't view application-specific libraries as an answer, mostly because they often overconstrain the developers to fit an application into a set of interfaces that were designed with a different purpose in mind. However, we acknowledge the value of optimized libraries, especially when they allow the use of an architecture feature that would otherwise not be exploited by an average programmer using minimal compiler optimizations.

proven and familiar toolchain, even if less optimized, is often a *must*. For example, the availability of ubiquitous *gcc* (the GNU compiler) already causes many developers to take a deep sigh of relief. *Backward compatibility* of tools is hence a plus: being able to replicate flags, dialect, and build environment is quickly becoming a necessity to ease the adoption of a new compiler and architecture.

3.6 THE EMBEDDED WORLD IS NOT IMMUNE TO LEGACY PROBLEMS

It is well known that the embedded world is largely immune to binary compatibility issues, which is one of the reasons for the success of VLIW processors in the embedded world. However other more subtle forms of compatibility get in the way.

Most embedded developers have strong allegiance to an existing embedded OS and will not switch easily. For real-time products, the interaction between the application and the real-time OS (RTOS) is closely guarded knowledge that requires many years of hard development to acquire.

Supporting several OSs on a customizable processor family can present significant engineering challenges. For example, the architecture-dependent code (such as context switching) depends on the state of the processor, which could be affected by customization. This adds complexity and risk, and increases the verification effort (and certification if needed) to ensure that a new customization is compatible with a given OS and version. Even PocketPC, which began life as Windows CE with support for multiple architectures, has narrowed its support over time to only the ARM platform.

The set of legacy issues does not stop at operating systems. Libraries, platform-specific quirks (e.g., how to program a real-time clock), endianness, assembler inserts, even specific compiler versions, are some of the other areas that need to be addressed when proposing a continuously changing processor instruction set.

Many of these issues can be solved through the notion of an architecture family sharing a base instruction set that is fixed across the multiple customized versions of the same processor (see Fig. 3.3). However, the temptation to use features that are in the currently targeted processor but are not part of the base ISA is strong. It requires much software engineering discipline to avoid getting trapped into using extended features that may be dropped in other variants, and unfortunately the time-to-market pressure is a very strong element working against this discipline.

A consequence of these legacy issues is that in many cases the presence of the "dominant" RISC processor (e.g., ARM) is mandatory for a given embedded product to address compatibility requirements. The good news is that the area of a simple scalar RISC is almost negligible in deep submicron geometries. The bad news is that this phenomenon even more strongly favors the "accelerator" model, where the customized CPU is used as a companion programmable accelerator in charge of offloading the main RISC of the most computing-intensive tasks.

If we look a bit further ahead, we can see dynamic binary translation techniques (such as the DELI dynamic execution system [22]) as a possible workaround for many of the legacy issues. Using a binary translator enables the older code base to be usable (perhaps at a small performance premium) right away. This gives more time to the software developers to gradually migrate to the new native ISA over an extended period of time without the need for an abrupt transition. We believe that the dynamic translation techniques that have matured in the general-purpose domain (e.g., Transmeta [23]) might gradually find their way into the embedded world to free the developers from some of the compatibility issues that may appear unsolvable today.

3.7   CUSTOMIZATION CAN BE TROUBLE

The success of an embedded product is often inversely proportional to its time to market, and hardware verification is by far the most domi-

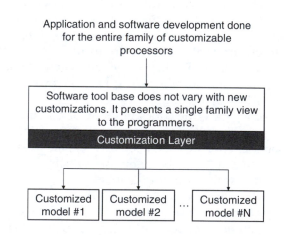

■ FIGURE 3.3

Processor families. Many issues about OS and application portability for customizable processors could be addressed by the notion of a "family" of processors. In a family of processors, there is a lot of commonality among members of the family that share a base instruction set. Application and OS development can be done once, with little or no change required other than recompiling to run the code correctly and with good performance on a new family member, even though the new core is object-code incompatible with the old one. Sometimes additional work is desirable in some application kernels to take advantage of the customization, however. This new work may take the form of changed compiler command-line switches, changes in unrolling amounts or the addition of pragmas, or, in the worst case, rewriting of some critical routines (derived from [1]).

nant component of time to market. Hence, anything that might increase verification risk will likely be shunned by product managers. Customizable processors unfortunately fall into this category, and the fact that they may even be generated by some obscure configuration tool makes them the first component to suspect as the cause of a system failure.

In other words, while engineers and designers love the concept of being able to tune a processor, product managers are usually horrified by that same concept. The whole idea of letting developers play around with the customization of a processor is perceived as fragile and incomplete, and something that is potentially a black-hole time sink. What the creators of a customizable family see as the most important differentiating feature might be interpreted as a liability by the marketplace.

Indeed, configuring a processor takes time, is a source of endless debates among the members of the engineering teams, and can make project deadlines slip. This reasoning is somewhat flawed, because ultimately you can end up with a better design that will make your life easier down the road. However, we have observed this reaction repeatedly, so it is important not to underestimate it, especially in the marketing and positioning messages for a customizable family.

As for the CPU vendor, the situation is not much better. As we mentioned, the entire issue of maintaining and supporting a family of customizable toolchains is an unsolved problem. Customers tend to be sensitive to these arguments and would look with suspicion on a vendor that does not address their legitimate support concerns. The basic problem is that—for the way in which a toolchain is maintained and verified today—every new customized CPU variant is practically a new processor, with a new toolchain. Now, if you know how many people (and how much CPU time) it takes to verify a single toolchain for a single CPU, it should be obvious that scaling that effort linearly is just not economically viable. Thus, until some new software engineering breakthrough happens that will allow quick verification and support of similar (but not equal) toolchains, this remains a major obstacle to overcome.

The consequences of these considerations are the usual ones: customization is constrained to be much less aggressive than one can envision and than architecture techniques allow. In the extreme, it is often the case that customization is normally hidden from the end user, and it is really only used in the product definition phase, long before any real design happens, as a guideline for the selection of a similar CPU.

3.8 CONCLUSIONS

In this chapter we have presented a critical, perhaps controversial, view of customizable processors. Through an analysis of our decade-long experience in developing, advocating, and using custom-fit VLIW cores,

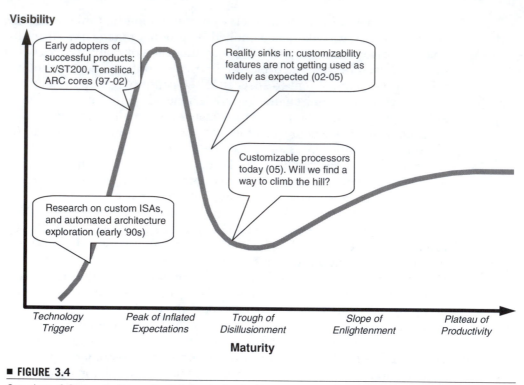

Visibility

Early adopters of
successful products:
Lx/ST200, Tensilica,
ARC cores (97-02)

Reality sinks in: customizability
features are not getting used as
widely as expected (02-05)

Customizable processors
today (05). Will we find a
way to climb the hill?

Research on custom ISAs,
and automated architecture
exploration (early '90s)

Technology   Peak of Inflated   Trough of   Slope of   Plateau of
Trigger   Expectations   Disillusionment   Enlightenment   Productivity

Maturity

■ **FIGURE 3.4**

Our view of Gartner's "Hype Cycle" for customizable processors. From the early successes of the late
1990s, we believe we are now quickly approaching the "trough of disillusionment" of the technology,
where many nontechnical barriers prevent benefits from being realized. Now that the major feasibility
questions have been answered, the big challenge for researchers and developers in the years to come
will be to address these other barriers.

we have shown where we think some of the barriers to customizations
are. It is important to observe that only some of these barriers can be
addressed through technical progress. Other barriers are deeply rooted
in the nature of the development cycle of embedded products and as
such may be much harder to overcome. While we arrived at these con-
clusions in the context of customizable VLIW cores, we believe they
apply much more widely and that many of them apply to all customizable hardware. We would advise anyone working in the field to be mind-
ful of these real-world effects.

We, like the other authors of this book, remain enthusiastic about the
elegant, efficient, and powerful technical solutions for scalability and
customizability. Our reservations are with the transfer from research
into products: in the financial, time-to-market, product lifecycle, and
marketing issues with customizable processors. The best research will
not only be technically sweet, it will also be practically engaged.
If we use Gartner's "Hype Cycle" [24] to illustrate the maturity of a

technology, we believe that customizable processors are now heading quickly toward the "trough of disillusionment" and that it will take a large effort to be able to climb back the hill toward maturity (Fig. 3.4).

We also believe that, in order to do so, it may be more effective to abandon the "one processor per application" idea, reaching a compromise of "one processor per application *domain*." Doing so will help to put aside some of the hardware engineering problems and will free up resources to attack the most fragile pieces of the technology. Compilers, tools, operating systems, and libraries for customizable platforms still need much work before the developers' community is ready to accept them. Making them more robust and amenable to coexist with other competing approaches (such as heterogeneous multiprocessing and nonprogrammable accelerators) will go a long way in the direction of making customizable processors a successful technology.

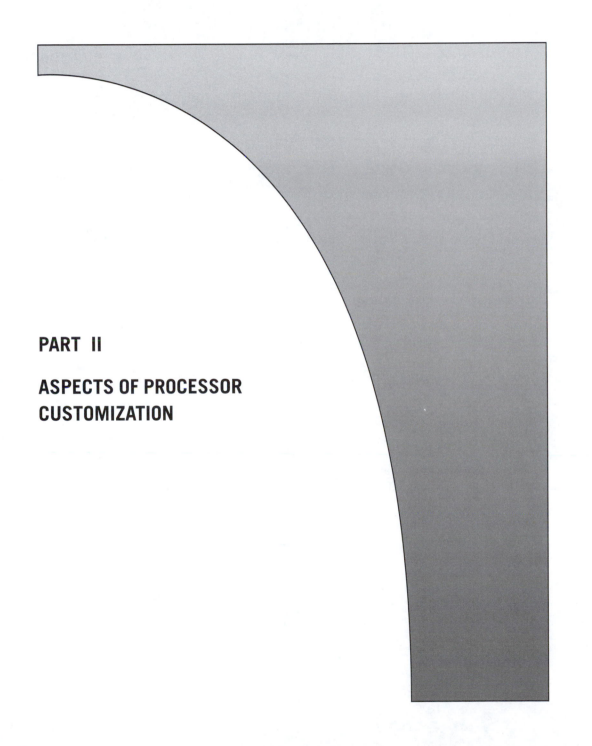

PART II

**ASPECTS OF PROCESSOR
CUSTOMIZATION**

ARCHITECTURE DESCRIPTION LANGUAGES

Prabhat Mishra and Nikil Dutt

Modeling plays a central role in the design automation of embedded processors. It is necessary to develop a specification language that can model complex processors at a higher level of abstraction and enable automatic analysis and generation of efficient tools and prototypes. The language should be powerful enough to capture high-level descriptions of the processor architectures. On the other hand, the language should be simple enough to allow correlation of the information between the specification and the architecture manual.

Architecture description languages (ADL) enable design automation of embedded processors as shown in Fig. 4.1. The ADL specification is used to generate various executable models, including simulator, compiler, and hardware implementation. The generated models enable various design automation tasks, including exploration, simulation, compilation, synthesis, test generation, and validation. Chapter 5 describes retargetable software tools for embedded processors. This chapter reviews the existing ADLs in terms of their capabilities in capturing a wide variety of

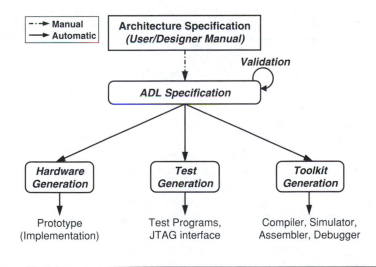

■ **FIGURE 4.1**

ADL-driven design automation of embedded processors.

embedded processors available today. Existing ADLs can be classified based on two aspects: content and objective. The content-oriented classification is based on the nature of the information an ADL can capture, whereas the objective-oriented classification is based on the purpose of an ADL. Existing ADLs can be classified into various content-based categories, such as structural, behavioral, and mixed ADLs. Similarly, contemporary ADLs can be classified into various objective-oriented categories, such as simulation oriented, synthesis oriented, test oriented, compilation oriented, and validation oriented. This chapter is organized as follows: Section 4.1 describes how ADLs differ from other modeling languages. Section 4.2 surveys the contemporary ADLs. Finally, Section 4.3 concludes this chapter with a discussion on expected features of future ADLs.

4.1   ADLs AND OTHER LANGUAGES

The phrase "architecture description language" has been used in the context of designing both software and hardware architectures. Software ADLs are used for representing and analyzing software architectures [1]. They capture the behavioral specifications of the components and their interactions that comprise the software architecture. However, hardware ADLs capture the structure (hardware components and their connectivity) and the behavior (instruction-set) of processor architectures. The concept of using machine description languages for specification of architectures has been around for a long time. Early ADLs such as ISPS [2] were used for simulation, evaluation, and synthesis of computers and other digital systems. This chapter surveys contemporary hardware ADLs.

How do ADLs differ from programming languages, hardware description languages, modeling languages, and the like? This section attempts to answer this question. However, it is not always possible to answer the following question: given a language for describing an architecture, what are the criteria for deciding whether or not it is an ADL? Specifications widely in use today are still written informally in natural languages such as English. Since natural language specifications are not amenable to automated analysis, there are possibilities of ambiguity, incompleteness, and contradiction—all problems that can lead to different interpretations of the specification. Clearly, formal specification languages are suitable for analysis and verification. Some have become popular because they are input languages for powerful verification tools such as a model checker. Such specifications are popular among verification engineers with expertise in formal languages. However, these specifications are not acceptable by designers and other tool developers. An ADL specification should have formal (unambiguous) semantics as well as easy correlation with the architecture manual.

In principle, ADLs differ from programming languages because the latter bind all architectural abstractions to specific point solutions, whereas ADLs intentionally suppress or vary such binding. In practice, architecture is embodied and recoverable from code by reverse-engineering methods. For example, it might be possible to analyze a piece of code written in C and figure out whether or not it corresponds to *Fetch* unit. Many languages provide architecture-level views of the system. For example, C++ offers the ability to describe the structure of a processor by instantiating objects for the components of the architecture. However, C++ offers little or no architecture-level analytical capabilities. Therefore, it is difficult to describe architecture at a level of abstraction suitable for early analysis and exploration. More importantly, traditional programming languages are not a natural choice for describing architectures, due to their inability to capture hardware features such as parallelism and synchronization. ADLs differ from modeling languages (such as UML) because the latter are more concerned with the behaviors of the whole rather than the parts, whereas ADLs concentrate on representation of components. In practice, many modeling languages allow the representation of cooperating components and can represent architectures reasonably well. However, the lack of an abstraction would make it harder to describe the instruction set of the architecture. Traditional hardware description languages (HDL), such as VHDL and Verilog, do not have sufficient abstraction to describe architectures and explore them at the system level. It is possible to perform reverse engineering to extract the structure of the architecture from the HDL description. However, it is hard to extract the instruction-set behavior of the architecture. In practice, some variants of HDLs work reasonably well as ADLs for specific classes of embedded processors.

There is no clear line between ADLs and non-ADLs. In principle, programming languages, modeling languages, and hardware description languages have aspects in common with ADLs, as shown in Fig. 4.2. Languages can, however, be discriminated from one another according to

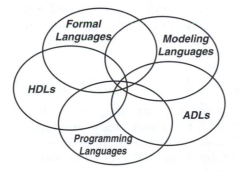

■ **FIGURE 4.2**

ADLs versus non-ADLs.

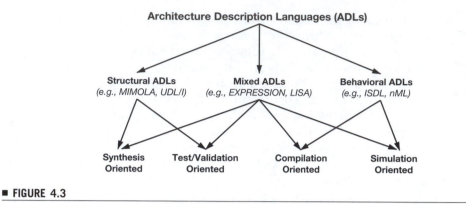

■ **FIGURE 4.3**

Taxonomy of ADLs.

how much architectural information they can capture and analyze. Languages that were born as ADLs show a clear advantage in this area over languages built for some other purpose and later co-opted to represent architectures.

4.2   SURVEY OF CONTEMPORARY ADLs

This section briefly surveys some of the contemporary ADLs in the context of designing customizable and configurable embedded processors. There are many comprehensive ADL surveys available in the literature, including ADLs for retargetable compilation [3] and SOC design [4]. Fig. 4.3 shows the classification of ADLs based on two aspects: *content* and *objective*. The content-oriented classification is based on the nature of the information an ADL can capture, whereas the objective-oriented classification is based on the purpose of an ADL. Contemporary ADLs can be classified into four categories based on the objective: simulation oriented, synthesis oriented, compilation oriented, and validation oriented. It is not always possible to establish a one-to-one correspondence between content-based and objective-based classification.

4.2.1   Content-Oriented Classification of ADLs

ADLs can be classified into three categories based on the nature of the information: structural, behavioral, and mixed. The structural ADLs capture the structure in terms of architectural components and their connectivity. The behavioral ADLs capture the instruction-set behavior of the processor architecture. The mixed ADLs capture both structure and behavior of the architecture. This section presents the survey using content-based classification of ADLs.

Structural ADLs

There are two important aspects to consider when designing an ADL: level of abstraction versus generality. It is very difficult to find an abstraction to capture the features of different types of processors. A common way to obtain generality is to lower the abstraction level. Register transfer level (RT level) is a popular abstraction level—low enough for detailed behavior modeling of digital systems and high enough to hide gate-level implementation details. Early ADLs are based on RT-level descriptions. This section briefly describes a structural ADL: MIMOLA [5].

MIMOLA

MIMOLA [5] is a structure-centric ADL developed at the University of Dortmund, Germany. It was originally proposed for microarchitecture design. One of the major advantages of MIMOLA is that the same description can be used for synthesis, simulation, test generation, and compilation. A tool chain that includes the MSSH hardware synthesizer, the MSSQ code generator, the MSST self-test program compiler, the MSSB functional simulator, and the MSSU RT-level simulator was developed based on the MIMOLA language [5]. MIMOLA has also been used by the RECORD [5] compiler. MIMOLA description contains three parts: the algorithm to be compiled, the target processor model, and additional linkage and transformation rules. The software part (algorithm description) describes application programs in a PASCAL-like syntax. The processor model describes microarchitecture in the form of a component netlist. The linkage information is used by the compiler to locate important modules, such as program counter and instruction memory. The following code segment specifies the program counter and instruction memory locations [5]:

```
LOCATION_FOR_PROGRAMCOUNTER PCReg;
LOCATION_FOR_INSTRUCTIONS IM[0..1023];
```

The algorithmic part of MIMOLA is an extension of PASCAL. Unlike other high-level languages, it allows references to physical registers and memories. It also allows use of hardware components using procedure calls. For example, if the processor description contains a component named MAC, programmers can write the following code segment to use the multiply-accumulate operation performed by MAC:

```
result := MAC(x, y, z);
```

The processor is modeled as a netlist of component modules. MIMOLA permits modeling of arbitrary (programmable or nonprogrammable) hardware structures. Similar to VHDL, a number of predefined, primitive operators exists. The basic entities of MIMOLA hardware models are modules and connections. Each module is specified by its port interface and its behavior. The following example shows the description of a multifunctional ALU module [5]:

```
MODULE ALU
  (IN inp1, inp2: (31:0);
  OUT outp: (31:0);
  IN ctrl;
  )
CONBEGIN
    outp <- CASE ctrl OF
      0: inp1 + inp2 ;
      1: inp1 - inp2 ;
      END;
CONEND;
```

The **CONBEGIN/CONEND** construct includes a set of concurrent assignments. In the example, a conditional assignment to output port *outp* is specified, which depends on the one-bit control input *ctrl*. The netlist structure is formed by connecting ports of module instances. For example, the following **MIMOLA** description connects two modules: *ALU* and accumulator *ACC*.

```
CONNECTIONS ALU.outp -> ACC.inp
            ACC.outp -> ALU.inp
```

The MSSQ code generator extracts instruction-set information from the module netlist. It uses two internal data structures: connection operation graph (COG) and instruction tree (I-tree). It is a very difficult task to extract the COG and I-trees, even in the presence of linkage information, due to the flexibility of an RT-level structural description. Extra constraints must be imposed for the MSSQ code generator to work properly. The constraints limit the architecture scope of MSSQ to microprogrammable controllers, in which all control signals originate directly from the instruction word. The lack of explicit description of processor pipelines or resource conflicts may result in poor code quality for some classes of VLIW or deeply pipelined processors.

Behavioral ADLs

The difficulty of instruction-set extraction can be avoided by abstracting behavioral information from the structural details. Behavioral ADLs explicitly specify the instruction semantics and ignore detailed hardware structures. Typically, there is a one-to-one correspondence between behavioral ADLs and instruction-set reference manual. This section briefly describes two behavioral ADLs: nML [6] and instruction set description language (ISDL) [7].

nML

nML is an instruction-set–oriented ADL proposed at the Technical University of Berlin, Germany. nML has been used by code generators CBC [8] and CHESS [9], and instruction-set simulators Sigh/Sim [10] and CHECKERS. Currently, the CHESS/CHECKERS environment is used for automatic and efficient software compilation and instruction-set simulation [11]. nML developers recognized the fact that several instructions share common properties. The final nML description would be compact and simple if the common properties were exploited. Consequently, nML designers used a hierarchical scheme to describe instruction sets. The instructions are the topmost elements in the hierarchy. The intermediate elements of the hierarchy are partial instructions (PI). The relationship between elements can be established using two composition rules: the AND rule and the OR rule. The AND rule groups several PIs into a larger PI, and the OR rule enumerates a set of alternatives for one PI. Therefore, instruction definitions in nML can be in the form of an AND-OR tree. Each possible derivation of the tree corresponds to an actual instruction. To achieve the goal of sharing instruction descriptions, the instruction set is enumerated by an attributed grammar [12]. Each element in the hierarchy has a few attributes. A nonleaf element's attribute values can be computed based on its children's attribute values. The following nML description shows an example of instruction specification [6]:

```
op numeric_instruction(a:num_action, src:SRC, dst:DST)
action }
  temp_src = src;
  temp_dst = dst;
  a.action;
  dst = temp_dst;
{
op num_action = add | sub
op add()
action = {
  temp_dst = temp_dst + temp_src
}
```

The definition of *numeric_instruction* combines three (PIs) with the AND rule: *num_action*, SRC, and DST. The first PI, *num_action*, uses the OR rule to describe the valid options for actions *add* or *sub*. The number of all possible derivations of *numeric_instruction* is the product of the size of *num_action*, *SRC*, and *DST*. The common behavior of all these options is defined in the *action* attribute of *numeric_instruction*. Each option for *num_action* should have its own action attribute defined as its specific behavior, which is referred by the *a.action* line. For example, the previous code segment has action description for *add* operation. Object code image and assembly syntax can also be specified in the same hierarchical manner.

nML also captures the structural information used by ISA. For example, storage units should be declared since they are visible to the instruction set. nML supports three types of storages: RAM, register, and transitory storage. Transitory storage refers to machine states that are retained only for a limited number of cycles (e.g., values on buses and latches). Computations have no delay in the nML timing model—only storage units have delay. Instruction delay slots are modeled by introducing storage units as pipeline registers. The result of the computation is propagated through the registers in the behavior specification. nML models constraints between operations by enumerating all valid combinations. The enumeration of valid cases can make nML descriptions lengthy. More complicated constraints, which often appear in DSPs with irregular ILP constraints or VLIW processors with multiple issue slots, are hard to model with nML. nML explicitly supports several addressing modes. However, it implicitly assumes an architecture model that restricts its generality. As a result, it is hard to model multicycle or pipelined units and multiword instructions explicitly.

ISDL

ISDL was developed at the Massachusetts Institute of Technology and used by the Aviv compiler [13] and GENSIM simulator generator [14]. The problem of constraint modeling is avoided by ISDL with explicit specification. ISDL is mainly targeted toward VLIW processors. Similar to nML, ISDL primarily describes the instruction set of processor architectures. ISDL mainly consists of five sections: instruction word format, global definitions, storage resources, assembly syntax, and constraints. It also contains an optimization information section that can be used to provide certain architecture-specific hints for the compiler to make better machine-dependent code optimizations. The instruction word format section defines fields of the instruction word. The instruction word is separated into multiple fields, each containing one or more subfields. The global definition section describes four main types: tokens, nonterminals, split functions, and macro definitions. Tokens are the primitive operands of instructions. For each token, assembly format and binary encoding information must be defined. An example token definition of a binary operand is:

```
Token X[0..1] X_R ival {yylval.ival = yytext[1]-'0';}
```

In this example, following the keyword *Token* is the assembly format of the operand. *X_R* is the symbolic name of the token used for reference. The *ival* is used to describe the value returned by the token. Finally, the last field describes the computation of the value. In this example, the assembly syntax allowed for the token *X_R* is *X0* or *X1*, and the values returned are 0 or 1, respectively. The value (last) field is used for behavioral definition and binary encoding assignment by nonterminals or instructions. Nonterminal is a mechanism provided to exploit commonalities among operations. The following code segment describes a nonterminal named *XYSRC*:

```
Non_Terminal ival XYSRC: X_D {$$ = 0;} |
                         Y_D {$$ = Y_D + 1;};
```

The definition of *XYSRC* consists of the keyword *Non_Terminal*, the type of the returned value, a symbolic name as it appears in the assembly, and an action that describes the possible token or nonterminal combinations and the return value associated with each of them. In this example, *XYSRC* refers to tokens *X_D* and *Y_D* as its two options. The second field (*ival*) describes the returned value type. It returns 0 for *X_D* or incremented value for *Y_D*. Similar to nML, storage resources are the only structural information modeled by ISDL. The storage section lists all storage resources visible to the programmer. It lists the names and sizes of the memory, register files, and special registers. This information is used by the compiler to determine the available resources and how they should be used.

The assembly syntax section is divided into fields corresponding to the separate operations that can be performed in parallel within a single instruction. For each field, a list of alternative operations can be described. Each operation description consists of a name, a list of tokens or nonterminals as parameters, a set of commands that manipulate the bitfields, RTL description, timing details, and costs. RTL description captures the effect of the operation on the storage resources. Multiple costs are allowed, including operation execution time, code size, and costs due to resource conflicts. The timing model of ISDL describes when the various effects of the operation take place (e.g., because of pipelining). In contrast to nML, which enumerates all valid combinations, ISDL defines invalid combinations in the form of Boolean expressions. This often leads to a simple constraint specification. It also enables ISDL to capture irregular ILP constraints. ISDL provides the means for compact and hierarchical instruction-set specification. However, it may not be possible to describe instruction sets with multiple encoding formats using the simple tree-like instruction structure of ISDL.

Mixed ADLs

Mixed languages captures both structural and behavioral details of the architecture. This section briefly describes three mixed ADLs: HMDES, EXPRESSION, and LISA.

HMDES

Machine description language HMDES was developed at University of Illinois at Urbana-Champaign for the IMPACT research compiler [15]. C-like preprocessing capabilities, such as file inclusion, macro expansion, and conditional inclusion, are supported in HMDES. An HMDES description is the input to the MDES machine description system of the Trimaran compiler infrastructure, which contains IMPACT as well as the Elcor research compiler from HP Labs. The description is first preprocessed, then optimized, and then translated to a low-level representation file. A machine database reads the low-level files and supplies information for the compiler backend through a predefined query interface. MDES captures both structure and behavior of target processors. Information is broken down into sections such as format, resource usage, latency, operation, and register. For example, the following code segment describes the register and register file. It describes 64 registers. The register file describes the width of each register and other optional fields such as generic register type (virtual field), speculative, static, and rotating registers. The value "1" implies speculative and "0" implies nonspeculative.

```
SECTION Register {
  R0(); R1(); ... R63();
  'R[0]'(); ... 'R[63]'();
  ...
}

SECTION Register_ File {
  RF_i(width(32) virtual(i) speculative(1)
      static(R0...R63) rotating('R[0]'...'R[63]'));
  ...
}
```

MDES allows only a restricted retargetability of the cycle-accurate simulator to the HPL-PD processor family [16]. MDES permits description of memory systems, but is limited to the traditional hierarchy (i.e., register files, caches, and main memory).

EXPRESSION

HMDES and other mixed ADLs require explicit description of reservation tables (RTs). Processors that contain complex pipelines, large amounts of parallelism, and complex storage subsystems typically contain a large number of operations and resources (and hence RTs). Manual specification of RTs on a per-operation basis thus becomes cumbersome and error-prone. The manual specification of RTs (for each configuration) becomes impractical during rapid architectural exploration. The EXPRESSION ADL [24] describes a processor as a netlist of units and storages to automatically generate RTs based on the netlist [17]. Unlike MIMOLA, the netlist representation of **EXPRESSION** is coarse grain. It uses a higher level of abstraction, similar to block-diagram level description in an architecture manual. **EXPRESSION ADL** was developed at the University of California, Irvine. The ADL has been used by the retargetable compiler and simulator generation framework [11]. An **EXPRESSION** description is composed of two main sections: behavior (instruction-set) and structure. The behavior section has three subsections: operations, instruction, and operation mappings. Similarly, the structure section consists of three subsections: components, pipeline/data-transfer paths, and memory subsystem.

The operation subsection describes the instruction-set of the processor. Each operation of the processor is described in terms of its opcode and operands. The types and possible destinations of each operand are also specified. A useful feature of **EXPRESSION** is the operation group, which groups similar operations together for the ease of later reference. For example, the following code segment shows an operation group (*alu_ops*) containing two ALU operations: *add* and *sub*.

```
(OP_GROUP alu_ops
  (OPCODE add
    (OPERANDS (SRC1 reg) (SRC2 reg/imm) (DEST reg))
    (BEHAVIOR DEST = SRC1 + SRC2)
  ...)
  (OPCODE sub
    (OPERANDS (SRC1 reg) (SRC2 reg/imm) (DEST reg))
    (BEHAVIOR DEST = SRC1 - SRC2)
  ...)
)
```

The instruction subsection captures the parallelism available in the architecture. Each instruction contains a list of slots (to be filled with operations), with each slot corresponding to a functional unit. The operation mapping subsection is used to specify the information needed by instruction selection and architecture-specific optimizations of the compiler. For example, it contains mapping between generic and target instructions. The component subsection describes each RT-level component in the architecture. The components can be pipeline units, functional units, storage elements, ports, and connections. For multicycle or pipelined units, the timing behavior is also specified.

The pipeline/data-transfer path subsection describes the netlist of the processor. The *pipeline path description* provides a mechanism to specify the units that comprise the pipeline stages, while the *data-transfer path description* provides a mechanism for specifying the valid data transfers. This information is used to both retarget the simulator and to generate RTs needed by the scheduler [17]. An example path declaration for the DLX architecture [18] is shown in Fig. 4.4. It illustrates that the processor has five pipeline stages. It also illustrates that the *Execute*

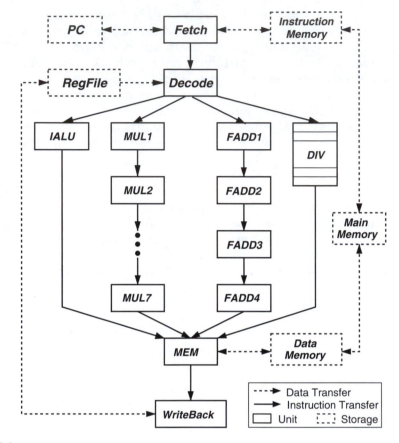

■ **FIGURE 4.4**

A VLIW DLX architecture.

stage has four parallel paths. Finally, it illustrates each path (e.g., it illustrates that the *FADD* path has four pipeline stages).

```
(PIPELINE Fetch Decode Execute MEM WriteBack)
(Execute (ALTERNATE IALU MULT FADD DIV))
(MULT (PIPELINE MUL1 MUL2 ... MUL7))
(FADD (PIPELINE FADD1 FADD2 FADD3 FADD4))
```

The memory subsection describes the types and attributes of various storage components (such as register files, SRAMs, DRAMs, and caches). The memory netlist information can be used to generate memory aware compilers and simulators [19]. EXPRESSION captures the data path information in the processor. The control path is not explicitly modeled. The instruction model requires an extension to capture interoperation constraints such as the sharing of common fields. Such constraints can be modeled by ISDL through cross-field encoding assignment.

LISA

LISA [20] was developed at Aachen University of Technology, Germany, with a simulator-centric view. The language has been used to produce production quality simulators [21]. An important aspect of LISA is its ability to capture control path explicitly. Explicit modeling of both datapath and control is necessary for cycle-accurate simulation. LISA has also been used to generate retargetable C compilers [22]. LISA descriptions are composed of two types of declarations: resource and operation. The resource declarations cover hardware resources such as registers, pipelines, and memories. The pipeline model defines all possible pipeline paths that operations can go through. An example pipeline description for the architecture shown in Fig. 4.4 is as follows:

```
PIPELINE int = {Fetch; Decode; IALU; MEM; WriteBack}
PIPELINE flt = {Fetch; Decode; FADD1; FADD2;
                FADD3; FADD4; MEM; WriteBack}
PIPELINE mul = {Fetch; Decode; MUL1; MUL2; MUL3; MUL4;
                MUL5; MUL6; MUL7; MEM; WriteBack}
PIPELINE div = {Fetch; Decode; DIV; MEM; WriteBack}
```

Operations are the basic objects in LISA. They represent the designer's view of the behavior, the structure, and the instruction set of the programmable architecture. Operation definitions capture the description of different properties of the system, such as operation behavior, instruction set information, and timing. These operation attributes are defined in several sections. LISA exploits the commonality of similar operations by grouping them into one. The following code segment describes the decoding behavior of two immediate-type (i_type) operations (ADDI and SUBI) in the DLX *Decode* stage. The complete behavior of an operation can be obtained by combining its behavior definitions in all the pipeline stages.

```
OPERATION i_type IN pipe_int.Decode {
  DECLARE {
    GROUP opcode = {ADDI || SUBI}
    GROUP rs1, rd = {fix_register};
  }
  CODING {opcode rs1 rd immediate}
  SYNTAX {opcode rd ``,'' rs1 ``,'' immediate}
  BEHAVIOR {reg_a = rs1; imm = immediate; cond = 0;
  }
  ACTIVATION {opcode, writeback}
}
```

A language similar to LISA is RADL. RADL [4] was developed at Rockwell, Inc., as an extension of the LISA approach that focuses on explicit support of detailed pipeline behavior to enable generation of production-quality, cycle-accurate, and phase-accurate simulators. Efficient software toolkit generation is also demonstrated using Mescal architecture description language (MADL) [23]. MADL uses the operation state machine (OSM) model to describe the operations at the cycle-accurate level. Due to OSM-based modeling of operations, MADL provides flexibility in describing a wide range of architectures, simulation efficiency, and ease of extracting model properties for efficient compilation.

4.2.2   Objective-Based Classification of ADLs

ADLs have been successfully used as a specification language for processor development. Rapid evaluation of candidate architectures is necessary to explore the vast design space and find the best possible design under various design constraints such as area, power, and performance. Fig. 4.5 shows a traditional ADL-based design space exploration flow. The application programs are compiled and simulated, and the feedback is used to modify the ADL specification. The generated simulator produces profiling data that can be used to evaluate an instruction set, performance of an algorithm, and required size of memory and registers. The generated hardware (synthesizable HDL) model can provide more accurate feedback related to required silicon area, clock frequency, and power consumption of the processor architecture. Contemporary ADLs can be classified into four categories based on the objective: compilation oriented, simulation oriented, synthesis oriented, or validation oriented.

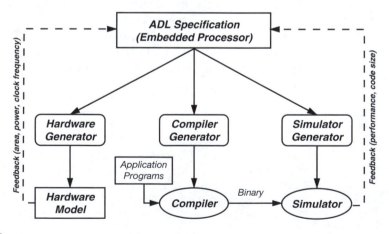

▪ **FIGURE 4.5**

ADL-driven design space exploration.

In this section we briefly describe the ADLs based on the objective-based classification. We primarily discuss the required capabilities of an ADL to perform the intended objective.

Compilation-Oriented ADLs

The goal of such an ADL is to enable automatic generation of retargetable compilers. A compiler is classified as retargetable if it can be adapted to generate code for different target processors with significant reuse of the compiler source code. Retargetability is typically achieved by providing target machine information in an ADL as input to the compiler along with the program corresponding to the application. Therefore, behavioral ADLs (e.g., ISDL [7] and nML [6]) are suitable for compiler generation. They capture the instruction set of the architecture, along with some structural information such as program counter and register details. Mixed ADLs are suitable for compiler generation since they capture both structure and behavior of the processor.

There is a balance between the information captured in an ADL and the information necessary for compiler optimizations. Certain ADLs (e.g., AVIV [13] using ISDL, CHESS [9] using nML, and Elcor [16] using MDES) explicitly capture all the necessary details such as instruction set and resource conflict information. Recognizing that the architecture information needed by the compiler is not always in a form that may be well suited for use by other tools (such as synthesis) or does not permit concise specification, some research has focused on extraction of such information from a more amenable specification. Examples include the MSSQ and RECORD compiler using MIMOLA [5], compiler optimizers using MADL [23], the retargetable C compiler based on LISA [22], and the EXPRESS compiler using EXPRESSION [24].

Simulation-Oriented ADLs

Simulation can be performed at various abstraction levels. At the highest level of abstraction, functional simulation (instruction set simulation) of the processor can be performed by modeling only the instruction set. Behavioral ADLs can enable generation of functional simulators. The cycle-accurate and phase-accurate simulation models yield more detailed timing information since they are at lower level of abstraction. Structural ADLs are good candidates for cycle-accurate simulator generation.

Retargetability (i.e., the ability to simulate a wide variety of target processors) is especially important in the context of customizable processor design. Simulators with limited retargetability are very fast but may not be useful in all aspects of the design process. Such simulators (e.g., HPL-PD [16] using MDES) typically incorporate a fixed architecture template and allow only limited retargetability in the form of parameters such as number of registers and ALUs. Due to OSM-based modeling of operations, MADL allows modeling and simulator generation for a wide range of architectures [23]. Based on the simulation model, simulators can be classified into three types: interpretive, compiled, and mixed. Interpretive simulators (e.g., GENSIM/XSIM [14] using ISDL) offer flexibility but are slow due to the need for a fetch, decode, and execution model for each instruction. Compilation-based approaches (e.g., [21] using LISA) reduce the runtime overhead by translating each target instruction into a series of host machine instructions that manipulate the simulated machine state. Recently proposed techniques (JIT-CCS [25] using LISA and IS-CS [26] using EXPRESSION) combine the flexibility of interpretive simulation with the speed of the compiled simulation.

Synthesis-Oriented ADLs

Structure-centric ADLs such as MIMOLA are suitable for hardware generation. Some of the behavioral languages (such as ISDL and nML) are also used for hardware generation. For example, the HDL generator HGEN [14] uses ISDL description, and the synthesis tool GO [11] is based on nML. Itoh et al. [27] have proposed a micro-operation description-based synthesizable HDL generation. Mixed languages such as LISA and EXPRESSION capture both the structure and behavior of the processor and enable HDL generation [17, 28, 29]. The synthesizable HDL generation approach based on LISA produces an HDL model of the architecture. The designer has the choice to generate a VHDL, Verilog, or SystemC representation of the target architecture [28].

Validation-Oriented ADLs

ADLs have been successfully used in both academia and industry to enable test generation for functional validation of embedded processors. Traditionally, structural ADLs such as MIMOLA [5] are suitable for test generation. Behavioral ADLs such as nML [11] have been used successfully for test generation. Mixed ADLs also enable test generation based on coverage of the ADL specification using EXPRESSION [30–32] as well as automated test generation for LISA processor models [33]. ADLs have been used in the context of functional verification of embedded processors [34] using a top-down validation methodology, as shown in Fig. 4.6. The first step in the methodology is to verify the ADL specification to ensure the correctness of the specified architecture [35]. The validated ADL specification can be used as a golden reference model for various validation tasks, including property checking, test generation, and equivalence checking. For example, the generated hardware model (reference) can be used to perform both property checking and equivalence checking of the implementation using EXPRESSION ADL [36].

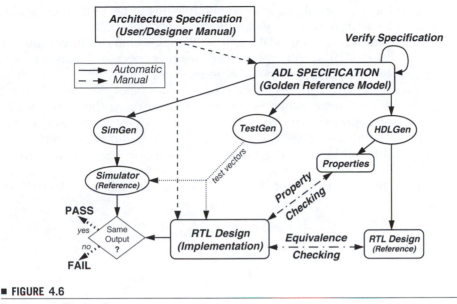

■ FIGURE 4.6

Top-down validation flow.

4.3 CONCLUSIONS

Design of customizable and configurable embedded processors requires the use of automated tools and techniques. ADLs have been used successfully in academic research as well as in industry for processor development. The early ADLs were either structure oriented (MIMOLA, UDL/I) or behavior oriented (nML, ISDL). As a result, each class of ADLs are suitable for specific tasks. For example, structure-oriented ADLs are suitable for hardware synthesis and unfit for compiler generation. Similarly, behavior-oriented ADLs are appropriate for generating compilers and simulators for instruction-set architectures and unsuited for generating cycle-accurate simulator or hardware implementation of the architecture. However, a behavioral ADL can be modified to perform the task of a structural ADL (and vice versa). For example, nML is extended by Target Compiler Technologies to perform hardware synthesis and test generation [11]. The later ADLs (LISA, HMDES, and EXPRESSION) adopted the mixed approach, where the language captures both structure and behavior of the architecture. ADLs designed for a specific domain (such as DSP or VLIW) or for a specific purpose (such as simulation or compilation) can be compact, and it is possible to automatically generate efficient (in terms of area, power, and performance) tools/hardware. However, it is difficult to design an ADL for a wide variety of architectures to perform different tasks using the same specification. Generic ADLs require the support of powerful methodologies to generate high-quality results, compared to domain-specific/task-specific ADLs.

In the future, existing ADLs will go through changes in two dimensions. First, ADLs will specify not only processor, memory, and coprocessor architectures, but also other components of the SoC architectures, including peripherals and external interfaces. Second, ADLs will be used for software toolkit generation, hardware synthesis, test generation, instruction-set synthesis, and validation of microprocessors. Furthermore, multiprocessor SoCs will be captured, and various attendant tasks will be addressed. The tasks include support for formal analysis, generation of RTOSs, exploration of communication architectures, and support for interface synthesis. The emerging ADLs will have these features.

CHAPTER 5

C COMPILER RETARGETING

Rainer Leupers

Compilers translate high-level language source code into machine-specific assembly code. For this task, any compiler uses a model of the target processor. This model captures the compiler-relevant machine resources, including the instruction set, register files, and instruction scheduling constraints. While in traditional target-specific compilers, this model is built-in (i.e., it is hard-coded and probably distributed within the compiler source code), a *retargetable compiler* uses an external processor model as an additional input that can be edited without the need to modify the compiler source code itself (Fig. 5.1). This concept provides retargetable compilers with high flexibility with regard to the target processor.

Retargetable compilers have been recognized as important tools in the context of embedded SoC design for several years. One reason is the trend toward increasing use of programmable processor cores as SoC platform building blocks, which provide the necessary flexibility for fast adoption of, for example, new media encoding or protocol standards and

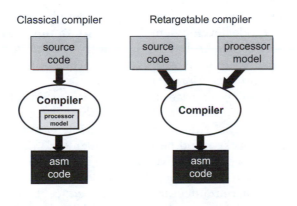

■ FIGURE 5.1

Classical versus retargetable compiler.

easy (software-based) product upgrading and debugging. While assembly language was predominant in embedded processor programming for quite some time, the increasing complexity of embedded application code now makes the use of high-level languages like C and C++ just as inevitable as in desktop application programming.

In contrast to desktop computers, embedded SoCs have to meet very high efficiency requirements in terms of MIPS per Watt, which makes the use of power-hungry, high-performance off-the-shelf processors from the desktop computer domain (together with their well developed compiler technology) impossible for many applications. As a consequence, hundreds of different domain- or even application-specific programmable processors have appeared in the semiconductor market, and this trend is expected to continue. Prominent examples include low-cost/low-energy microcontrollers (e.g., for wireless sensor networks), number-crunching digital signal processors (e.g., for audio and video codecs), as well as network processors (e.g., for Internet traffic management).

All these devices demand their own programming environment, obviously including a high-level language (mostly ANSI C) compiler. This requires the capability to quickly design compilers for new processors, or variations of existing ones, without starting from scratch each time. While compiler design traditionally has been considered a very tedious and hour-intensive task, contemporary retargetable compiler technology makes it possible to build operational (not heavily optimizing) C compilers within a few weeks and more decent ones approximately within a single person-year. Naturally, the exact effort heavily depends on the complexity of the target processor, the required code optimization and robustness level, and the engineering skills. However, compiler construction for new embedded processors is now certainly much more feasible than a decade ago. This permits us to employ compilers not only for application code development, but also for optimizing an embedded processor architecture itself, leading to a true "compiler/architecture codesign" technology that helps to avoid hardware-software mismatches long before silicon fabrication.

This chapter summarizes the state-of the art in retargetable compilers for embedded processors and outlines their design and use by means of examples and case studies. In Section 5.1, we provide some compiler construction background needed to understand the different retargeting technologies. Section 5.2 gives an overview of some existing retargetable compiler systems. In Section 5.3, we describe how the previously mentioned "compiler/architecture codesign" concept can be implemented in a processor architecture exploration environment. A detailed example of an industrial retargetable C compiler system is discussed in Section 5.4. Finally, Section 5.5 concludes and takes a look at potential future developments in the area.

5.1 COMPILER CONSTRUCTION BACKGROUND

The general structure of retargetable compilers follows that of well proven classical compiler technology, which is described in textbooks such as [1–4]. First, there is a language *frontend* for source code analysis. The frontend produces an *intermediate representation* in which a number of machine-independent code optimizations are performed. Finally, the *backend* translates the intermediate representation into assembly code while performing additional machine-specific code optimizations.

5.1.1 Source Language Frontend

The standard organization of a frontend comprises a *scanner*, a *parser*, and a *semantic analyzer* (Fig. 5.2). The scanner performs lexical analysis on the input source file, which is first considered just as a stream of ASCII characters. During lexical analysis, the scanner forms substrings of the input string to groups (represented by *tokens*), each of which corresponds to a primitive syntactic entity of the source language, (e.g., identifiers, numerical constants, keywords, or operators). These entities can be represented by regular expressions, for which in turn finite automata can be constructed and implemented that accept the formal languages generated by the regular expressions. Scanner implementation is strongly facilitated by tools like lex [5].

The scanner passes the tokenized input file to the parser, which performs syntax analysis with regard to the context-free grammar underlying the source language. The parser recognizes syntax errors and, in case of a correct input, builds up a tree data structure that represents the syntactic structure of the input program.

Parsers can be constructed manually based on the LL(k) and LR(k) theory [2]. An LL(k) parser is a top-down parser, (i.e., it tries to generate a derivation of the input program from the grammar start symbol according to the grammar rules). In each step, it replaces a non-terminal by the right-hand side of a grammar rule. To decide which rule to apply out of possibly many alternatives, it uses a lookahead

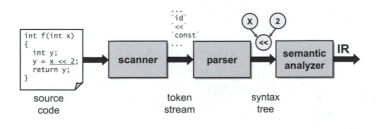

■ **FIGURE 5.2**

Source language frontend structure.

of k symbols on the input token stream. If the context-free grammar shows certain properties, this selection is unique, so that the parser can complete its job in linear time in the input size. The same also holds for LR(k) parsers, which can process a broader range of context-free grammars, though. They work bottom-up, i.e., the input token stream is step-by-step reduced until finally reaching the start symbol. Instead of making a reduction step solely based on the knowledge of the k lookahead symbols, the parser additionally stores input symbols temporarily on a stack until enough symbols for an entire right-hand side of a grammar rule have been read. Due to this, the implementation of an LR(k) parser is less intuitive and requires some more effort than for LL(k).

Constructing LL(k) and LR(k) parsers manually provides some advantage in parsing speed. However, in most practical cases, tools like yacc [5] (which generates a variant of LR(k) parsers) are employed for semi-automatic parser implementation.

Finally, the semantic analyzer performs correctness checks not covered by syntax analysis, e.g., forward declaration of identifiers and type compatibility of operands. It also builds up a symbol table that stores identifier information and visibility scopes. In contrast to scanners and parsers, there are no widespread standard tools like lex and yacc for generating semantic analyzers. Frequently, attribute grammars [1] are used, though, for capturing the semantic actions in a syntax-directed fashion, and special tools like ox [6] can extend lex and yacc to handle attribute grammars.

5.1.2   Intermediate Representation and Optimization

In most cases, the output of the frontend is an intermediate representation (IR) of the source code that represents the input program as assembly-like, yet machine-independent, low-level code. Three-address code (Fig. 5.3 and 5.4) is a common IR format.

There is no standard format for three-address code, but usually all high-level control flow constructs and complex expressions are

```
int fib (int m )
{ int f0 = 0 , f1 = 1 , f2 , i;
  if (m < = 1) return m;
  else
  for ( i=2; i<=m;   i++ ) {
  f2 = f0 + f1 ;
  f0 = f1 ;
  f1 = f2 ; {
  return  f2 ;
}
```

▪ **FIGURE 5.3**

Sample C source file fib.c (Fibonacci numbers).

```
int fib (int m_2)
{
  int f0_4, f1_5, f2_6, i_7, t1, t2, t3, t4, t6, t5;

            f0_4 = 0;
            f1_5 = 1;
            t1 = m_2 <= 1;
            if (t1) goto LL4;
            i_7 = 2;
            t2 = i_7 <= m_2;
            t6 = ! t2;
            if (t6) goto LL1;
   LL3:     t5 = f0_4 + f1_5;
            f2_6 = t5;
            f0_4 = f1_5;
            f1_5 = f2_6;
   LL2:     t3 = i_7;
            t4 = t3 + 1;
            i_7 = t4;
            t2 = i_7 <= m_2;
            if (t2) goto LL3;
   LL1:     goto LL5;
   LL4:     return m_2;
   LL5:     return f2_6;

}
```

■ **FIGURE 5.4**

Three address code IR for source file fib.c. Temporary variable identifiers inserted by the frontend start with letter "t". All local identifiers have a unique numerical suffix. This particular IR format is generated by the LANCE C frontend [7].

decomposed into simple statement sequences consisting of three-operand assignments and gotos. The IR generator inserts *temporary variables* to store intermediate results of computations.

Three-address code is a suitable format for performing different types of *flow analysis*, i.e., control and data flow analysis. Control flow analysis first identifies the basic block[1] structure of the IR and detects the possible control transfers between basic blocks. The results are captured in a control flow graph (CFG). Based on the CFG, more advanced control flow analyses can be performed, e.g., in order to identify program loops. Fig. 5.5 shows the CFG generated for the example from Fig. 5.3 and 5.4.

---

[1] A basic block is a sequence of IR statements with unique control flow entry and exit points.

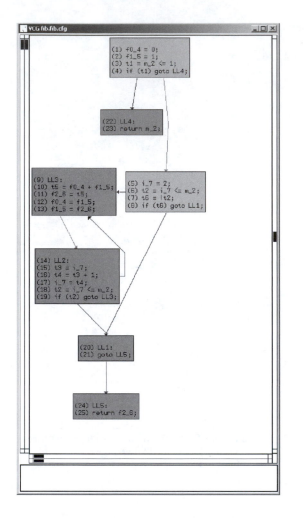

■ **FIGURE 5.5**

Control flow graph for fib.c.

Data flow analysis works on the statement level and determines interdependencies between computations. For instance, the data flow graph (DFG) from Fig. 5.6 shows relations of the form "statement X computes a value used as an argument in statement Y."

Both the CFG and the DFG form the basis for many code optimization passes at the IR level. These include common subexpression elimination, jump optimization, loop invariant code motion, dead code elimination, and other "Dragon Book" [1] techniques. Due to their target machine independence, these IR optimizations are generally considered complementary to machine code generation in the backend and are supposed to be useful "on the average" for any type of target. However, care must be taken to select an appropriate IR optimization sequence or

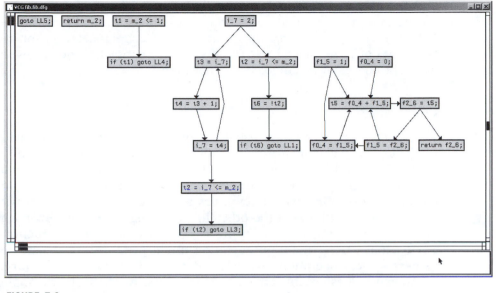

■ **FIGURE 5.6**

Data flow graph for fib.c.

script for each particular target, since certain (sometimes quite subtle) machine dependencies do exist. For instance, common subexpression elimination removes redundant computations to save execution time and code size. At the assembly level, however, this effect might be over-compensated by the higher register pressure that increases the amount of spill code. Moreover, there are many interdependencies between the IR optimizations themselves. For instance, constant propagation generally creates new optimization opportunities for constant folding, and vice versa, and dead code elimination is frequently required as a "cleanup" phase in between other IR optimizations. A poor choice of IR optimizations can have a dramatic effect on final code quality. Thus, it is important that IR optimizations be organized in a modular fashion, so as to permit enabling and disabling of particular passes during fine-tuning of a new compiler.

5.1.3 Machine Code Generation

During this final compilation phase, the IR is mapped to target assembly code. Since for a given IR an infinite number of mappings as well as numerous constraints exist, this is clearly a complex optimization problem. In fact, even many optimization subproblems in code generation are NP-hard, i.e., they require exponential runtime for optimal solutions. As a divide-and-conquer approach, the backend is thus generally organized into three main phases: *code selection*, *register allocation*, and *scheduling*, which are implemented with a variety of heuristic algorithms. Dependent on the exact problem definition, all of these phases may be considered NP-hard, e.g., [8] analyzes the complexity of code generation for certain types of target machines. The rest of this subsection describes the different phases of machine code generation in detail.

Code Selection

The IR of an application is usually constructed using primitive arithmetic, logical, and comparison operations. The target architecture might combine several such primitive operations into a single instruction. A classical example is the multiply and accumulate (MAC) instruction found in most DSPs, which combines a multiplication operation with a successive addition. On the other hand, a single primitive operation might have to be implemented using a sequence of machine instructions. For example, the C negation operation might need to be implemented using a logical operation not followed by an increment, if the target architecture does not implement negation directly. The task of mapping a sequence of primitive IR operations to a sequence of machine instructions is performed by the *code selector* or *instruction selector*.

In particular for target architectures with complex instruction sets, such as CISCs and DSPs, careful code selection is key for good code quality. Due to complexity reasons, most code selectors work only on trees [9], even though generalized code selection for arbitrary DFGs can yield higher code quality for certain architectures [10, 11]. The computational effort for solving the NP-hard generalized code selection problem is normally considered too high in practice, whereas optimum code selection for trees can be efficiently accomplished using *dynamic programming* techniques as described later.

For the purpose of code selection, the optimized IR is usually converted into a sequence of tree-shaped DFGs or *data flow trees* (DFTs). Each instruction is represented as a *tree-pattern* that can partially cover a subtree of a DFT and is associated with a *cost*. Fig. 5.7 shows a set of instruction patterns for the Motorola 68K CPU where each pattern has the same cost. As can be seen in Fig. 5.8, the same DFT can be covered in multiple ways using the given instruction patterns. Using the cost metric for the machine instructions, the code selector aims at a minimum-cost covering of the DFTs by the given instruction patterns.

Generally the code-selection algorithms for DFTs use special context-free grammars to represent the instruction patterns. In such grammars, each rule is annotated with the cost of the corresponding pattern (Fig. 5.9). The nonterminals in the grammar rules usually correspond to the different register classes or memory-addressing mechanisms of the target architecture and are placeholders for the results produced by the corresponding instructions. In each step of code selection, a subtree of the entire DFT is matched to a grammar rule, and the subtree is

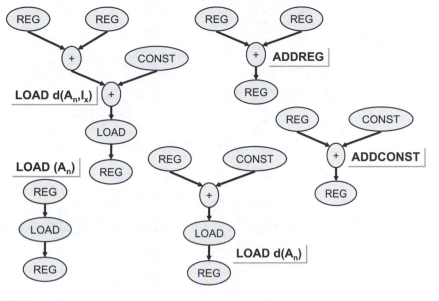

FIGURE 5.7

Five instruction patterns available for a Motorola 68K CPU.

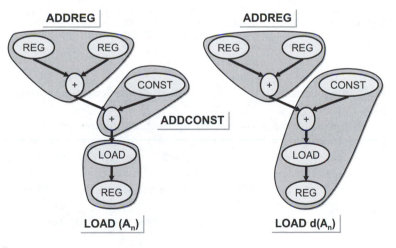

FIGURE 5.8

Two possible coverings of a DFG using the instruction patterns from Fig. 5.7.

replaced by the nonterminal in the corresponding rule (Fig. 5.10). Since the context-free grammars representing instruction patterns are generally very ambiguous, each subtree, potentially, can be matched using different candidate rules. The optimum code selection, therefore, must

Terminals: {MEM, +, −, *}
Non Terminals: {reg1, reg2}
Start symbol: reg1

Instruction Syntax	Cost	Rule
add reg1, reg2, reg1	2	reg1 → +(reg1, reg2)
add reg1, MEM, reg1	2	reg2 → +(reg1, MEM)
sub MEM, reg1, reg1	2	reg1 → -(MEM, reg1)
mul reg1, reg2, reg1	2	reg1 → *(reg1, reg2)
mul reg1, MEM, reg2	2	reg2 → *(reg1, MEM)
mac reg1, reg2, MEM	2	reg1 → +(*(reg1, reg2), MEM)
mov reg1, reg2	1	reg2 → reg1
mov reg2, reg1	1	reg1 → reg2
load MEM, reg2	1	reg2 → MEM

▪ **FIGURE 5.9**

A context-free grammar representing different instruction patterns. The two nonterminals (reg1, reg2) in the grammar represent two different register classes. The right-hand side of each grammar rule is generated by preorder traversal of the corresponding instruction pattern; the left-hand side refers to where the instruction produces its result.

take into account the costs of the different candidate rules to resolve these ambiguities.

The generalized dynamic-programming algorithm of code selection for DFTs applies a bottom-up technique to enumerate the costs of different possible covers using the available grammar rules. It calculates the minimum-cost match at each node of the DFT for each *nonterminal* of the grammar (Fig. 5.11). After the costs of different matches are computed, the tree is traversed in a top-down fashion selecting the minimum-cost nonterminal (and the corresponding instruction pattern) for each node.

Register Allocation

The code selector generally assumes that there is an infinite number of *virtual registers* to hold temporary values. Subsequent to code selection, the register allocator decides which of these temporaries are to be kept in machine registers to ensure efficient access. Careful register allocation is key for target machines with RISC-like load-store architectures and large register files. Frequently, there are many more simultaneously live

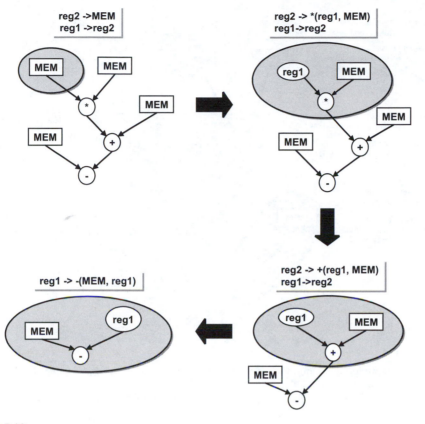

■ **FIGURE 5.10**

DFT covering using the grammar rules of Fig. 5.9. In each step a subtree is replaced by the nonterminal(s) corresponding to one or more matching grammar rule(s).

variables than physically available machine registers. In such cases, the register allocator inserts *spill code* to temporarily store register variables to main memory. Obviously, spill code needs to be minimized in order to optimize program performance and code size.

Many register allocators use a graph coloring approach [12, 13] to accomplish this task. Although graph coloring is an NP-complete problem, an efficient approximation algorithm exists that produces fairly good results. A register allocator first creates an *interference graph* by examining the CFG and DFG of an application. Each node of the interference graph represents a temporary value. An edge between two nodes indicates that the corresponding temporaries are simultaneously *live* and therefore cannot be assigned to the same register (Fig. 5.12). If the target processor has K registers, then coloring the graph with K colors is a valid register assignment. Spills are generated when K-coloring of the graph is not possible.

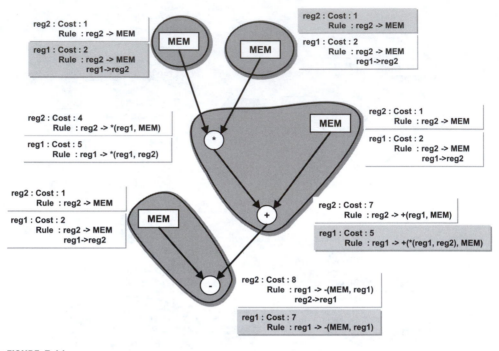

▪ **FIGURE 5.11**

Dynamic programming technique for code selection on the DFT of Fig. 5.10 using the grammar rules of Fig. 5.9. For each intermediate node of the DFT, the minimum cost match for each nonterminal is calculated in a bottom up fashion. The optimum cover consists of the rules in the grey boxes.

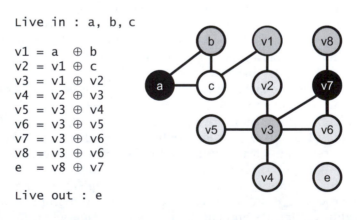

▪ **FIGURE 5.12**

A code fragment and the corresponding interference graph. Since the graph is 3-colorable, a valid register assignment with only three registers can be generated.

Instruction Scheduling

For target processors with instruction *pipeline hazards* and/or *instruction-level parallelism (ILP)*, such as VLIW machines, instruction scheduling is an absolute necessity. For pipelined architectures without any hardware interlocks or data forwarding mechanisms, instruction scheduling is required to ensure correct program semantics. For processors with high ILP, instruction scheduling is necessary to exploit the fine-grained parallelism in the applications. Like other backend phases, optimal scheduling is an intractable problem. However, there exist a number of powerful scheduling heuristics, such as list scheduling and trace scheduling [3].

Instruction scheduling is usually employed to resolve the (potentially) conflicting accesses to the same hardware resource(s) by a pair of instructions. The scheduler must take into account the *structural*, *data*, or *control* hazards that usually result from pipelining. The schedulability of any two instructions, X and Y, under such hazards can usually be described using *latency* and *reservation* tables.

For the instruction X, the latency table specifies the number of cycles X takes to produce its result. This value places a lower bound on the number of instructions that Y, if it uses the result of X, must a wait before it starts executing. In such cases, the scheduler must insert the required number of NOPs (or other instructions that are not dependent on X) between X and Y to ensure correct program semantics. The reservation table specifies the number of hardware resources used by an instruction and the durations of such usages. The scheduler must ensure that X and Y do not make conflicting accesses to any hardware resource during execution. For example, if hardware multiplication takes three cycles to complete, then the scheduler must always keep two multiplication instructions at least three cycles apart (again by inserting NOPs or other nonmultiplication instructions).

Scheduling can be done locally (for each basic block) or globally (across basic blocks). Most of the local schedulers employ the list scheduling technique. In contrast, the global scheduling techniques (such as trace scheduling and percolation scheduling) are more varied and out of the scope of the current discussion.

In the list scheduling approach, the DFG of a basic block is topologically sorted on the basis of data dependencies. The *independent nodes* of the topologically sorted graph are entered into a *ready* (for scheduling) set. Then, one member of the ready set is removed, along with the edges emanating from it, and is scheduled. The process is then recursively applied on the resulting graph.

An example of list scheduling for a four-slot VLIW machine, with three ALUs and one division unit, is presented in Fig. 5.13. The division operation has a latency of four, and all other operations have a latency of one. Initially, the ready set for the topologically sorted DFG consists of four nodes: two left shift and two division operations (nodes 1, 2, 3, 4).

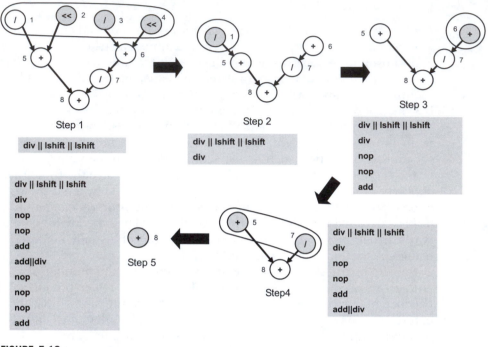

▪ FIGURE 5.13

List scheduling applied to a DFG for a four-slot VLIW machine. For each step of the scheduling algorithm, nodes in the ready set are encircled, and the nodes finally selected by the algorithm are shown in grey. The partially scheduled code after each step is shown in grey boxes. Each node has been given a number for a better understanding of the explanation provided in the text.

The left shift operations (nodes 2 and 3) can be scheduled simultaneously in two of the three available ALUs. However, only one division can be scheduled in the available division unit. The tie between the two division operations is broken by selecting node 4, which lies on the *critical path* of the graph.

In the next scheduling step, only node 1 is available for scheduling. Although node 6 has no predecessor in the graph after removal of nodes 3 and 4, it can only be scheduled in the fifth cycle, i.e., after the division operation of node 3 produces its result. The scheduler schedules node 6 in step 3 after inserting two NOPs to take care of the latency of the division operation. The last two steps schedule the rest of the graph as shown in the figure.

Note that selecting node 1 in the first step would have resulted in a longer schedule than the current one. In general, the effectiveness of list scheduling lies in selecting the *most eligible* member from the ready set that should result in an optimal or near-optimal schedule. A number of heuristics [3], such as the one used in this example (i.e., selecting nodes on the critical path), have been proposed for this task.

Other Backend Optimizations

Besides these three standard code-generation phases, a backend frequently also incorporates different target-specific code optimization passes. For instance, address code optimization [14] is useful for a class of DSPs, so as to fully utilize dedicated address generation hardware for pointer arithmetic. Many VLIW compilers employ loop unrolling and software pipelining [15] for increasing instruction-level parallelism in the hot spots of application code. Loop unrolling generates larger basic blocks inside loop bodies and hence provides better opportunities for keeping the VLIW functional units busy most of the time. Software pipelining rearranges the loop iterations so as to remove intraloop data dependencies that otherwise would obstruct instruction-level parallelism. Finally, NPU architectures for efficient protocol processing require yet a different set of machine-specific techniques [16] that exploit bit-level manipulation instructions.

The separation of the backend into multiple phases is frequently needed to achieve sufficient compilation speed but tends to compromise code quality due to interdependencies between the phases. In particular, this holds for irregular "non-RISC" instruction sets, where the phase interdependencies are sometimes very tight. Although there have been attempts to solve the code generation problem in its entirety, e.g., based on ILP [17], such "phase-coupled" code-generation techniques are still far from widespread use in real-world compilers.

5.2   APPROACHES TO RETARGETABLE COMPILATION

From the previous discussions it is obvious that compiler retargeting mainly requires adaptations of the backend, even though IR optimization issues certainly should not be neglected. To provide a retargetable compiler with a processor model, as sketched in Fig. 5.1, a formal machine description language is required. For this purpose, dozens of different approaches exist. These can be classified with regard to the intended target processor class (e.g., RISC versus VLIW) and the modeling abstraction level, e.g., purely behavioral, compiler-oriented versus more structural, architecture-oriented modeling styles.

Behavioral modeling languages make the task of retargeting easier, because they explicitly capture compiler-related information about the target machine, i.e., instruction set, register architecture, and scheduling constraints. On the other hand, they usually require good understanding of compiler technology. In contrast, *architectural modeling languages* follow a more hardware design–oriented approach and describe the target machine in more detail. This is convenient for users not familiar with compiler technology. However, automatic retargeting gets more difficult, because a "compiler view" needs to be extracted from the architecture model while eliminating unnecessary details.

In the following, we will briefly discuss a few representative examples of retargetable compiler systems. For a comprehensive overview of existing systems, see [18].

5.2.1   MIMOLA

MIMOLA denotes both a mixed programming and hardware HDL and a hardware design system. As the MIMOLA HDL serves multiple purposes, e.g., RTL simulation and synthesis, the retargetable compiler MSSQ [19, 20] within the MIMOLA design system follows the previously mentioned architecture-oriented approach. The target processor is described as an RTL netlist, consisting of components and interconnect. Fig. 5.14 gives an example of such an RTL model.

Since the HDL model comprises all RTL information about the target machine's controller and data path, it is clear that all information relevant for the compiler backend of MSSQ is present, too. However, this information is only implicitly available, and consequently the lookup of this information is more complicated than in a behavioral model.

MSSQ compiles an "extended subset" of the PASCAL programming language directly into binary machine code. Due to its early introduction, MSSQ employs only few advanced code optimization techniques (e.g., there is no graph-based global register allocation) but performs the source-to-architecture mapping in a straightforward fashion, on a statement-by-statement basis. Each statement is represented by a DFG, for which an isomorphic subgraph is searched in the target data path. If this matching fails, the DFG is partitioned into simpler components, for which graph matching is invoked recursively.

In spite of this simple approach, MSSQ is capable of exploiting ILP in VLIW-like architectures very well, due to the use of a flexible instruction scheduler. However, code quality is generally not acceptable in case of complex instruction sets and load/store data paths. In addition, it shows comparatively high compilation times, due to the need for exhaustive graph matching.

The MIMOLA approach shows very high flexibility in compiler retargeting, since in principle any target processor can be represented as an RTL HDL model. In addition, it avoids the need to consistently maintain multiple different models of the same machine for different design phases, e.g., simulation and synthesis, as all phases can use the same "golden" reference model. MSSQ demonstrates that retargetable compilation is possible with such unified models, even though it does not handle architectures with complex instruction pipelining constraints well (which is a limitation of the tool, though, rather than of the approach

```
MODULE SimpleProcessor (IN inp:(7:0); OUT outp:(7:0); STRUCTURE
IS TYPE InstrFormat = FIELDS        -- 21-bit horizontal instruction word
          imm:        (20:13);
          RAMadr:     (12:5);
          RAMctr:      (4);
          mux:         (3:2);
          alu:         (1:0);
        END;
      Byte = (7:0); Bit = (0);   -- scalar types


PARTS                             -- instantiate behavioral modules
 IM: MODULE InstrROM (IN adr: Byte; OUT ins: InstrFormat);
      VAR storage: ARRAY[0..255] OF InstrFormat;
      BEGIN ins <- storage[adr]; END;
 PC, REG: MODULE Reg8bit (IN data: Byte; OUT outp: Byte);
         VAR R: Byte;
         BEGIN R := data; outp <- R; END;
 PCIncr: MODULE IncrementByte (IN data: Byte; OUT inc: Byte);
         BEGIN outp <- INCR data; END;
 RAM: MODULE Memory (IN data, adr: Byte; OUT outp: Byte; FCT c: Bit);
      VAR storage: ARRAY[0..255] OF Byte;
      BEGIN
       CASE c OF: 0: NOLOAD storage; 1: storage[adr]:= data; END;
       outp <- storage[adr];
      END;
 ALU: MODULE AddSub (IN d0, d1: Byte; OUT outp: Byte; FCT c: (1:0));
      BEGIN                  -- "%" denotes binary numbers
       outp <- CASE c OF %00: d0+d1; %01: d0-d1; %1x: d0; END;
      END;
 MUX: MODULE Mux3x8 (IN d0,d1,d2: Byte; OUT outp: Byte; FCT c: (1:0));
      BEGIN outp <- CASE c OF 0: d0;  1: d1; ELSE: d2; END; END;

CONNECTIONS
 -- controller:                    -- data path:
 PC.outp        -> IM.adr;         IM.ins.imm   -> MUX.d0;
 PC.outp        -> PCIncr.data;    inp          -> MUX.d1;   -- primary input
 PCIncr.outp    -> PC.data;        RAM.outp  -> MUX.d2;
 IM.ins.RAMadr -> RAM.adr;         MUX.outp  -> ALU.d1;
 IM.ins.RAMctr -> RAM.c;           ALU.outp  -> REG.data;
 IM.ins.alu    -> ALU.c;           REG.outp  -> ALU.d0;
 IM.ins.mux    -> MUX.c;           REG.outp  -> outp;       -- primary output
END; -- STRUCTURE
LOCATION_FOR_PROGRAMCOUNTER PC;
LOCATION_FOR_INSTRUCTIONS IM;
END; -- STRUCTURE
```

■ **FIGURE 5.14**

MIMOLA HDL model of a simple processor.

itself). The disadvantage, however, is that the comparatively detailed modeling level makes it more difficult to develop the model and to understand its interaction with the retargetable compiler, since, e.g., the instruction set is "hidden" in the model.

Some of the limitations have been removed in RECORD, another MIMOLA HDL-based retargetable compiler that comprises dedicated code optimizations for DSPs. To optimize compilation speed, RECORD uses an *instruction set extraction technique* [21] that bridges the gap between RTL models and behavioral processor models. Key ideas of MSSQ, e.g., the representation of scheduling constraints by binary partial instructions, have also been adopted in the CHESS compiler [22, 23], one of the first commercial tool offerings in that area. In the Expression compiler [24], the concept of structural architecture modeling has been further refined to increase the reuse opportunities for model components.

5.2.2   GNU C Compiler

The widespread GNU C compiler gcc [25] can be retargeted by means of a machine description file that captures the compiler view of a target processor in a behavioral fashion. In contrast to MIMOLA, this file format is heterogeneous and solely designed for compiler retargeting. The gcc compiler is organized into a fixed number of different passes. The frontend generates a three-address code-like intermediate representation (IR). There are multiple built-in Dragon Book IR optimization passes, and the backend is driven by a specification of instruction patterns, register classes, and scheduler tables. In addition, retargeting gcc requires C code specification of numerous support functions, macros, and parameters.

The gcc compiler is robust and well supported, it includes multiple source language frontends, and it has been ported to dozens of different target machines, including typical embedded processor architectures like ARM, ARC, MIPS, and Xtensa. However, it is very complex and hard to customize. It is primarily designed for "compiler-friendly" 32-bit RISC-like load-store architectures. While porting to more irregular architectures, such as DSPs, is possible as well, this generally results in huge retargeting effort and/or insufficient code quality.

5.2.3   Little C Compiler

Like gcc, retargeting the "little C compiler" lcc [26, 27] is enabled via a machine description file. In contrast to gcc, lcc is a "lightweight" compiler that comes with much less source code and only a few built-in optimizations; hence, lcc can be used to design compilers for certain architectures very quickly. The preferred range of target processors is similar to that of gcc, with some further restrictions on irregular architectures.

To retarget lcc, the designer has to specify the available machine registers, as well as the translation of C operations (or IR operations, respectively) to machine instructions by means of *mapping rules*. The following excerpt from lcc's Sparc machine description file [27] exemplifies two typical mapping rules:

```
addr:    ADDP4(reg,reg)    "%%%0+%%%1"
reg:     INDIRI1(addr)     "ldsb [%0],%%%c\n"
```

The first line instructs the code selector how to cover address computations ("addr") that consist of adding two 4-byte pointers ("ADDP4") stored in registers ("reg"). The string "%%%0+%%%1" denotes the assembly code to be emitted, where "%0" and "%1" serve as placeholders for the register numbers to be filled later by the register allocator (and "%%" simply emits the register identifier symbol "%"). Since "addr" is only used in context with memory accesses, here only a substring without assembly mnemonics is generated.

The second line shows the covering of a 1-byte signed integer load from memory ("INDIRI1"), which can be implemented by assembly mnemonic "ldsb," followed by arguments referring to the load address ("%0", returned from the "addr" mapping rule) and the destination register ("%c").

By specifying such mapping rules for all C/IR operations, plus around 20 relatively short C support functions, lcc can be retargeted quite efficiently. However, lcc is very limited in the context of non-RISC embedded processor architectures. For instance, it is impossible to model certain irregular register architectures (as, e.g., in DSPs) and there is no instruction scheduler, which is a major limitation for targets with ILP. Therefore, lcc has not found wide use in code generation for embedded processors so far.

5.2.4 CoSy

The CoSy system from ACE [28] is a retargetable compiler for multiple source languages, including C and C++. Like gcc, it includes several Dragon Book optimizations but shows a more modular, extensible software architecture, which permits us to add IR optimization passes through well defined interfaces.

For retargeting, CoSy comprises a backend generator that is driven by the *CGD machine description format*. Similar to gcc and lcc, this format is full-custom and designed for use only in compilation. Hence, retargeting CoSy requires significant compiler know-how, particularly with regard to code selection and scheduling. Although it generates the backend automatically from the CGD specification, including standard algorithms for code selection, register allocation, and scheduling, the designer has to fully understand the IR-to-assembly mapping and how the architecture constrains the instruction scheduler.

The CGD format follows the classical backend organization. It includes mapping rules, a register specification, and scheduler tables. The register specification is a straightforward listing of the different register classes and their availability for the register allocator (Fig. 5.15).

Mapping rules are the key element of CGD (Fig. 5.16). Each rule describes the assembly code to be emitted for a certain C/IR operation, depending on matching conditions and cost metric attributes. Similar to

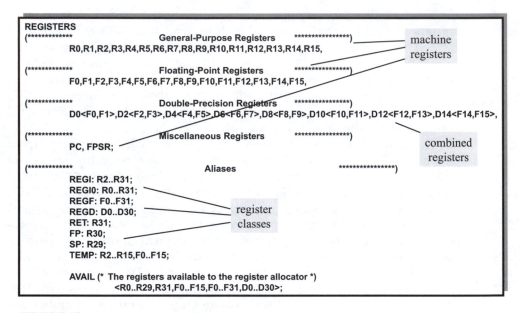

▪ **FIGURE 5.15**

CGD specification of processor registers.

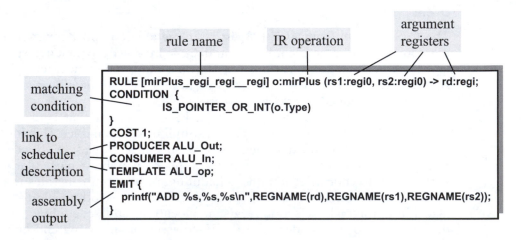

▪ **FIGURE 5.16**

CGD specification of mapping rules.

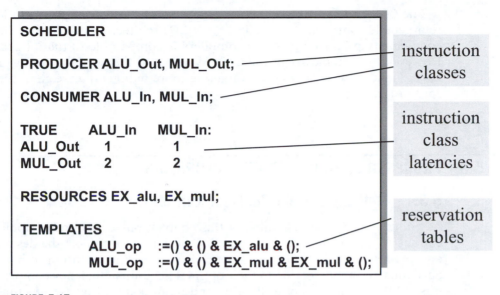

```
SCHEDULER

PRODUCER ALU_Out, MUL_Out;

CONSUMER ALU_In, MUL_In;

TRUE      ALU_In     MUL_In:
ALU_Out    1          1
MUL_Out    2          2

RESOURCES EX_alu, EX_mul;

TEMPLATES
          ALU_op   :=() & () & EX_alu & ();
          MUL_op   :=() & () & EX_mul & EX_mul & ();
```

instruction
classes

instruction
class
latencies

reservation
tables

■ **FIGURE 5.17**

CGD specification of scheduler tables.

gcc and lcc, the register allocator later replaces symbolic registers with physical registers in the generated code.

Mapping rules also contain a link to the CGD scheduler description. By means of the keywords "PRODUCER" and "CONSUMER," the instructions can be classified into groups, so as to make the scheduler description more compact. For instance, arithmetic instructions performed on a certain ALU generally have the same latency values. In the scheduler description itself (Fig. 5.17), the latencies for pairs of instruction groups are listed as a table of numerical values. As explained in Section 5.1.3, these values instruct the scheduler to arrange instructions a minimum amount of cycles apart from each other. Different types of interinstruction dependencies are permitted; here the keyword "TRUE" denotes data dependency.[2]

Via the "TEMPLATE" keyword, a reservation table entry is referenced. The "&" symbol separates the resource use of an instruction group over the different cycles during its processing in the pipeline. For instance, in the last line of Fig. 5.17, instruction group "MUL_op" occupies resource "EX_mul" for two subsequent cycles.

_____

[2] Data dependencies are sometimes called "true," since they are induced by the source program itself. Hence, they cannot be removed by the compiler. In contrast, there are "false" or antidependencies that are only introduced by code generation via reuse of registers for different variables. The compiler should aim at minimizing the amount of false dependencies, in order to maximize the instruction scheduling freedom.

The CGD processor modeling formalism makes CoSy quite a versatile retargetable compiler. Case studies for RISC architectures show that the code quality produced by CoSy compilers is comparable to that of gcc. However, the complexity of the CoSy system and the need for compiler background knowledge make retargeting more tedious than, e.g., in the case of lcc.

5.3   PROCESSOR ARCHITECTURE EXPLORATION

5.3.1   Methodology and Tools for ASIP Design

As pointed out in the beginning of this chapter, one of the major applications of retargetable compilers in SoC design is to support the design and programming of *ASIPs*. ASIPs receive increasing attention in both academia and industry due to their optimal flexibility/efficiency compromise [29]. The process of evaluating and refining an initial architecture model step by step to optimize the architecture for a given application is commonly called *architecture exploration*. Given that the ASIP application software is written in a high-level language like C, it is obvious that compilers play a major role in architecture exploration. Moreover, to permit frequent changes of the architecture during the exploration phase, compilers have to be retargetable.

Today's most widespread architecture exploration methodology is sketched in Fig. 5.18. It is an iterative approach that requires multiple remappings of the application code to the target architecture. In

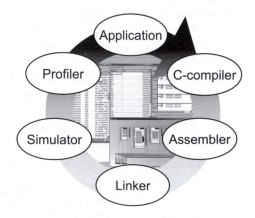

■ FIGURE 5.18

Processor architecture exploration loop.

each iteration, the usual software development tool chain (C compiler, assembler, linker) is used for this mapping. Since exploration is performed with a virtual prototype of the architecture, an instruction set simulator together with a profiler are used to measure the efficiency and cost of the current architecture with regard to the given (range of) applications, e.g., in terms of performance and area requirements.

We say that hardware (processor architecture and instruction set) and software (application code) "match" if the hardware meets the performance and cost goals and there is no over- or underutilization of HW resources. For instance, if the HW is not capable of executing the "hot spots" of the application code under the given timing constraints, e.g., due to insufficient function units, too much spill code, or too many pipeline stalls, then more resources need to be provided. On the other hand, if many function units are idle most of the time or half of the register file remains unused, this indicates an underutilization. Fine-grained profiling tools make such data available to the processor designer. However, it is still a highly creative process to determine the exact source of bottlenecks (application code, C compiler, processor instruction set, or microarchitecture) and to remove them by corresponding modifications while simultaneously considering their potential side effects.

If the HW/SW match is initially unsatisfactory, the ASIP architecture is further optimized, dependent on the bottlenecks detected during simulation and profiling. This optimization naturally requires hardware design knowledge and may comprise, e.g., addition of application-specific custom machine instructions, varying register file sizes, modifying the pipeline architecture, adding more function units to the data path, or simply removing unused instructions. The exact consequences of such modifications are hard to predict, so that usually multiple iterations are required in order to arrive at an optimal ASIP architecture that can be handed over to synthesis and fabrication.

With the research foundations of this methodology laid in the 1980s and 1990s (see [18] for a summary of early tools), several commercial offerings are available now in the EDA industry, and more and more startup companies are entering the market in that area. While ASIPs offer many advantages over off-the-shelf processor cores (e.g., higher efficiency, reduced royalty payments, and better product differentiation), a major obstacle is still the potentially costly design and verification process, particularly concerning the software tools shown in Fig. 5.18. In order to minimize these costs and to make the exploration loop efficient, all approaches to processor architecture exploration aim at automating the retargeting of these tools as much as possible. In addition, a link to hardware design has to be available in order to accurately estimate area, cycle time, and power consumption of a new ASIP. In most cases this is enabled by automatic HDL generation capabilities for processor models, which provide a direct entry to gate-true estimations via traditional synthesis flows and tools.

One of the most prominent examples of an industrial ASIP is the Tensilica Xtensa processor [30] (see also Chapter 6). It provides a basic RISC core that can be extended and customized by adding new machine instructions and adjusting parameters, e.g., for the memory and register file sizes. Software development tools and an HDL synthesis model can be automatically generated. Application programming is supported via the gcc compiler and a more optimizing in-house C compiler variant. The Tensilica Xtensa, together with its design environment, completely implement the exploration methodology from Fig. 5.18. On the other hand, the use of a largely predefined RISC core as the basic component poses limitations on the flexibility and the permissible design space. An important new entry to the ASIP market is Stretch [31]. Their configurable S5000 processor is based on the Xtensa core but includes an embedded field programmable gate array (FPGA) for processor customization. While FPGA vendors have combined processors and configurable logic on a single chip for some time, the S5000 "instruction set extension fabric" is optimized for implementation of custom instructions, thus providing a closer coupling between processor and FPGA. In this way, the ASIP becomes purely software-configurable and field-programmable, which reduces the design effort, but at the expense of reduced flexibility.

5.3.2   ADL-Based Approach

More flexibility is offered by the tool suite from Target Compiler Technologies [23], which focuses on the design of ASIPs for signal processing applications. In this approach, the target processor can be freely defined by the user in the nML ADL. In contrast to a purely compiler-specific machine model, such as in the case of gcc or CoSy's CGD, an ADL such as nML also captures information relevant for the generation of other software development tools, e.g., simulator, assembler, and debugger, and hence covers a greater level of detail. On the other hand, in contrast to HDL-based approaches to retargetable compilation, such as MIMOLA, the abstraction level is still higher than RTL and usually allows for a concise explicit modeling of the instruction set. The transition to RTL only takes place once the ADL model is refined to an HDL model for synthesis.

LISATek is another ASIP design tool suite that originated at Aachen University [32]. It has first been commercialized by LISATek, Inc., and is now available as a part of CoWare's SoC design tool suite [33]. LISATek uses the LISA 2.0 ADL for processor modeling. A LISA model captures the processor resources, such as registers, memories, and instruction pipelines, as well as the machine's ISA. The ISA model is composed of *operations* (Fig. 5.19), consisting of *sections* that describe the binary coding, timing, assembly syntax, and behavior of machine operations at

```
OPERATION ADD IN pipe.EX {
  // declarations
  DECLARE {
   INSTANCE writeback;
   GROUP src1, dst = {reg};
   GROUP src2 = {reg || imm};}

  // assembly syntax
  SYNTAX {"addc" dst "," src1 "," src2}

  // binary encoding
  CODING { 0b0101 dst src1 src2 }

  // behavior (C code)
  BEHAVIOR {
   u32 op1, op2, result, carry;
   if (forward) {
    op1 = PIPELINE_REGISTER(pipe,EX/WB).result;}
   else {
    op1 = PIPELINE_REGISTER(pipe,DC/EX).op1;}
   result = op1 + op2;
   carry = compute_carry(op1, op2, result);
   PIPELINE_REGISTER(EX/WB).result = result;
   PIPELINE_REGISTER(EX/WB).carry = carry;}

  // pipeline timing
  ACTIVATION {writeback, carry_update}
}
```

■ **FIGURE 5.19**

LISA operation example: execute stage of an ADD instruction in a cycle-true model with forwarding hardware modeling.

different abstraction levels. In an instruction-accurate model (typically used for early architecture exploration), no pipeline information is present, and each operation corresponds to one instruction. In a more fine-grained, cycle-accurate model, each operation represents a single pipeline stage of one instruction. LISATek permits the generation of software development tools (compiler, simulator, assembler, linker, debugger) from a LISA model, and embeds all tools into an integrated graphical user interface (GUI) environment for application and architecture profiling. In addition, it supports the translation of LISA models to synthesizable VHDL and Verilog RTL models. Fig. 5.20 shows the intended ASIP design flow with LISATek. In addition to an implementation of the exploration loop from Fig. 5.18, the flow also comprises the synthesis path via HDL models, which enables back-annotation of gate-level hardware metrics.

In [34] it has been exemplified how the LISATek architecture exploration methodology can be used to optimize the performance of an ASIP for an IPv6 security application. In this case study, the goal was to

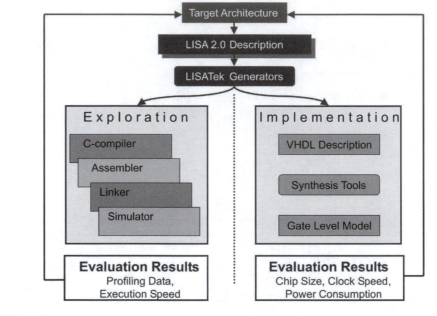

LISATek-based ASIP design flow.

enhance a given processor architecture (MIPS32) by means of dedicated machine instructions and a microarchitecture for fast execution of the computing-intensive Blowfish encryption algorithm (Fig. 5.21) in IPsec. Based on initial application C code profiling, hot spots were identified that provided the first hints on appropriate custom instructions. The custom instructions were implemented as a coprocessor (Fig. 5.22) that communicates with the MIPS main processor via shared memory. The coprocessor instructions were accessed from the C compiler generated from the LISA model via compiler intrinsics. This approach was feasible due to the small number of custom instructions required, which can be easily utilized with small modifications of the initial Blowfish C source code. LISATek-generated instruction-set simulators embedded into a SystemC-based cosimulation environment were used to evaluate candidate instructions and to optimize the coprocessor's pipeline microarchitecture on a cycle-accurate level.

Finally, the architecture implementation path via LISA-to-VHDL model translation and gate-level synthesis was used for further architecture fine-tuning. The net result was that Blowfish execution over the original MIPS was increased by a factor of five, at the expense of an additional coprocessor area of 22K gates. This case study demonstrates that ASIPs can

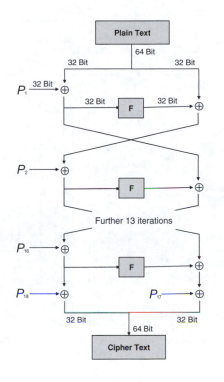

Blowfish encryption algorithm for IPsec: P_i denotes a 32-bit subkey, F denotes the core subroutine consisting of substitutions and add/xor operations.

provide excellent efficiency combined with **IP** reuse opportunities for similar applications from the same domain. Simultaneously, the iterative, profiling-based exploration methodology permits us to achieve such results quickly, i.e., typically within a few personnel weeks.

The capability of modeling the ISA behavior in arbitrary C/C++ code makes LISA very flexible with regard to different target architectures and enables the generation of high-speed ISA simulators based on the JITCC technology [35]. As in the MIMOLA and Target approaches, LISATek follows the "single golden model" paradigm, i.e., only one ADL model (or automatically generated variants of it) is used throughout the design flow in order to avoid consistency problems and to guarantee "correct-by-construction" software tools during architecture exploration. Under this paradigm, the construction of retargetable compilers is a challenging problem, since in contrast to special-purpose languages like CGD, the ADL model is not tailored toward compiler support only. Instead, similar to **MIMOLA/MSSQ** (see Section 5.2.1), the compiler-relevant information needs to be extracted with special techniques. This is discussed in more detail in the next section.

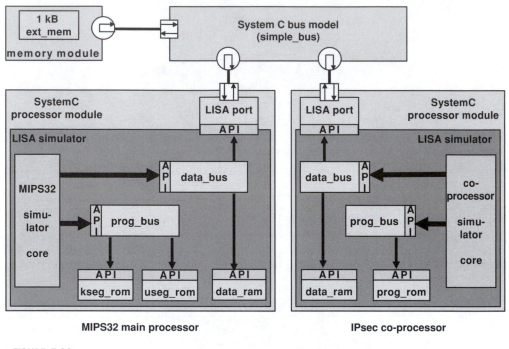

■ FIGURE 5.22

MIPS32/coprocessor system resulting from Blowfish architecture exploration (simulation view).

5.4   C COMPILER RETARGETING IN THE LISATek PLATFORM

5.4.1   Concept

The design goals for the retargetable C compiler within the LISATek environment were to achieve high flexibility and good code quality at the same time. Normally, these goals are contradictory, since the more the compiler can exploit knowledge of the range of target machines, the better the code quality is, and vice versa. In fact, this inherent tradeoff has been a major obstacle for the successful introduction of retargetable compilers for quite some time.

However, a closer look reveals that this only holds for "push-button" approaches to retargetable compilers, where the compiler is expected to be retargeted fully automatically once the ADL model is available. If compiler retargeting follows a more pragmatic user-guided approach (naturally at the cost of a slightly longer design time), then one can escape from this dilemma. In the case of the LISA ADL, an additional constraint is the unrestricted use of C/C++ for operation behavior descriptions. Due to the need for flexibility and high simulation speed, it is impossible to sacrifice this description vehicle. On the other hand, this makes it very difficult to automatically derive the compiler semantics of operations, due to large syntactic variances in operation descriptions. In addition, hardware-oriented languages like ADLs do not at all contain certain types of compiler-related information, such as C-type bit widths and function calling conventions, which makes an interactive, GUI-based retargeting environment useful, anyway.

To maximize the reuse of existing, well tried compiler technology and to achieve robustness for real-life applications, the LISATek C compiler builds on the CoSy system (Section 5.2.4) as a backbone. Since CoSy is capable of generating the major backend components (code selector, register allocator, scheduler) automatically, it is sufficient to generate the corresponding CGD fragments (see Figs. 5.15–5.17) from a LISA model in order to implement an entire retargetable compiler tool chain.

5.4.2 Register Allocator and Scheduler

Out of the three backend components, the register allocator is the easiest one to retarget, since the register information is explicit in the ADL model. As shown in Fig. 5.15, essentially only a list of register names is required, which can be largely copied from the resource declaration in the LISA model. Special cases (e.g., combined registers, aliases, special purpose registers such as the stack pointer) can be covered by a few one-time user interactions in the GUI.

As explained in Section 5.1.3, generation of the instruction scheduler is driven by two types of tables: latency tables and reservation tables. Both are only implicit in the ADL model. Reservation tables model interinstruction conflicts. Similar to the MSSQ compiler (Section 5.2.1), it is assumed that all such conflicts are represented by instruction encoding conflicts.[3] Therefore, reservation tables can be generated by examining the instruction encoding formats in a LISA model.

Fig. 5.23 and 5.24 exemplify the approach for two possible instruction formats or *compositions*. Composition 0 is VLIW-like and allows us to encode two parallel 8-bit instructions. In composition 1, all 16 instruction bits are required due to an 8-bit immediate constant that needs to be encoded. In the corresponding LISA model, these two formats are modeled by means of a switch/case language construct.

The consequences of this instruction format for the scheduler are that instructions that fit into one of the 8-bit slots of composition 0 can be scheduled in either of the two, while an instruction that requires an immediate operand blocks other instructions from being scheduled in

---

[3] This means that parallel scheduling of instructions with conflicting resource usage is already prohibited by the instruction encoding itself. Architectures for which this assumption is not valid appear to be rare in practice; if necessary, there are still simple workarounds via user interaction, e.g., through manual addition of artificial resources to the generated reservation tables.

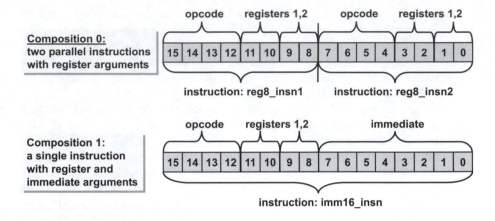

■ FIGURE 5.23

Two instruction encoding formats (compositions).

```
OPERATION decode_op
{
 DECLARE
 {
   ENUM  composition = {composition0,
                        composition1};
   GROUP  reg8_insn1, reg8_insn2 = {reg8_op};
   GROUP imm16_insn = {imm16_op};
 }
 SWITCH(compositions)
 {
   CASE composition0:
   {
    CODING AT (progam_counter)
       {insn_reg == reg8_insn1 | | reg8_insn2}
    SYNTAX {reg8_insn1 "," reg_insn2}
   }
   CASE composition1:
   {
    CODING AT (progam_counter)
       {insn_reg == imm16_insn}
    SYNTAX {imm16_insn}
   }
 }
}
```

■ FIGURE 5.24

LISA model fragment for instruction format from Fig. 5.23.

parallel. The scheduler generator analyzes these constraints and constructs *virtual resources* to represent the interinstruction conflicts. Naturally, this concept can be generalized to handle more complex, realistic cases for wider VLIW instruction formats. Finally, a reservation table in CGD format (see Fig. 5.17) is emitted for further processing with CoSy.[4]

The second class of scheduler tables, latency tables, depends on the resource access of instructions as they run through the different instruction pipeline stages. In LISA, cycle-accurate instruction timing is described via *activation sections* inside the LISA operations. One operation can invoke the simulation of other operations downstream in the pipeline during subsequent cycles, e.g., an instruction fetch stage would typically be followed by a decode stage and so forth. This explicit modeling of pipeline stages makes it possible to analyze the reads and writes to registers at a cycle-true level. In turn, this information permits us to extract the different types of latencies, e.g., due to a data dependency.

An example is shown in Fig. 5.25, where there is a typical four-stage pipeline (fetch, decode, execute, writeback) for a load/store architecture with a central register file.[5] By tracing the operation activation chain, one can see that a given instruction makes two read accesses in stage "decode" and a write access in stage "writeback." For arithmetic instructions executed on the ALU, for instance, this implies a latency value of 2 (cycles) in case of a data dependency. This information can again be translated into CGD scheduler latency tables (Fig. 5.17). The current version of the scheduler generator, however, is not capable of automatically analyzing forwarding/bypassing hardware, which is frequently used to minimize latencies due to pipelining. Hence, the fine-tuning of latency tables is performed via user interaction in the compiler retargeting GUI, which allows the user to add specific knowledge in order to override potentially too conservative scheduling constraints, thereby improving code quality.

5.4.3 Code Selector

As sketched in Section 5.1.3, retargeting the code selector requires specification of instruction patterns (or mapping rules) used to cover a data flow graph representation of the IR. Since it is difficult to extract instruction semantics from an arbitrary C/C++ specification as in the LISA behavior models, this part of backend retargeting is least automated. Instead, the GUI offers the designer a mapping dialog (Fig. 5.26;

---

[4] We also generate a custom scheduler as an optional bypass of the CoSy scheduler. The custom one achieves better scheduling results for certain architectures [36].
[5] The first stage in any LISA model is the "main" operation that is called for every new simulation cycle, similar to the built-in semantics of the "main" function in ANSI C.

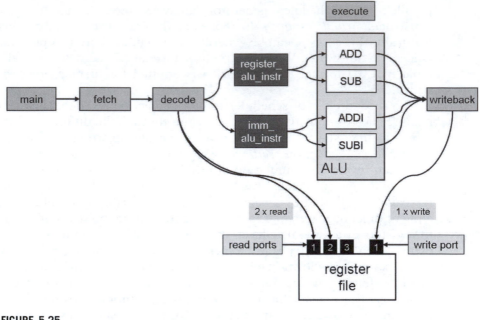

Register file accesses of an instruction during its processing over different pipeline stages.

see also [37]) that allows for a manual specification of mapping rules. This dialog enables the "drag-and-drop" composition of mapping rules, based on (a) the IR operations needed to be covered for a minimal operational compiler and (b) the available LISA operations. In the example from Fig. 5.26, an address computation at the IR level is implemented with two target-specific instructions (LDI and ADDI) at the assembly level.

Although significant manual retargeting effort is required with this approach, it is much more comfortable than working with a plain compiler generator such as CoSy (Section 5.2.4), since the GUI hides many compiler internals from the user and takes the underlying LISA processor model explicitly into account, e.g., concerning the correct assembly syntax of instructions. Moreover, it ensures very high flexibility with regard to different target processor classes, and the user gets immediate feedback on consistency and completeness of the mapping specification.

The major drawback, however, of this approach is a potential *model consistency* problem, since the LISA model is essentially overlaid with a (partially independent) code selector specification. To eliminate this problem, yet retain flexibility, the LISA language has recently been enhanced with *semantic sections* [38]. These describe the behavior of

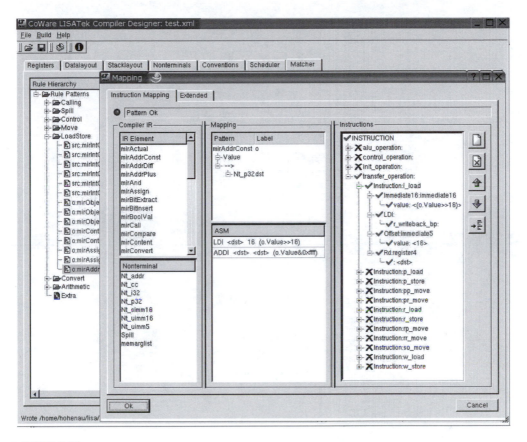

■ **FIGURE 5.26**

GUI for interactive compiler retargeting (code selector view).

operations from a pure compiler perspective and in a canonical fashion. In this way, semantic sections eliminate syntactic variances and abstract from details such as internal pipeline register accesses or certain side effects that are only important for synthesis and simulation.

Fig. 5.27 shows a LISA code fragment for an ADD instruction that generates a carry flag. The core operation ("dst = src1 + src2") could be analyzed easily in this example, but in reality, more C code lines might be required to capture this behavior precisely in a pipelined model. The carry flag computation is modeled with a separate if-statement, but the detailed modeling style might obviously vary. On the other hand, the compiler only needs to know that (a) the operations adds two registers, and (b) it generates a carry, independent of the concrete implementation.

The corresponding semantics model (Fig. 5.28) makes this information explicit. Semantics models rely on a small set of precisely defined *micro-operations* ("_ADDI" for "integer add" in this example) and capture

```
OPERATION ADD {
 DECLARE {
  GROUP src1, dst = {reg};
  GROUP src2 = { reg || imm };}
 SYNTAX { "add" dst "," src1 "," src2 }
 CODING { 0b0000 src1 src2 dst }
 BEHAVIOR {
  dst = src1 + src2;
  if (((src1 < 0) && (src2 < 0)) ||
      ((src1 > 0) && (src2 > 0) &&
       (dst < 0)) ||
      ((src1 > 0) && (src2 < 0) &&
       (src1 > -src2)) ||
      ((src1 < 0) && (src2 > 0) &&
       (-src1 < src2)))
  { carry = 1; }}}
```

▪ **FIGURE 5.27**

Modeling of an ADD operation in LISA with carry flag generation as a side effect.

```
OPERATION ADD {
 DECLARE {
  GROUP src1, dst = { reg };
  GROUP src2 = { reg || imm };}
 SYNTAX { "add" dst "," src1 "," src2 }
 CODING { 0b0000 src1 src2 dst }
 SEMANTICS { _ADDI[_C] ( src1, src2 ) -> dst; }}

OPERATION reg {
 DECLARE {
  LABEL index; }
 SYNTAX { "R" index = #U4 }
 CODING { index = 0bxxxx }
 SEMANTICS { _REGI(R[index])<0..31> }}
```

▪ **FIGURE 5.28**

Compiler semantics modeling of the ADD operation from Fig. 5.27 and a micro-operation for register file access (micro-operation "_REGI").

compiler-relevant side effects with special attributes (e.g., "_C" for carry generation). This is feasible, since the meaning of generating a carry flag (and similar for other flags like zero or sign) in instructions like ADD does not vary between different target processors.

Frequently, there is no one-to-one correspondence between IR operations (compiler dependent) and micro-operations (processor dependent). Therefore, the code selector generator that works with the semantic sections must be capable of implementing complex IR patterns by sequences of micro-operations. For instance, it might be needed to implement a 32-bit ADD on a 16-bit processor by a sequence of an ADD followed by an ADD-with-carry. For this "lowering," the code selector generator relies on an extensible default library of *transformation rules*. Vice versa, some LISA operations may have complex semantics (e.g., a DSP-like multiply-accumulate) that cover multiple IR operations at a time. These complex instructions are normally not needed for an operational compiler but should be utilized to optimize code quality. Therefore, the code selector generator analyzes the LISA processor model for such instructions and automatically emits mapping rules for them.

The use of semantic sections in LISA enables a much higher degree of automation in code selector retargeting, since the user only has to provide the semantics per LISA operation, while mapping rule generation is completely automated (except for user interactions possibly required to extend the transformation rule library for a new target processor).

The semantics approach eliminates the previously mentioned model consistency problem at the expense of introducing a potential *redundancy* problem. This redundancy is due to the coexistence of separate behavior (C/C++) and semantics (micro-operations) descriptions. The user has to ensure that behavior and semantics do not contradict. However, this redundancy is easily to deal with in practice, since behavior and semantics are local to each single LISA operation. Moreover, as outlined in [39], coexistence of both descriptions can even be avoided in some cases, since one can generate the behavior from the semantics for certain applications.

5.4.4 Results

The retargetable LISATek C compiler has been applied to numerous different processor architectures, including RISC, VLIW, and network processors. Most importantly, it has been possible to generate compilers for all architectures with limited effort, on the order of some person-weeks, depending on the processor complexity. This indicates that the semi-automatic approach outlined in Section 5.4 works for a large variety of processor architectures commonly found in the domain of embedded systems.

While this flexibility is a must for retargetable compilers, code quality is an equally important goal. Experimental results confirm that the code quality is generally acceptable. Fig. 5.29 shows a comparison between the gcc compiler (Section 5.2.2) and the CoSy based LISATek C compiler for a MIPS32 core and some benchmarks programs. The latter one is an "out-of-the-box" compiler that was designed within two person-weeks, while the gcc compiler, due to its wide use, most likely incorporates significantly more personnel hours. On the average, the LISATek compiler shows an overhead of 10% in performance and 17% in code size. With specific compiler optimizations added to the generated backend, this gap could certainly be further narrowed.

Further results for a different target (Infineon PP32 network processor) show that the LISATek compiler generates better code (40% in performance, 10% in code size) than a retargeted lcc compiler (Section 5.2.3),

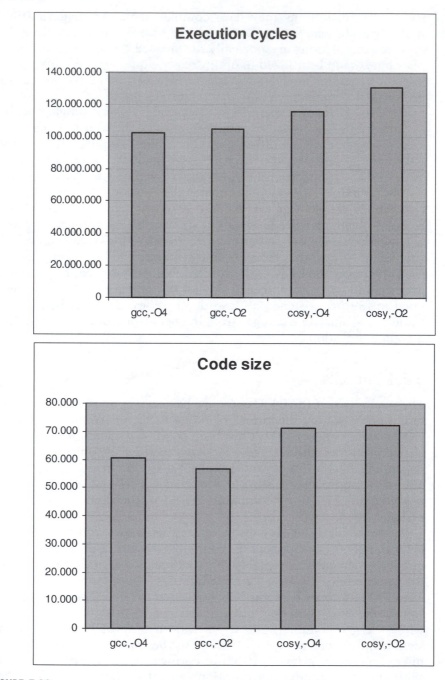

■ **FIGURE 5.29**

Code-quality comparison for a MIPS32 processor core.

due to more built-in code optimization techniques in the CoSy platform. Another data point is the ST200 VLIW processor, where the LISATek compiler has been compared to the ST Multiflow, a heavily optimizing target-specific compiler. In this case, the measured overhead has been 73% in performance and 90% in code size, which is acceptable for an "out-of-the-box" compiler that was designed with at least an order of magnitude less time than the Multiflow. Closing this code quality gap would require adding special optimization techniques, e.g., to utilize predicated instructions, which are currently ignored during automatic compiler retargeting. Additional optimization techniques are also expected to be required for highly irregular DSP architectures, where the classical backend techniques (Section 5.1.3) tend to produce unsatisfactory results. From our experience, we conclude that such irregular architectures can hardly be handled in a completely retargetable fashion, but will mostly require custom optimization engines for highest code quality. The LISATek/CoSy approach enables this by means of a modular, extensible compiler software architecture, naturally at the expense of an increased design effort.

5.5 SUMMARY AND OUTLOOK

Motivated by the growing use of ASIPs in embedded SoCs, retargetable compilers have made their way from academic research to EDA industry and application by system and semiconductor houses. While still being far from perfect, they increase design productivity and help to obtain better quality results. The flexibility of today's retargetable compilers for embedded systems can be considered satisfactory, but more research is required on how to make code optimization more retargetable.

We envision a pragmatic solution where optimization techniques are coarsely classified with regard to different target processor families, (e.g., RISCs, DSPs, NPUs, and VLIWs), each of which shows typical hardware characteristics and optimization requirements. For instance, software pipelining and utilization of SIMD instructions are mostly useful for VLIW architectures, while DSPs require address code optimization and a closer coupling of different backend phases. Based on a target processor classification given by the user with regard to these categories, an appropriate subset of optimization techniques would be selected, each of which is retargetable only within its family of processors.

Apart from this, we expect a growing research interest in the following areas of compiler-related EDA technology:

- *Compilation for low power and energy:* Low power and/or low energy consumption have become primary design goals for embedded systems. As such systems are more and more dominated by software executed by programmable embedded processors, it is obvious that compilers, also, may play an important role, since they control the code efficiency. At first glance, it appears that program energy minimization is identical to performance optimization, assuming that power consumption is approximately constant over the execution time. However, this is only a rule of thumb, and the use of fine-grained instruction-level energy models [40, 41] shows that there can be a trade-off between the two optimization goals, which can be explored with special code generation techniques. The effect is somewhat limited, though, when neglecting the memory subsystem, which is a major source of energy consumption in SoCs. More optimization potential is offered by exploitation of small on-chip (scratchpad) memories, which can be treated as entirely compiler-controlled, energy efficient caches. Dedicated compiler techniques, such as those described in [42, 43], are required to ensure an optimum use of scratchpads for program code and/or data segments.

▪ *Source-level code optimization:* Despite powerful optimizing code transformations at the IR or assembly level, the resulting code can be only as efficient as the source code passed to the compiler. For a given application algorithm, an infinite number of C code implementations exist, each possibly resulting in different code quality after compilation. For instance, downloadable reference C implementations of new algorithms are mostly optimized for readability rather than performance, and high-level design tools that generate C as an output format usually do not pay much attention to code quality. This motivates the need for code optimizations at the source level (e.g., C-to-C transformations) that complement the optimizations performed by the compiler while retaining the program semantics. Moreover, such C-to-C transformations are inherently retargetable, since the entire compiler is used as a backend in this case. Techniques like those described in [44–46] exploit the implementation space at the source level to significantly optimize code quality for certain applications, while tools like PowerEscape [47] focus on efficient exploitation of the memory hierarchy in order to minimize power consumption of C programs.

▪ *Complex application-specific machine instructions:* Recent results in ASIP design automation indicate that a high performance gain is best achieved with complex application-specific instructions that go well beyond the classical custom instructions, like multiply-accumulate for DSPs. While there are approaches to synthesizing such custom instructions based on application code analysis [48–50], the interaction with compilers is not yet well understood. In particular, tedious manual rewriting of the source code is still required in many cases to make the compiler aware of new instructions. This slows down the ASIP design process considerably, and in an ideal environment the compiler would automatically exploit custom instruction set extensions. This will require generalization of classical code selection techniques to cover more complex constructs like directed acyclic graphs or even entire program loops.

ACKNOWLEDGMENTS

The author gratefully acknowledges the contributions to the work described in this chapter made by M. Hohenauer, J. Ceng, K. Karuri, and O. Wahlen from the Institute for Integrated Signal Processing Systems at RWTH Aachen University, Germany. Research on the tools described in this chapter has been partially supported by CoWare, Inc.

CHAPTER 6

AUTOMATED PROCESSOR CONFIGURATION AND INSTRUCTION EXTENSION

David Goodwin, Steve Leibson, and Grant Martin

The idea of an ASIP—application-specific instruction-set processor—is central to every chapter in the book. Nevertheless, we will briefly review the ASIP concept to put the idea of automated processor configuration into context.

General-purpose processors are exactly that: processors that implement a relatively generic instruction set applicable to a wide variety of applications that cannot be easily predicted in advance or that are of a diverse nature. The trend in general-purpose microprocessor design over the last three decades has been to create faster and faster processors with deeper pipelines (to achieve high clock rates) and performance accelerators such as speculative execution, all focused on maximizing performance without particular concern for energy consumption, peak power usage, or cost. Recently this trend has hit a wall, and there has been a subsequent shift in emphasis to multicore processors, each running several threads to maximize general-purpose performance at lower clock rates.

Embedded processors generally have much more severe constraints on energy consumption and cost than do general-purpose processors. Over time, embedded processors based on classical RISC concepts have adopted many of the same clock-rate escalation and superscalar design approaches used by general-purpose microprocessors, gradually pushing up energy consumption and cost in gates per unit of useful computation.

There is a profound factor motivating the move to ASIPs: waste. Every unnecessary instruction executed wastes time and, more importantly, power. Every 32×32-bit multiplication of quantities that only require 20 bits of precision wastes time and more than 50% of the energy consumed in that computation. Every inefficient mapping of an application kernel into less-than-optimal code on a general-purpose processor consumes resources (user's time, battery life, energy) better spent on other uses. This is well described in Chapter 2.

But how can an ASIP be created? Tailoring a processor to an application has been more of an art than an exact science, and, as described in Chapter 3, the process demands a considerable amount of effort if done on a manual ad hoc basis. Many existing approaches to ASIP creation require the in-depth knowledge of a processor architect, the software knowledge of applications specialists, and the hardware-implementation skills of a team of experienced digital designers. Both structural, coarse-grained configuration parameters (for example, the inclusion or exclusion of functional units, the width of processor-to-memory or bus interfaces, the number and size of local and system memories) and fine-grained instruction extensions (the addition of application-specific tuned instructions that accelerate the processing of major functional application kernels by a factor of $2\times$, $10\times$, and more) are possible in ASIP configuration. Deciding on the specific configuration parameters and extended instructions can be akin to finding the optimal needle in the proverbial haystack—and requires years of broad experience in a host of design disciplines.

With this in mind, the design and use of ASIPs on a widespread basis across multiple application domains demand a more automated process for creating these processors from high-level configuration specifications.

6.1 AUTOMATION IS ESSENTIAL FOR ASIP PROLIFERATION

An automated process to create an ASIP, driven by an application specification, has proven feasible through use in dozens of taped-out and fabricated SoC designs [1, 2]. This result is somewhat contrary to the pessimism shown by Fisher, Faraboschi, and Young. However, as they discuss, the issue is not creating one or more customized processors for an application; it is creating a brand new microarchitecture and toolchain for each design. As our experience has shown, this level of effort is not necessary in many practical applications, because customizing a well designed and proven microarchitecture via an automated process has been proven to work effectively. Such an automated process generates both the hardware implementation of the configured processor core and the relevant software tooling, including the instruction-set simulator, compiler, debugger, assembler, and all related tools for creating applications for the processor and evaluating their performance. It is also essential that such a creation process be fast. To effectively explore the design space for the processor configuration, developers must be able to transform a specification into an implementation in minutes or hours. Because the process of ASIP configuration and instruction extension is usually iterative and incremental, it is a prerequisite, at least with manual specification methods, that a design team be able to evaluate several configurations a day, and in the course of a 2- or 3-week period, examine as many as 50 or 60 different configurations to find a reasonably optimal choice. Such design-space exploration is only possible with an automated ASIP configuration process and a disciplined methodology (such as the MESCAL methodology discussed in [3]).

But even with an automated methodology and toolflow for generating both the processor hardware and a software tool chain from a configuration and instruction-extension specification, the question remains: How best to define that specification? When considering in particular the fine-grained specification of instruction extensions, the skills required of a designer or design team are often difficult to come by. Certainly, a design team developing such a specification for a specific application domain needs to understand aspects of hardware microarchitecture design and the implications of concurrent processing. In addition, members of the team need to understand the application from the sequential programming point of view and understand how to structure new instructions to allow relative ease of programming. Thus a combination of hardware and software skills coupled with deep application understanding often is required to build a successful specification for an ASIP. Unfortunately, this is not an easy skill set for a team to acquire, especially at the beginning of an ASIP development process.

So what is the next step in the evolution of an ASIP methodology? Jumping from the idea of a general-purpose processor to the idea of an ASIP was the first big step. The development of a highly automated, low-risk, and reliable ASIP-generation process and methodology, driven by a compact and efficient specification, was the second step. The third step is automating the creation of the ASIP specification, based on an automated analysis and abstraction of the underlying application source code.

Chapter 5, by Leupers, gives an overview of the various techniques used to configure and extend processors. Chapter 7, by Pozzi and Ienne discusses some research into the extension process. Additional research into automated configuration exploration is discussed in Chapter 11, by Parameswaran, Henkel, and Cheung. This chapter discusses a practical, commercial example of such techniques—and one that has been used on many real-life product designs.

6.2 THE TENSILICA XTENSA LX CONFIGURABLE PROCESSOR

The most recent generation of Tensilica configurable processor architectures is the Xtensa LX architecture introduced in the summer of 2004. The Xtensa LX architecture has a number of configuration options, as illustrated in Fig. 6.1. In addition, instruction extensions can be described in an instruction set architecture (ISA) description language, the Tensilica instruction extension (TIE) language. We can regard these

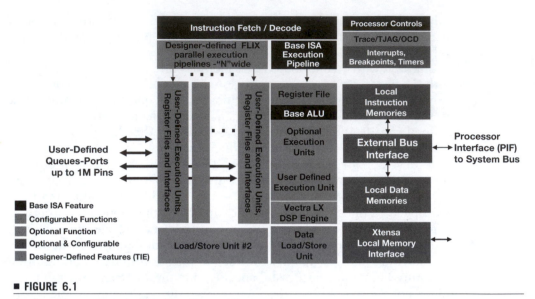

■ FIGURE 6.1

Configuration options for Tensilica Xtensa LX processors.

configuration/extension options as being coarse grained (structural) and fine grained (instruction extensions via TIE).

For the Xtensa LX processors, the configuration options include:

- Single or multi-issue architecture with an opportunity for multiple execution units to be active simultaneously, configured by the user

- Flexible-length instruction-set extensions (FLIX) for efficient code size and an ability to intermix instruction widths and create the multioperation instructions to control the multi-issue architecture

- An optional second load/store unit to increase the classical ISA bandwidth where desired (for example, in DSP-type applications)

- Either a 5- or 7-stage pipeline, the latter to improve the matching between processor performance and on-chip memory speed

- Size of the register file

- Optional inclusion of specialized functional units, including 16- and 32-bit multipliers, multiplier-accumulator (MAC), single-precision floating-point unit, and DSP unit

- Configurable on-chip debug, trace, and JTAG ports

- A wide variety of local memory interfaces, including instruction and data cache size and associativity, local instruction and data

RAM and ROM, general-purpose local memory interface (XLMI), and memory-management capabilities including protection and address translation (these interfaces are configurable as to size and address-space mapping)

- Timers, interrupts, and exception vectors

- Processor bus interface (PIF), including width, protocol, and address decoding, for linking to classical on-chip buses

- TIE instruction extensions including special-purpose functions, dedicated registers, and wire interfaces mapped into instructions

- Options for communications directly into and out of the processor's execution units via TIE ports and queues that allow FIFO communications channels (either unbuffered, or n-deep), memories, and peripheral devices to be directly hooked into instruction extensions defined in the TIE language (as a result, multiprocessor systems using direct FIFO-based communications between execution units in the processor datapath are now possible)

6.3 GENERATING ASIPs USING XTENSA

TIE is a Verilog-like specification language that allows one to define instruction extensions that are directly implemented by an automated processor generator as hardware in the processor datapath and are then controlled by software in user code. The TIE language allows one to specify the input and output arguments of the new instructions, the storage elements/registers used by the instructions, and the syntax and semantics of the instructions. Tensilica's TIE compiler transforms these instruction extensions into both hardware implementations and modifications to the software tool flow, so that compilers, debuggers, simulators, and related software tools are aware of them. As an example of the impact of configuration and instruction extensions, an ASIP designed for the Data Encryption Standard (DES) application can execute that application more than 50 times faster than a general-purpose processor with the addition of very few instructions to accelerate the algorithm's inner loops. These extra dimensions of configurability and extensibility significantly open up the design space for embedded processors.

Beyond the base ISA, the Tensilica processor can be configured with multiple execution units. With clever compilation and instruction encoding, these multiple execution units can be controlled by embedded software in an efficient and highly application-specific way. Multiple configured processors can be assembled into a subsystem using a variety of intertask communications semantics to accomplish an embedded application. In particular, the ability to hook FIFO queues directly into the execution pipeline so that data produced by a task running on one processor can be directly passed to a task running on a second opens up the possibility of building high-performance, programmable subsystems that can replace fixed-hardware implementations of functions in many application areas.

It is essential that the configurable processor have a highly automated generation process to speed development and reduce design errors. In Tensilica's case, this capability is offered via an Internet-based configuration service accessible to designers through client-based software. The automated process allows designers to enter configuration options via a series of entry screens, where the feasibility of the set of configuration parameters is cross-checked during entry, and the designer is given warnings or errors for erroneous parametric combinations. Instruction extensions are created in the TIE language using an editor that is part of the Tensilica integrated development environment (IDE), Xtensa Xplorer, which is based on Eclipse (www.eclipse.org). The TIE compiler checks the designer-defined instruction-extension code and generates a local copy of the relevant software tools (compiler, instruction-set simulator, and so forth), which can be used for rapid software development and experimentation long before chip fabrication.

At some point, the designer will want to generate the complete set of software tools and the HDL hardware description for the processor. The generation process allows the configuration options and the designer-specified TIE files to be uploaded to Tensilica's secure server environment. Within one to two hours, the complete ASIP will be generated

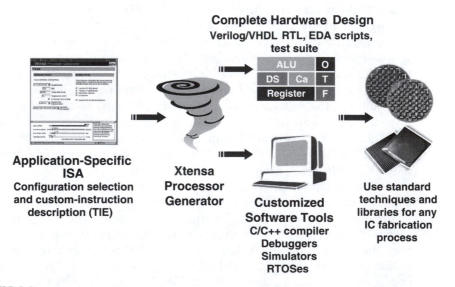

Application-Specific ISA
Configuration selection and custom-instruction description (TIE)

Xtensa Processor Generator

Complete Hardware Design
Verilog/VHDL RTL, EDA scripts, test suite

Customized Software Tools
C/C++ compiler
Debuggers
Simulators
RTOSes

Use standard techniques and libraries for any IC fabrication process

▪ **FIGURE 6.2**

The Tensilica processor generation process.

and verified. The software tools and RTL for the ASIP can then be downloaded to a local workstation for further work on software development, instruction extension, and performance analysis and optimization. This automated ASIP generation process is extensively verified [4] in the process of developing each new release of Tensilica software and hardware. It has been validated in field use by more than 70 licensees on many different processor instantiations over the last several years. Fig. 6.2 illustrates the Tensilica processor configuration and generation process.

This very brief overview of the Tensilica processor architecture, configuration, and extension possibilities is elaborated in much more detail in [1]. Experience with the approach has produced performance-optimized ASIP implementations for a large variety of applications [3]. The impact of configurable extensible ASIPs on power requirements and energy consumption is described in [5].

6.4 AUTOMATIC GENERATION OF ASIP SPECIFICATIONS

The next major step in ASIP development is to move from automatic generation based on manually generated ASIP specifications to the automatic generation of ASIP specifications. The XPRES compiler [6] represents both a major advance in the art of automated configurable-processor specification and a first step in moving processor-based design to a higher level of abstraction.

XPRES enhances the Xtensa processor generator (XPG) by eliminating the need to manually create ISA extensions. As shown in Fig. 6.3(a), XPRES analyzes a C/C++ application and generates sets of configuration options and TIE descriptions that specify a family of ASIPs.

Each set of configuration options and TIE code is processed by the XPG to produce an ASIP implementation and corresponding software tool chain. To target an application to a particular ASIP, the developer simply recompiles the original application or a related application using the C/C++ compiler generated by the XPG. The customized C/C++ compiler automatically exploits the ASIP's application-specific ISA extensions to increase application performance, as shown in Fig. 6.3(b).

ASIP generation using XPRES has four phases:

1. XPRES profiles the application, identifies performance-critical functions and loops, and collects the operation mix and dependence graph for each.

2. XPRES uses the application profile to generate a family of potential ASIPs, each providing a performance improvement at the cost of some additional hardware. The family of potential ASIPs may contain millions of candidates, representing combinations of custom

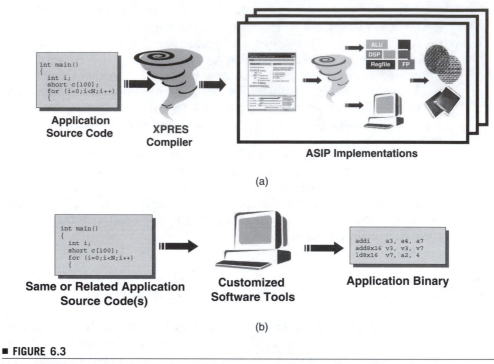

(a)

(b)

■ **FIGURE 6.3**

Using XPRES to automatically create a family of ASIPs.

ISA extensions that exploit the instruction, data, and pipeline parallelism present in the application.

3. Each potential ASIP is evaluated based on estimated hardware gate count and application performance impact to produce a Pareto-style curve of the lowest cost ASIP at every achievable performance level. The family of ASIPs range from those providing a modest performance improvement at small cost (for example, 10% to 30% speedup at a cost of a few hundred gates) to those providing a dramatic performance increase at a larger cost (for example, 20 times speedup at a cost of 100,000 gates).

4. For each Pareto-optimal ASIP, the corresponding set of configuration options and TIE description is created.

The result of this process is a set of ASIPs with a range of performance/cost characteristics (cost translates into silicon area on the SoC) presented on a Pareto-style curve, as shown in Fig. 6.4. The development team needs only to pick the ASIP with the right performance and cost characteristics for the target application and then submit that ASIP to the XPG for implementation.

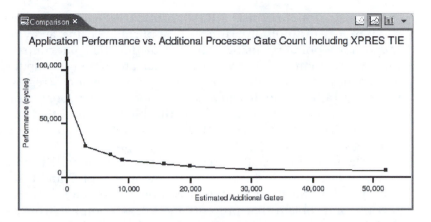

■ **FIGURE 6.4**

The XPRES Compiler presents the designer with a series of ASIPs that provide increasing amounts of application-specific performance for an increasing amount of silicon area.

6.5 CODING AN APPLICATION FOR AUTOMATIC ASIP GENERATION

To get maximum performance from a traditional ASIP, the developer must write an application that exploits the specific abilities of the ASIP. The developer may need to use intrinsics or assembly code to access ASIP operations that cannot be inferred automatically by the compiler. The developer may also need to structure an application's algorithms and data specifically for the ASIP. Then, when the application is ported to a new ASIP, the developer must repeat the optimization task because the intrinsics, assembly, algorithm structure, and data structures required for the new ASIP are likely to differ significantly from the existing ASIP.

XPRES eliminates these problems by exposing a meta-ISA. The meta-ISA allows the developer to write just one version of the application in a high-level language. The meta-ISA represents a large space of ASIPs capable of exploiting varying degrees of instruction, data, and pipeline parallelism. So, by writing the application to a meta-ISA, the developer is targeting a range of ASIP implementations with widely varying performance and cost characteristics.

To target the meta-ISA, the developer simply implements the application's algorithms in a clean manner that exposes the maximum amount of parallelism, without intrinsics or assembly code. Often, the algorithm's meta-ISA implementation will closely resemble the mathematical or standards-body definition of the algorithm, simplifying maintenance and debugging.

The *meta-ISA* concept is a little abstract. Perhaps the easiest way to think of it is to think of writing a C application using array references rather than pointers, using standard arithmetic and control instructions rather than intrinsics, using regular and simply computed access expressions to data rather than complex expressions (regular strides in arrays rather than irregular ones), and nesting loops to a limited degree in a natural order. Indeed, the kind of code that works well is often very similar to code that works well in hardware behavioral synthesis.

As the following results show, exploiting data parallelism often provides the largest performance improvement. Use of C and C++ pointers can make it difficult for the C/C++ compiler to extract data parallelism. A common problem that prevents the C/C++ compiler from extracting data parallelism is aliasing, the potential that two different pointers will access the same data element(s). Another common problem that prevents the compiler from producing the most efficient code is alignment. The Xtensa C/C++ compiler supports additional directives that allow the developer to express the aliasing and alignment properties of application pointers. Using these directives, the developer can annotate an application to expose data parallelism.

6.6 XPRES BENCHMARKING RESULTS

Before discussing in more detail how XPRES works, we will summarize some benchmark results that show how ASIP generation can be completely automated for realistic applications. Tensilica has extensively benchmarked its Xtensa LX processor both out of the box (unaugmented) and using manual processes for processor configuration and extension. These tests have been carried out using benchmarks from the Embedded Microprocessor Consortium (EEMBC) and BDTI, for example. The XPRES technology offers a new wrinkle in benchmarking because it automates the process of creating ASIPs that achieve the EEMBC "full-fury" level of optimization without manual effort.

Under EEMBC's out-of-the-box benchmarking rules, results from code compiled with the Tensilica C/C++ compiler and run on an XPRES-generated processor qualify as out-of-the-box (not optimized) results. The results of an EEMBC consumer benchmark run using such a processor appear in Fig. 6.5. This figure compares the performance of a stock Xtensa V processor core with that of an Xtensa LX processor core that has been automatically extended using the XPRES Compiler. (Note: Fig. 6.5 compares the Xtensa V and Xtensa LX microprocessor cores running at 260 and 330 MHz, respectively, but the scores are reported on a per-megahertz basis, canceling out performance differences attributable to clock rate. The standard versions of the Xtensa V and Xtensa LX

Benchmark Score
Iterations Comparison

Consumer Benchmarks

Processor Name-Clock	Tensilica Xtensa T1050 -260	Tensilica Xtensa LX 330 MHz
ConsumermarkTM	.08696	.51997
Compress JPEG	.04796	0.056
Decompress JPEG	.06837	0.083
High Pass Grey-scale filter	0.45136	8.661
RGB to CYMK Conversion	0.50251	7.572
RGB to YIQ Conversion	0.33830	6.310

Processor Name-Clock	Tensilica Xtensa T1050 -260	Tensilica Xtensa LX 330 MHz
ConsumermarkTM	1.00	5.98
Compress JPEG	1.00	1.17
Decompress JPEG	1.00	1.21
High Pass Grey-scale filter	1.00	19.19
RGB to CYMK Conversion	1.00	15.07
RGB to YIQ Conversion	1.00	18.65

■ **FIGURE 6.5**

Tensilica's XPRES Compiler Benchmarking Results. Copyright EEMBC. Reproduced by permission.

processor cores have the same ISA and produce the same benchmark results.)

The XPRES Compiler boosted the stock Xtensa processor's performance by 1.17 times to 18.7 times, as reported under EEMBC's out-of-the-box rules. The time required to create this performance boost was a few hours. These performance results demonstrate the sort of performance gains designers can expect from automatic processor configuration. However, these results also suggest that benchmark performance comparisons become even more complicated with the introduction of configurable-processor technology and that the designers making comparisons of processors using such benchmarks must be especially careful to understand how the benchmark tests for each processor are conducted.

6.7 TECHNIQUES FOR ASIP GENERATION

When creating an application-specific ISA, XPRES employs a number of techniques to increase application performance. These techniques include very long instruction word (VLIW) instruction formats to exploit instruction parallelism, single instruction multiple data (SIMD) or vectorized operations to exploit data parallelism, and fusion operations to exploit pipeline parallelism. Each technique differs in the cost of the hardware required for its implementation. To demonstrate these techniques, we use the following example loop. In this example, a, b, and c are arrays of 8-bit data.

```
for (i = 0; i < n; i++)
  c[i] = (a[i] * b[i]) >> 4;
```

The base Xtensa LX processor is a single-issue RISC that does not contain VLIW instructions, SIMD operations, or fused operations. For comparison with XPRES-generated ASIPs, we configure the base Xtensa LX to have a 32-bit multiplier. The following assembly code implements our example loop. On a base Xtensa LX, the example loop requires 8 cycles per iteration.

```
loop:
    l8ui     a11,a10,0     (load 8 bit unsigned)
    l8ui     a12,a9,0
    addi     a9,a9,1       (add immediate)
    mul16u   a11,a11,a12   (multiply 16 bit unsigned)
    addi     a10,a10,1
    srai     a11,a11,4     (shift right arithmetic immediate)
    s8i      a11,a14,0     (store 8 bit)
    addi     a14,a14,1
```

The XPRES Compiler technology starts with a fully functional, 32-bit Xtensa LX microprocessor core and then adds hardware to it in the form of additional execution units and corresponding machine instructions to speed processor execution for the target application. Analysis of the source application code guides the generation of processor-configuration alternatives, which are evaluated using sophisticated performance and area estimators. Using characterized and tuned estimators allows XPRES to evaluate millions of possible microarchitectures as part of its design-space exploration in a short period of time—a few seconds to a few hours, depending on the size and characteristics of the application.

6.7.1 Reference Examples For Evaluating XPRES

To demonstrate how XPRES automatically generates a wide range of customized ASIPs, we present results for a variety of reference examples.

TABLE 6.1 ■ Reference examples

Example	Description
idwt	Inverse discrete wavelet transform
autocor	Auto-correlation
rgbcmyk	Color conversion
fft	Radix-4 fast Fourier transform
gsm-encode	GSM audio encoder
gsm-decode	GSM audio decoder
compress	Data compression
djpeg	Image compression
m88ksim	Processor simulation
li	Lisp interpreter

The results show the types of parallelism that can best be exploited for each example and the relative application speedup and hardware cost provided by each type of parallelism. Table 6.1 describes the examples used to illustrate automatic ASIP generation. All applications are compiled using the Xtensa C/C++ compiler, using profile-feedback and full interprocedural optimization. Based on information derived from the XPRES-generated TIE file describing an ASIP, the Xtensa C/C++ compiler automatically exploits the custom ISA extensions to optimize each application without source-code changes or assembly-language coding. Actual application speedup is determined using a cycle-accurate simulator, assuming all memory accesses hit in the cache.

6.7.2   VLIW-FLIX: Exploiting Instruction Parallelism

Including VLIW instructions in an ISA allows a single instruction to contain multiple independent operations. A VLIW *format* partitions an instruction into a number of *slots*, each of which may contain a set of operations. A C/C++ compiler can increase application performance by using software-pipelining and instruction-scheduling techniques to pack multiple operations into one VLIW instruction.

For example, an ASIP designed for the example loop can be created by extending the base Xtensa ISA with a VLIW format. The complete TIE description for one possible VLIW format is shown here. This format exploits the instruction parallelism of the example loop by allowing three operations to be executed in parallel. The f format specifies a 64-bit instruction composed of three slots; s0, s1, and s2. The operations allowed in each slot are specified by the slot_opcodes statements.

```
format f 64 { s0, s1, s2 }
slot_opcodes s0 { L8UI, S8I }
slot_opcodes s1 { SRAI, MUL16U }
slot_opcodes s2 { ADDI }
```

When the example loop is compiled to target this ASIP, the following assembly code is produced. The loop executes in just 3 cycles, 2.6 times the speedup compared to the base Xtensa LX processor.

```
loop:
   l8ui  a8,a9,0;      mul16u  a12,a10,a8;  addi  a9,a9,1
   l8ui  a10,a11,0;    nop;                 addi  a11,a11,1
   s8i   a13,a14,508;  srai    a13,a12,a4;  addi  a14,a14,1
```

XPRES considers several factors when estimating the hardware cost of a VLIW format. Because a VLIW format issues and executes multiple independent operations in parallel, the hardware must contain multiple parallel instruction decoders for these operations. Also, if multiple operations in the same VLIW format access the same register file, that register file must contain enough read and write ports to satisfy all possible accesses. In addition, if the VLIW format allows multiple instances of an operation to appear in the format, the execution hardware required to implement that operation must be duplicated so that the multiple instances of the operation can execute in parallel.

Fig. 6.6 shows the performance improvement and hardware cost that result when extending a base Xtensa processor with VLIW formats. Each

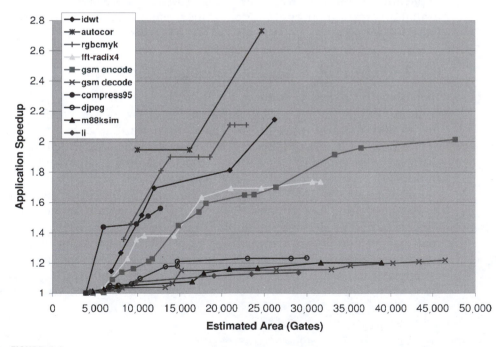

■ FIGURE 6.6

Application speedup and hardware cost for ASIPs exploiting instruction parallelism (VLIW-FLIX).

line in the chart represents an application, and each point on a line represents an ASIP generated by XPRES for that application. To exploit instruction parallelism, XPRES generates 32-bit and 64-bit instruction formats, each containing two, three, or four slots. Each slot contains a mix of the base Xtensa operations.

As the results show, the applications contain a range of instruction parallelism that can be exploited by the ASIPs. As a point of reference, the area occupied by 125,000 gates in a 130-nm process technology is about 1 mm^2, so the area occupied by the base Xtensa processor in the same process is about 0.25 mm^2. Thus the total ASIP area (not including memories) is $0.25 + (\text{gates}/125,000)\text{mm}^2$. For example, for the `autocor` benchmark, XPRES creates three ASIPs by adding VLIW extensions to the base Xtensa. For one ASIP, these extensions improve performance by nearly two times using about 10,000 extra gates. The total size of the ASIP is $0.25 + (10,000/125,000) = 0.33\text{mm}^2$.

For the Xtensa LX processor, VLIW multioperation instructions are implemented using Tensilica's FLIX technology. FLIX instructions consist of multiple independent operations, in contrast with the dependent multiple operations of fused and SIMD instructions. Each operation in a FLIX instruction is independent of the others, and the Xtensa C/C++ compiler can pack independent operations into a FLIX-format instruction as needed to accelerate code. While the native Xtensa processor instructions are 16 or 24 bits wide, FLIX instructions are 32 or 64 bits wide, to allow the needed flexibility to fully describe multiple independent operations. FLIX instructions can be seamlessly and modelessly intermixed with native Xtensa instructions. Thus FLIX avoids the code bloat that is often a side effect of classical VLIW instruction encoding.

6.7.3 SIMD (Vectorization): Exploiting Data Parallelism

Including SIMD operations in an ISA allows a single operation to perform a computation on more than one data element at a time. A SIMD operation is characterized by the operation it performs on each data element and by the *vector length*, the number of data elements that it operates on in parallel. A C/C++ compiler can increase application performance by using automatic parallelization and vectorization techniques to transform application loops to use SIMD operations.

For example, by extending the base Xtensa ISA with the set of SIMD operations required by the example loop, an ASIP can be designed to exploit the data parallelism available in the loop. A partial TIE description for one possible set of SIMD operations appears next. The TIE code describes two vector register files. Each register in the `v` register file holds a vector of eight 8-bit data elements. Each register in the `x` register file holds a vector of eight 16-bit elements. One SIMD operation is shown, `ashri16x8`, which arithmetically shifts each 16-bit element of an eight data element vector to the right by the amount specified by an immediate operand.

```
regfile vr8x8 64 4 v
regfile vr16x8 128 2 x
operation ashri16x8 { out vr16x8 a, in vr16x8 b, in vimm i } {
   assign a = { { {16{b[127]}}, b[127:112] } >> i[3:0],
                { {16{b[111]}}, b[111:96] } >> i[3:0],
                ...};
}
...
```

Compiling the example loop to target this ASIP produces a loop composed of SIMD operations that perform eight computations in parallel. The C/C++ compiler has used software pipelining to hide the latency of the multiply operation. Thus, the loop performs eight original iterations in just six cycles; 10.6 times the speedup compared to the base Xtensa processor.

```
loop:
   liu8x8          v0,a8,8
   ashri16x8       x0,x0,4
   liu8x8          v1,a9,8
   cvt16x8sr8x8s   v2,x0
   mpy8r16x8u      x0,v0,v1
   siu8x8          v2,a10,8
```

To estimate the hardware cost required to implement SIMD operations, XPRES considers the additional logic required to perform operations on multiple data elements in parallel and the vector register file(s) used to hold the vectors of data elements.

Fig. 6.7 shows the performance improvement and hardware cost that result when extending a base Xtensa processor with SIMD operations. The chart shows only those benchmarks that have a significant amount of data parallelism exploitable by XPRES. To exploit data parallelism, XPRES can generate SIMD operations that operate on vectors of 2, 4, 8, or 16 elements.

The Xtensa C/C++ compiler automatically vectorizes application loops using the XPRES-generated SIMD operations. Loop acceleration from vectorization is usually on the order of the number of SIMD units within the enhanced instruction. Thus, a 2-operation SIMD instruction approximately doubles loop performance, and an 8-operation SIMD instruction speeds up loop execution by about eight times.

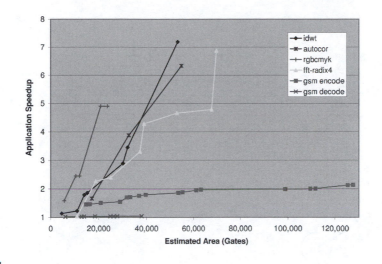

■ **FIGURE 6.7**

Application speedup and hardware cost for ASIPs exploiting data parallelism (SIMD).

6.7.4 Operator Fusion: Exploiting Pipeline Parallelism

A fused operation is created by combining two or more operations. Using a fused operation reduces the code size of the application and reduces the number of instructions that must be executed to complete the application. Also, the latency of the fused operation may be less than the combined latency of the operations it replaces.

Fig. 6.8(a) shows the dataflow graph for the body of the example loop. By extending the base Xtensa ISA with a fused operation that implements the portion of the dataflow graph highlighted in Fig. 6.8(b), the example loop can be implemented with the following assembly code.

```
loop:
    l8ui     a12,a10,0
    l8ui     a11,a9,0
    addi     a10,a10,1
    addi     a9,a9,1
    fusion   a14,a11,a12
```

With the fused operation, the example loop requires five cycles per iteration; this is a 1.6 times speedup compared to the base Xtensa processor.

The hardware cost of a fused operation is determined by the additional logic required to implement the fusion. The fusion may also require the addition of one or more register file ports to access its operands. However, fused operations that have constant inputs often have low hardware cost

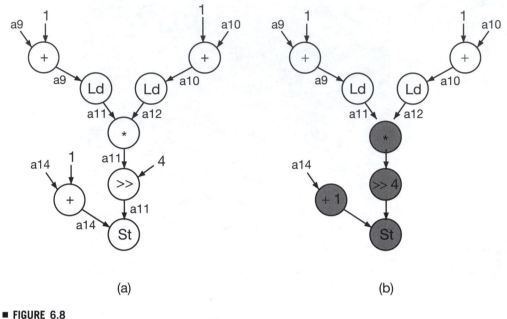

(a)   (b)

▪ **FIGURE 6.8**

Dataflow graph for the example loop and subgraph representing the fused operation.

(the shift-by-four in the example fusion is effectively free) while potentially providing significant performance benefit.

Fig. 6.9 shows the performance improvement and hardware cost that result when extending a base Xtensa processor with fused operations. For these results, XPRES generates fused operations with up to three input operands, one output operand, and a latency of one or two cycles.

6.7.5 Combining Techniques

To get the maximum performance improvement for a given hardware cost, the designer must consider an ISA that can contain any combination of SIMD operations, fused operations, and operations that combine those techniques (for example, a single operation that can perform four parallel multiply-by-three-accumulate computations on two vectors of four integers, producing a result vector of four integers). In addition, the designer must consider the use of VLIW to allow multiple independent operations to be executed in parallel.

For example, by extending the base Xtensa ISA with SIMD and fused operations, and with a VLIW format, an ASIP can be designed to exploit the instruction, data, and pipeline parallelism available in the example

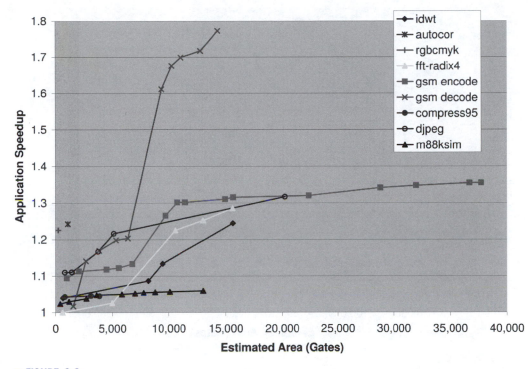

■ **FIGURE 6.9**

Application speedup and hardware cost for ASIPs exploiting pipeline parallelism (fusion).

loop. For this ASIP, the example loop can be implemented with the following assembly code.

```
loop:
si8x8   v2,a10,8;   liu8x8 v0,a8,8; fusion.mpy.ashrix8 v2,v0,v1
siu8x8  v5,a10,16;  liu8x8 v1,a9,8; fusion.mpy.ashrix8 v5,v3,v4
liu8x8  v3,a8,8;    liu8x8 v4,a9,8; nop
```

The loop requires three cycles to perform 16 iterations of the original loop; a 42.6 times speedup compared to the base Xtensa processor.

Fig. 6.10 shows the performance improvement and hardware cost that result when extending a base Xtensa processor with VLIW formats, SIMD operations, and fused operations. By combining the techniques, XPRES is able to significantly increase the performance of many of the benchmark applications. By selectively using some or all of the techniques presented earlier, these results also show how XPRES delivers a range of ASIPs for each application. These ASIPs span a wide range of

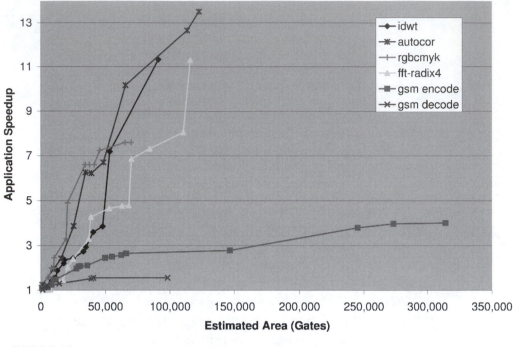

Application speedup and hardware cost for ASIPs exploiting instruction, data, and pipeline parallelism.

performance improvement and hardware cost, allowing the designer to optimize the design for the specific application and system requirements.

6.8   EXPLORING THE DESIGN SPACE

The number of configurations that XPRES can generate is, of course, vast. Therefore, it is important to use techniques to constrain design-space exploration to reasonable limits. XPRES essentially deals with the performance-critical loops found during application profiling. For each loop, several steps are performed that define the configuration space to be explored. If a loop can be vectorized, then configurations are generated for vector lengths ranging between a user-settable minimum and maximum; otherwise, only scalar operations are used.

Configuration generation then creates a set of seed configurations for a loop. These configurations include ones with no fused instructions and ones with candidate fusions that cover the loop dependence graph and improve the cycle count. The operations in the seed configurations are adjusted to support the current vector length under consideration.

The next step is to expand the configuration with additional resources (functional units or other hardware structures) that might increase performance. Here, XPRES looks at critical resources that most limit performance. Expansion of configurations stops, depending on the user-specified slot limits (multioperation limits) and the processor's memory interface limits.

The last step assigns operations and resources to issue slots, assuming that multiple-issue slots are available in the search space. This step has the objective of maximizing performance while minimizing resource cost. XPRES enumerates all resource-to-issue slot assignments and operation allocations are then explored for each possible assignment using heuristics that further try to lower hardware costs.

Each assigned, or slotted, configuration can then be estimated for overall hardware cost based on a series of estimators using factors such as register-file, functional-unit, memory-interface, instruction-fetch, and instruction-decode costs. The set of configurations is pruned based on minimum cycle count for the loop. A heuristic combines configurations together based on optimizing the overall cycle count for the set of loops in the application.

This procedure is explained in considerably more detail in [6].

6.9   EVALUATING XPRES ESTIMATION METHODS

XPRES estimates application performance and ASIP area using the following techniques. Fig. 6.11 shows the assumed idealized performance and area for a hypothetical family of ASIPS. As more ISA extensions are added to an ASIP, the area increases and the number of cycles required to execute the application decreases, as shown by the "Estimated" dataset. XPRES does not directly estimate the maximum-clock-frequency slowdown that likely results from increased ASIP area. The maximum clock frequency slowdown typically causes the actual ASIP performance to be less than the estimated performance, as shown by the "Real" dataset. XPRES assumes that the performance degradation caused by the decreased clock frequency will not offset the performance improvement achieved from the cycle count decrease.

An important metric in evaluating the accuracy of the XPRES estimation heuristics is the monotonicity of the real results. Let $E_{area}(n)$ represent the estimated hardware area for ASIP n, and $E_{perf}(n)$ represent the estimated application speedup for ASIP n. Similarly, let $R_{area}(n)$ and $R_{perf}(n)$ represent the real hardware area and application speedup for ASIP n. XPRES generates a Pareto-optimal family of ASIPs such that estimated performance monotonically increases with increasing estimated area; that is, for ASIPs x and y, $E_{area}(x) > E_{area}(y) \Rightarrow E_{perf}(x) > E_{perf}(y)$. To ensure that the design space is accurately explored, the real results

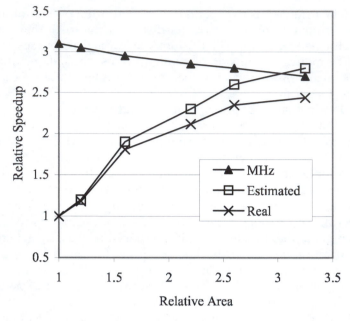

Idealized XPRES estimation results.

should have similar monotonic behavior; that is, for ASIPs x and y, $R_{area}(x) > R_{area}(y) \Rightarrow R_{perf}(x) > R_{perf}(y)$. In addition, an increase in estimated area or estimated performance should result in an increase in real area or real performance:

$$E_{area}(x) > E_{area}(y) \Rightarrow R_{area}(x) > R_{area}(y)$$
$$E_{perf}(x) > E_{perf}(y) \Rightarrow R_{perf}(x) > R_{perf}(y)$$

An example of a hypothetical family of ASIPs where the real performance results are not monotonically increasing is shown in Fig. 6.12. The two points marked x indicate the estimated and real performance for the ASIP with nonmonotonic real performance. Assuming that all other ASIP estimations result in monotonically increasing real performance, the shaded region represents the design space of ASIPs that is not accurately explored by the tool. The shaded region likely contains a large number of ASIPs that could have been created instead of ASIP x and that would potentially have had monotonically increasing real performance. The inaccurate estimates cause these ASIPs to be discarded.

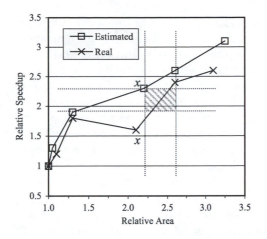

Example showing nonmonotonic ASIP performance due to inaccurate estimation.

6.9.1 Application Performance Estimation

XPRES estimates the performance of an application on a particular ASIP by estimating the decrease in cycles obtained by exploiting the ISA extensions provided by the ASIP. For each application region, r, XPRES first estimates $C_{\text{ref}}(r)$, the number of cycles required to execute the region's base operations on a reference Xtensa LX processor. The reference Xtensa LX implements a general-purpose RISC ISA with no FLIX formats, no SIMD operations, and no fused operations. XPRES then estimates $C_{\text{ASIP}}(r)$, the number of cycles required to execute the region using the FLIX formats, SIMD operations, and fused operations implemented by the ASIP. Finally, XPRES estimates application performance on the ASIP as a speedup relative to the reference processor. Given the set of application regions, R, and the total application cycle count on the reference processor, P, estimated application speedup is:

$$S_{\text{ASIP}} = \frac{P}{P - \sum_{r \in R}(C_{\text{ref}}(r) - C_{\text{ASIP}}(r))}$$

Using the earlier estimation, XPRES generates a family of potential ASIPs. To generate a more accurate performance estimate, the software toolchain created by the processor generator for each ASIP is used to compile and simulate the application. The resulting cycle count is used as the performance estimate for the ASIP.

6.9.2 ASIP Area Estimation

The XPRES area estimator relies on an *area characterization library* (ACL). The ACL accurately characterizes the area of primitive hardware building blocks. The building blocks include arithmetic operators such as adders, multipliers, and shifters; logical operators such as exclusive-or; storage elements such as flip-flops and latches; and basic hardware elements such as multiplexors and comparators. Each building block is parameterized by the bit-width of the corresponding logic. For certain building blocks, the number of parameters and range of allowed values for each parameter leads to a combinatorial explosion in the number of possible area characterizations. For these building blocks, a subset of allowed values is characterized and interpolation is used to estimate the remaining values.

To estimate the area required for an entire ASIP, XPRES uses the ACL to estimate the area of the data path logic that implements the operations, the area of the register files and state required by the operations, and the area of additional ASIP logic. To estimate the area of the data path logic, XPRES sums the estimated area of the arithmetic and logic operators that compose the data path and the estimated area of the storage elements used for pipelining the data path. XPRES's estimate of the area of each ASIP register file is based on the width and depth of the register file, the number of register-file read and write ports, and the pipeline stages that read and write each port. XPRES also estimates ASIP area required for operation-decode logic, instruction-fetch logic, and data-memory-interface logic. Together these estimates form the area estimate for the entire ASIP.

6.9.3 Characterization Benchmarks

Table 6.2 shows the three benchmarks used to evaluate the accuracy of the estimation heuristics used in XPRES. Each benchmark is compiled using the Xtensa C/C++ compiler at the highest optimization level, using interprocedural optimization and profile-feedback information. All performance and area results are shown relative to a reference Xtensa processor implementing a general-purpose ISA. To provide a more meaningful comparison, the reference Xtensa includes a number of ISA options such as a 32-bit multiplier (and thus has an area of about 62,000 gates, significantly larger than the base Xtensa processor area of 25,000 gates). The cycles required to execute an iteration of each benchmark on the reference Xtensa is shown in Table 6.2. For `autocor`, the benchmark operates on one frame of 500 16-bit samples with 32 lags. For `fft`,

TABLE 6.2 ■ Characterization benchmark descriptions

Benchmark	Ref. Cycles	Description
autocor	124,660	Auto-correlation
fft	25,141	Fast Fourier transform
gsm_encode	178,923	GSM encoder

the benchmark performs a 256-point radix-4 FFT. For `gsm_encode`, the benchmark operates on one frame of 160 16-bit samples.

6.9.4 Performance and Area Estimation

Fig. 6.13 shows the real and estimated performance and area for each benchmark, normalized to the performance of the reference Xtensa processor. From the family of ASIPs generated by XPRES using all

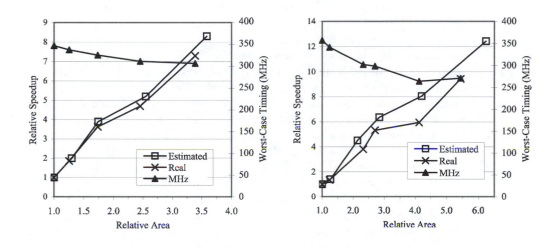

(a) autocor

(b) fft

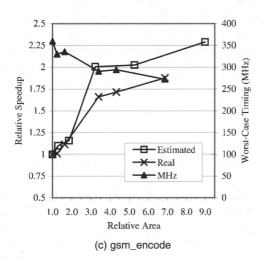

(c) gsm_encode

FIGURE 6.13

Timing and area characteristics.

possible types of ISA extensions, four ASIPs generated for `autocor`, five ASIPs generated for `fft`, and five ASIPs generated for `gsm_encode` are shown because they represent interesting tradeoffs between performance and area. The "Estimated" results show the performance and area for each ASIP as estimated by XPRES. As described at the beginning of Section 6.9, XPRES does not account for the potential maximum-clock-frequency reduction for larger ASIPs when estimating performance. The "Real" results show the actual ASIP performance, taking into account the difference in ASIP clock frequency shown in the "MHz" results. To determine the real performance and area, each ASIP is synthesized with a physical synthesis tool using a 130-nm low-Vt library and a frequency target of 400 MHz.

When the decrease in clock frequency is monotonic and in relative proportion to the area increase, the XPRES performance estimation increases monotonically and closely matches the real performance, as demonstrated by the `autocor` benchmark. For the `fft` benchmark, the clock frequency decrease is monotonic, but not in relative proportion to the area increase. As a result, the real performance increases monotonically but does not closely match the estimated performance for the two largest ASIPs. For the `gsm_encode` benchmark, the clock frequency decrease is not monotonic, but the nonmonotonic variations are small. So, the real performance still increases monotonically but does not closely match the estimated performance for several ASIPs.

Fig. 6.13 also shows the real and estimated ASIP area for each benchmark, normalized to the area of the reference Xtensa processor. For all benchmarks, the real area increases monotonically with increasing performance. For some ASIPs of the `fft` and `gsm_encode` benchmarks, there is significant difference between the estimated and real areas, especially for the larger ASIPs. This is likely caused by physical synthesis optimizations that are not accounted for by area estimation heuristics.

The estimation inaccuracies exposed by the `gsm_encode` and `fft` benchmarks are being explored to determine whether or not the estimators can be improved by accounting for how physical synthesis optimizations affect real timing and area.

6.10   CONCLUSIONS AND FUTURE OF THE TECHNOLOGY

ASIPs provide the time-to-market and programmability benefits of general-purpose processors while satisfying the demanding performance, size, and power requirements of embedded SoC designs. The XPRES compiler, by providing an easily targeted meta-ISA, removes the need for the application developer to target applications using the peculiarities of any particular ASIP. The meta-ISA represents a range of potential ASIPs that exploit instruction, data, and pipeline parallelism to increase application performance. Based on performance and hardware cost requirements, XPRES automatically generates a family of ASIPs from a single version of the application source code, allowing the developer to quickly explore performance and cost tradeoffs.

Benchmark application results demonstrate that XPRES is effective in generating ASIPs that exploit the wide range of instruction, data, and pipeline parallelism available in the applications. By exploiting all three types of parallelism, XPRES generates ASIPs that execute applications many times faster than a general-purpose processor. Calibration of XPRES technology shows a reasonable match between its area and performance estimators and real results, but there is room for improvement. These areas are being explored.

Another area of interest in further developing the technology is to look at a wider range of instruction fusions. XPRES currently aims to generate fairly generic fused instructions rather than highly specific ones, to some extent trading off ultimate performance for generality and future flexibility. By offering a wider range of instruction fusions including very application-specific ones within the design space being explored, XPRES can offer a wider range of ASIPs that can further improve performance at the cost of additional gates for highly application-specific instructions. In addition, a wider range of multicycle instructions can be automatically explored. In theory, with a wide range of deep, multicycle, highly application-specific fused instructions, combined with the SIMD and VLIW capabilities, XPRES should come very close to the optimized performance of application-specific hardware behavioral synthesis while retaining the flexibility, programmability, and generality of a processor-based design approach.

The XPRES compiler's ability to automatically generate a range of ASIP implementations and software tools, and the ease of use by ordinary software, hardware, and system designers, heralds a new era in automated embedded system design.

CHAPTER 7

AUTOMATIC INSTRUCTION-SET EXTENSIONS

Laura Pozzi and Paolo Ienne

Modern synthesizable embedded processors can typically be customized in two different senses: on one side, they often have optional or parametrizable features (e.g., optional load/store units or variable-size register files); on the other side, they are prepared for extensions of the instruction-set with application-specific or domain-specific instructions. In Chapter 6, the authors have concentrated on an example of a commercially available customizable processor and have primarily discussed how the large design space corresponding to the various options and parameters can be explored *automatically* by a tool. Complementarily, in this chapter we focus on understanding the potentials of current techniques to identify *automatically* the best instruction-set extensions for a given application. Note that instruction-set extensibility is often available also in ASIC processors distributed as hard-macros, as well as in processors implemented as hard cores in FPGA devices, and it therefore has a rather wide applicability. Specifically, we present here in succession a number of techniques and refinements that have been proposed in the last few years and that address increasingly complex instructions. After understanding existing techniques to automatically identify *instruction-set extensions (ISE)*, Chapter 8 will discuss with examples some remaining issues that still limit the capabilities of fully automatic ISE identification.

7.1 BEYOND TRADITIONAL COMPILERS

Many customizable processors are now on the market, and they come in a variety of forms and technologies. Some are synthesizable cores, while others are generally available as hard-macros; most target ASIC implementations, but an increasing number are meant for implementation on FPGAs. A list of the best known ones, certainly not exhaustive, includes ARC ARCtangent, Tensilica Xtensa, the STMicroelectronics ST200 family, MIPS CorExtend, Xilinx Virtex 4 and MicroBlaze, and Altera NIOS I and II. Automation is well developed in the toolchains offered with such devices: typically the toolchains include retargetable compilers, simulators, synthesizers, and debuggers; most of the time, these tools can be easily adapted to fit user-defined instruction set extensions.

This chapter focuses on a missing link in today's automation of customizable embedded processor design: as already mentioned, we address here the *automatic* identification of most profitable ISEs for a given application, directly from its source code. This represents a very interesting and actual challenge, where the goal is to devise techniques that are able to mimic the choices of an expert designer, but cut the long time to market traditionally necessary for many personnel-months of manual investigations and refinements. Automation can also help reduce risk by generating complex testbenches and/or property files to be used in verification.

Raising the level of processor design automation in such a direction—i.e., defining automatically its instruction set (or at least its extensions beyond a basic set)—represents a significant departure from the traditional concept of a compiler. This reasoning can be followed in Fig. 7.1. At first, a compiler was simply a translator from source-code to machine-code, possibly embodying a great sophistication in producing optimized code for such machine—see Fig. 7.1(a). Therefore, the role of a compiler was to *produce code for a fixed machine*. Then, research in retargetable compilers became prominent, recognizing the advantage of maintaining a single piece of code for compiling to different machine targets. This scenario is depicted in Fig. 7.1(b), where the machine description is an input to the compiler—possibly an input to the compiler compilation process, as in `gcc`, as opposed to an input to the process of running the

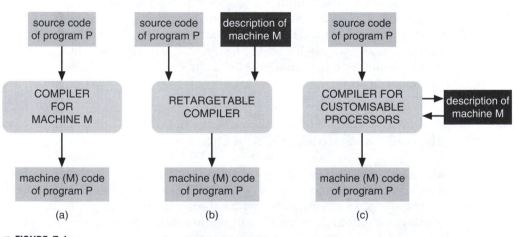

▪ **FIGURE 7.1**

(a) A "standard" compiler for a specific machine. (b) A retargetable compiler reads a machine description and generates code for it. (c) A compiler for a customizable processor can generate by itself the description of the best machine for a given application and then produce code for it.

compiler. Therefore, a retargetable compiler, at least conceptually, first *reads the underlying machine description*, then *produces code for it*. By raising the level of automation of embedded processor design, the techniques overviewed in this chapter revisit the role of the compiler and its importance. In fact, if it is possible to automate the process of tuning the machine's instruction set, then the compiler tends to become a tool that first *defines the underlying machine* and then *produces code for it*—as depicted in Fig. 7.1(c). In this chapter we will see how this is in fact already possible today—at least for a specific subset of the machine description, namely significant extensions of the instruction set.

7.1.1 Structure of the Chapter

We will begin by looking at the simplest problem: identifying clusters of elementary operations that would get the maximum advantage when performed in hardware as a single new instruction; in Section 7.2 we will show the basic approach to identifying new instructions for a processor, given the high-level application code. Since this approach could incur, in the worst case, a computational complexity that would grow exponentially with the problem size, we will briefly discuss in Section 7.3 some heuristics that can help coping with especially large instances of the problem. The idea of Section 7.2, and therefore of Section 7.3 too, is limited to additional instructions that are rather simple in nature: pure combinations of ALU operations. On the other hand, especially in ad hoc embedded systems, designers very often add some local registers and memories to the dedicated functional units, because in most cases these are extremely efficient and effective ways of accelerating operation. In Section 7.4, we will illustrate the challenges of discovering automatically which are the best values or data structures to store locally in the functional units. Finally, in Section 7.5, we will show how to kill two birds with one stone while addressing typical constraints of real microarchitectures: on one side, one has to pipeline complex functional units so as not to slow down the cycle time of the whole processor; on the other, the bandwidth between the register file and the functional units is generally limited, and large datasets must be sequentialized. It turns out that these two problems can be solved together, to mutual advantage.

7.2 BUILDING BLOCK FOR INSTRUCTION SET EXTENSION

Loosely stated, the problem of identifying instruction-set extensions consists of detecting clusters of operations that, when implemented as a single complex instruction, maximize some metric—typically performance. Such clusters must invariably satisfy some constraint; for instance, they must produce a single result or use no more than four input values. In this section, we will present an algorithm for identifying the best such clusters [1] for a given application code; it represents an improvement in some respects on a previously published one [2]. This algorithm will serve as a base for most of the extensions and advanced techniques detailed in the following sections.

It should be noted that it is an exact algorithm in that it completely explores the design space for the best solution (in the sense precised later). Other exact algorithms have been presented in recent years [3, 4], and all reach, essentially, the same solution with different degrees of efficiency.

7.2.1   Motivation

Fig. 7.2 shows the dataflow graph of the basic block most frequently executed in a typical embedded processor benchmark, *adpcmdecode*; nodes represent atomic operations (typically corresponding to standard processor instructions) and edges represent data dependencies. The first observation is that identification based on recurrence of clusters would hardly find candidates of more than three to four operations. Additionally, one should notice that recurring clusters such as M0 have several inputs and could be often prohibitive. In fact, choosing larger, albeit non-recurrent, clusters might ultimately reduce the number of inputs and/or outputs: subgraph M1 satisfies even the most stringent constraints of two operands and one result. An inspection of the original code suggests that this subgraph represents an approximate 16×4–bit multiplication and is therefore the most likely manual choice of a designer even under severe area constraints. For different reasons, most previous algorithms would bail out before identifying such large subgraphs, and yet, despite the apparently large cluster size, the resulting functional unit can be implemented as a very small piece of hardware.

Availability of a further input would also include the following accumulation and saturation operations (subgraph M2 in Fig. 7.2). Furthermore, if additional inputs and outputs are available, one would like to implement *both* M2 and M3 as part of the same instruction—thus exploiting the parallelism of the two disconnected graphs. The algorithm presented here is exact and identifies all of the previously mentioned instructions depending on the given user constraints [1].

7.2.2   Problem Statement: Identification and Selection

We call $G(V, E)$ the *direct acyclic graphs* (DAGs) representing the *data flow graph* (DFG) of each basic block; the nodes V represent primitive operations, and the edges E represent data dependencies. Each graph G is associated to a graph $G^+(V \cup V^+, E \cup E^+)$, which contains additional nodes V^+ and edges E^+. The additional nodes V^+ represent input and

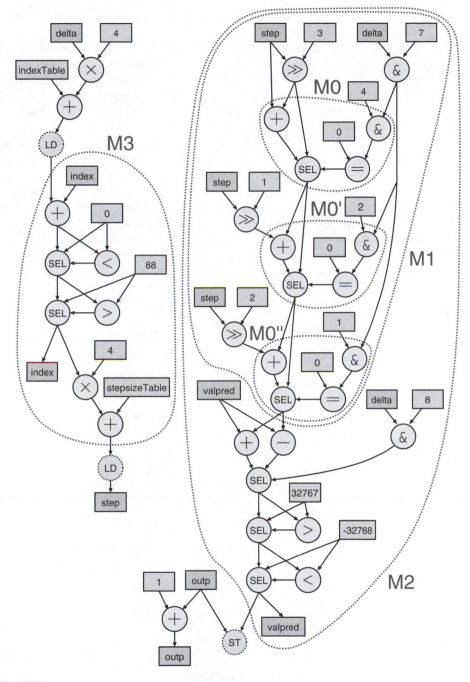

■ **FIGURE 7.2**

Motivational example from the *adpcmdecode* benchmark [5]. **SEL** represents a selector node and results from applying an if-conversion pass to the code.

output variables of the basic block. The additional edges E^+ connect nodes V^+ to V, and nodes V to V^+.

A *cut* S is an induced subgraph of G: $S \subseteq G$. There are $2^{|V|}$ possible *cuts*, where $|V|$ is the number of nodes in G. An arbitrary function $\mathrm{M}(S)$ measures the merit of a cut S. It is the objective function of the optimization problem introduced next and typically represents an estimation of the speedup achievable by implementing S as a special instruction.

We call IN (S) the number of predecessor nodes of those edges that enter the cut S from the rest of the graph G^+. They represent the number of input values used by the operations in S. Similarly, OUT (S) is the number of predecessor nodes in S of edges exiting the cut S. They represent the number of values produced by S and used by other operations, either in G or in other basic blocks. We call the cut S *convex* if there exists no path from a node $u \in S$ to another node $v \in S$ that involves a node $t \notin S$. Fig. 7.3(b) shows an example of nonconvex cut.

Finally, the values N_{in} and N_{out} express some features of the microarchitecture and indicate the register-file read and write ports, respectively,

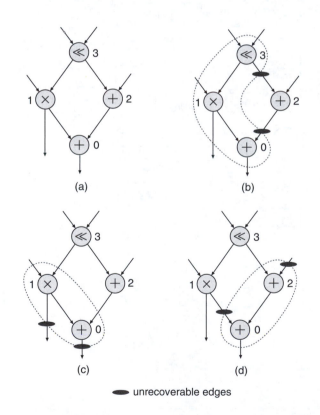

▪ **FIGURE 7.3**

(a) A topologically sorted graph. (b) A nonconvex cut. (c) A cut violating the output check, for N_{out} = 1. (d) A cut violating the permanent input check, for N_{in} = 1.

that can be used by the special instruction. Also, due to microarchitectural constraints, some operation types might not be allowed in a special instruction, depending on the underlying organization chosen. This may reflect, for instance, the existence or absence of memory ports in the functional units: when no memory ports are desirable, load and store nodes must be always excluded from possible cuts, as reflected in the example of Fig. 7.2 (nodes labeled LD and ST represent memory loads and stores, respectively). We call F (with $F \subseteq V$) the set of *forbidden* nodes, which should never be part of S. This constraint can be somehow relaxed, as it was shown by Biswas et al. [6, 7]; we will discuss such extensions in Section 7.4.

Considering each basic block independently, the identification problem can now be formally stated as follows:

Problem 1 (single-cut identification) *Given a graph G^+ and the microarchitectural features N_{in}, N_{out}, and F, find the cut S that maximizes M(S) under the following constraints:*

  1. IN $(S) \leq N_{in}$,

  2. OUT $(S) \leq N_{out}$,

  3. $F \cap S = \emptyset$, and

  4. S is convex.

The convexity constraint is a legality check on the cut S and is needed to ensure that a feasible schedule exists: as Fig. 7.3(b) shows, if all inputs of an instruction are supposed to be available at issue time and all results are produced at the end of the instruction execution, no possible schedule can respect the dependences of this graph once S is collapsed into a single instruction.

We call N_{bb} the number of basic blocks in an application. Since we will allow several special instructions from all N_{bb} basic blocks, we will need to find up to N_{instr} disjoint cuts, that together, give the maximum advantage. This problem, referred here as *selection*, could be solved non-optimally by repeatedly solving Problem 1 on all basic blocks and by simply selecting the N_{instr} best ones. Formally, the problem that we want to solve is:

Problem 2 (selection) *Given the graphs G_i^+ of all basic blocks, the respective sets F_i, and the microarchitectural features N_{in} and N_{out}, find up to N_{instr} cuts S_j that maximize $\sum_j M(S_j)$ under the same constraints of Problem 1 for each cut S_j.*

One should notice that this formulation implicitly assumes that the benefits of multiple instructions S_j—represented by M (S_j)—are perfectly additive. In some architectures, such as RISC and single-issue processors, this is practically exact. In other architectures, such as VLIW, some secondary effects might make the merit function not exactly additive; however, the detailed computation of the advantage of several cuts simultaneously would require scheduling the remaining operations and would therefore be infeasible in this context.

This formulation of the problems addresses ISE exclusively limited to data flow operations. We only indirectly attack the control flow by applying typical compiler transformations aimed at maximizing the exploitable parallelism, such as if-conversion. We believe that a comprehensive solution of the identification of dataflow clusters is a precondition to the exploration of the control flow. Similarly, the approach presented here does not address the storage of some variables in the functional unit implementing an instruction—in other words, the functional units we describe at this point may contain pipeline registers but not architecturally visible ones: as mentioned already, Section 7.4 addresses and relaxes this issue.

7.2.3   Identification Algorithm

Enumerating all possible cuts within a basic block exhaustively is not computationally feasible. We describe here an exact algorithm that explores the complete search space but effectively detects and prunes infeasible regions during the search. It exactly solves Problem 1 and is a building block for the solution of Problem 2; the reader interested in the latter is referred to the original paper [1].

The algorithm starts with a topological sort on G: nodes of G are ordered such that if G contains an edge (u, v) then u appears after v in the ordering. We write that $u \succ v$. Fig. 7.3(a) shows a topologically sorted graph. The algorithm uses a recursive search function based on this ordering to explore an abstract search tree.

The search tree is a binary tree of nodes representing possible cuts. It is built from a root representing the empty cut, and each couple of 1- and 0-branches at level i represents the addition or not of the node of G having topological order i, to the cut represented by the parent node. Nodes of the search tree immediately following a 0-branch represent the same cut as their parent node and can be ignored in the search. Fig. 7.4 shows the search tree for the example of Fig. 7.3(a), with some tree nodes labeled with their cut values. The search proceeds as a preorder traversal of the search tree. It can be shown that in some cases there is no need to branch toward lower levels; therefore, the search space is pruned.

A trivial case when it is useless to explore the subtree of a particular node in the search tree is when such a node represents a cut S, which contains a forbidden node in F. Clearly, such S cannot be a solution of Problem 1, nor can it be any solution that adds further nodes.

But the real usefulness of this traversal order can be understood from the following discussion. Suppose, for instance, that the output port constraint has already been violated by the cut defined by a certain tree

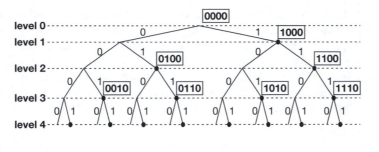

■ **FIGURE 7.4**

The search tree corresponding to the graph shown in Fig. 7.3(a).

node: adding nodes that appear later in the topological ordering (i.e., predecessors in the DAG) cannot reduce the number of outputs of the cut. An example of this situation is given in Fig. 7.3(c), where an output constraint of 1 is violated at inclusion of node 1 and cannot be recovered. Similarly, if the convexity constraint is violated at a certain tree node, there is no way of regaining the feasibility by considering the insertion of nodes of G that appear later in the topological ordering. Considering, for instance, Fig. 7.3(b) after inclusion of node 3, the only ways to regain convexity are to either include node 2 or remove from the cut nodes 0 or 3: due to the use of a topological ordering, both solutions are impossible in a search step subsequent to insertion of node 3. As a consequence, when the output-port or the convexity constraints are violated when reaching a certain search tree node, the subtree rooted at that node can be eliminated from the search space.

Additionally, we can also limit the search space when specific violations of the input-port constraint happen. For this we note that there are cases when edges entering a cut S from G^+ cannot be removed from IN (S) by further adding to the cut nodes of G that appear later in the topological ordering. We call $\text{IN}_f (S)$ the number of predecessor nodes of those edges that enter S from the rest of G^+ and either (1) belong to V^+ or F—that is, they come from either primary input variables or forbidden nodes, or (2) are nodes that have been already considered in the tree traversal and have been assigned a 0 in the cut—i.e., they have been excluded from the cut. In the first case, there is no node in the rest of G that can remove this input value. In the second case, nodes that could remove the inputs exist, but they have been already permanently excluded from the cut. Of course, it is always $\text{IN}_f (S) \leq \text{IN} (S)$, and we call such inputs *permanent*. As an example, consider Fig. 7.3(d); after inclusion of node 2, the cut has accumulated two permanent inputs: one is the external input of node 2, and another is the input of node 0, which was made permanent when node 1 was excluded from the cut. Similarly to what happens for the output-port and the convexity constraints, when the number of permanent inputs $\text{IN}_f (S)$ violates the input-port constraint when reaching a particular search-tree node, the subtree rooted at that node can be eliminated from the search space.

Fig. 7.5 gives the algorithm in pseudo C notation. The search tree is implemented implicitly, by use of the recursive `search` function. The parameter `current_choice` defines the direction of the branch, and the parameter `current_index` defines the index of the graph node and the level of the tree on which the branch is taken. The function `forbidden` returns true if S contains a node in F. The functions `input_port_check`, `permanent_input_port_check`, and `output_port_ check` return a true value when, respectively, $IN (S) \leq N_{in}$, $IN_f (S) \leq N_{in}$, and $OUT (S) \leq N_{out}$. The function `convexity_check` returns a true value when S is convex. When either the output port check, the permanent-input port check, or the convexity check fail, or when a leaf is reached during the search, the algorithm backtracks. The best solution is updated only if all constraints are satisfied by the current cut.

Fig. 7.6 shows the application of the algorithm to the graph given in Fig. 7.3(a) with $N_{out} = 1$ and $N_{in} = 1$. Only 4 cuts pass the output port check, the permanent input port check, and the convexity check, while 6 cuts are found to violate a constraint, resulting in elimination of 5 more cuts. Among 16 possible cuts, only 10 are therefore actually considered

```
identification() {
    for (i = 0; i < NODES; i++) cut[i] = 0;
    topological_sort();
    search(1, 0);
    search(0, 0);
}
search(current_choice, current_index) {
    cut[current_index] = current_choice;
    if (current_choice == 1) {
        if (forbidden()) return;
        if (!output_port_check()) return;
        if (!permanent_input_port_check()) return;
        if (!convexity_check()) return;
        if (input_port_check()) {
            calculate_speedup();
            update_best_solution();
        }
    }
    if ((current_index + 1) == NODES) return;
    current_index = current_index + 1;
    search(1, current_index);
    search(0, current_index);
}
```

▪ **FIGURE 7.5**

The single cut identification algorithm.

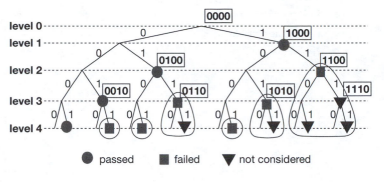

passed   **failed**   **not considered**

■ **FIGURE 7.6**

The execution trace of the single cut algorithm for the graph given in Fig. 7.3(a), for $N_{out} = 1$ and $N_{out} = 1$. Note that cut 1100 corresponds to the subgraph shown in Fig. 7.3(c)—it violates output constraints—and cut 1010 corresponds to the subgraph shown in Fig. 7.3(d)—it fails the permanent input check.

at all (the empty cut is never considered), and only 4 are effectively possible solutions of the problem.

The graph nodes contain $O(1)$ entries in their adjacency lists on average, since the number of inputs for a graph node is limited in every practical case. Combined with a single node insertion per algorithm step, the `input_port_check`, `permanent_input_port_check`, `output_port_check`, `convexity_check`, and `calculate_speedup` functions can be implemented in $O(1)$ time using appropriate data structures. The overall complexity of the algorithm is therefore $O(2^{|V|})$, but, although still exponential, the algorithm reduces in practice the search space very tangibly. Fig. 7.7 shows the run time performance of the algorithm using an input port constraint of four and an output port constraint of two on basic blocks extracted from several benchmarks. The actual performance follows rather a polynomial trend in all practical cases considered; however, an exponential tendency is also visible. Constraint-based subtree elimination plays a key role in the algorithm performance: the tighter the constraints are, the faster the algorithm is.

7.2.4 Results

Speedup is measured by means of a particular function M $(\cdot)$, assumed to express the merit of a specific cut. M (S) represents an estimation of the speedup achievable by executing the cut S as a single instruction in a specialized datapath. It should be noticed that the algorithm itself does not depend on the actual function M $(\cdot)$—it only requires the function to be evaluated efficiently.

In software, we estimate the latency in the execution stage of each instruction; in hardware, we evaluate the latency of each operation by synthesising arithmetic and logic operators on a common 0.18-μm CMOS

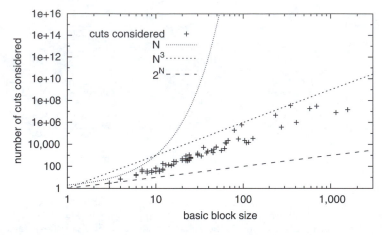

■ **FIGURE 7.7**

Number of cuts considered by the single cut algorithm with $N_{in} = 4$ and $N_{out} = 2$ for several graphs (basic blocks) with size varying between 2 and 1,500 nodes.

process and normalize to the delay of a 32-bit multiply-accumulate. The accumulated software values of a cut estimate its execution time in a single-issue processor. The latency of a cut as a single instruction is approximated by a number of cycles equal to the ceiling of the sum of hardware latencies over the cut critical path. The difference between the software and hardware latency is used to estimate the speedup.

The described algorithm was implemented within the MachSUIF framework and tested on several MediaBench [5], EEMBC [8], and cryptography benchmarks. In order to show the potentials of the algorithms described here with respect to the state of the art, a comparison is shown with a generalization of Clubbing [9], and with the MaxMISO [10] identification algorithms. The first is a greedy linear-complexity algorithm that can detect n-input m-output graphs, where n and m are user-given parameters. The second is a linear complexity algorithm that identifies maximal-size single-output and unbounded-input graphs. This comparison shows the power of the approach presented in this section when in search for large, possibly nonrecurrent, ISEs. Note that since the spirit of the presented approach is to find as large ISEs as possible, close to those found by experienced designers, recurrence of the identified subgraphs is not checked and is considered to be equal to one by default.

Furthermore, a simple algorithm was implemented, denoted with Sequences, that can identify recurrent sequences of two instructions only; this makes possible a comparison with works that propose as ISEs small, recurrent subgraphs, such as shift-add or multiply-accumulate. Sequences traverses the dataflow graph once and selects an ISE every time it identifies a positive-gain two-instruction sequence. Recurrence of the identified sequences is checked, and the gain is modified accordingly—i.e., it is multiplied by recurrence. Of course the algorithm thus constructed is very simple and less sophisticated than others presented in literature, but this simple implementation is meant to give the order of magnitude of the potential gain of such small ISEs, even when recurrence is considered.

Results for Iterative, Clubbing, MaxMISO, and Sequences can be observed in Fig. 7.8. Three main points should be noted. First, Iterative constantly outperforms other algorithms; in particular, for low input/output constraints, all algorithms have generally similar performances (see, for example, benchmarks *adpcmencode* and *viterbi*), but in the case of higher, and yet still reasonable, constraints, Iterative excels. Second, a large performance potential lays in multiple output, and therefore possibly disconnected, graphs, and the algorithms presented here are among the first ones to exploit it. Disconnected graphs are indeed chosen by Iterative in most benchmarks; an example is given by *adpcmdecode* and *aes*, discussed in detail later. Last, note that the potential in the large and nonrecurrent graphs chosen by Iterative is constantly much greater than that of small recurrent sequences, especially once the I/O constraints are loosened. This can be seen by observing the performance of Sequences, for example in benchmarks *fft* and *autocor*. Recall that current VLIW architectures like ST200 and TMS320 can commit 4 values per cycle and per cluster: it is therefore reasonable to loosen the constraints. Furthermore, Section 7.5 shows a method for exploiting large ISEs, requiring large I/O, even in presence of architectures that allow only a small number of read/write ports to the register file.

In the light of the motivation expressed with the help of Fig. 7.2, it is useful to analyze the case of *adpcmdecode*: (a) Clubbing is generally limited in the size of the instructions identified, and Sequences selects the couples *and-compare* and *shift-add* of subgraphs M0. (b) MaxMISO finds the correct solution (corresponding to M2 in the figure) with a constraint of more than two inputs. Yet, when given only two input ports, it cannot find M1 because M1 is part of the larger 3-input MaxMISO M2. (c) Iterative manages to increase the speedup further when multiple outputs are available; in such cases, it chooses a disconnected subgraph at once, consisting in M2+M3. Iterative *is the only algorithm that truly adapts to the available microarchitectural constraints.*

Of course, the worst-case complexity of Iterative is much higher than that of Clubbing, MaxMISO, or Sequences, but it is on average well below exponential complexity, as Fig. 7.7 shows. The run times of Iterative are shown in Fig. 7.9, for input/output constraints of 2-1, 4-2, 4-3, and 8-4, and for a single ISE. It can be seen that run times are almost exclusively in the order of seconds or lower; only benchmarks *md5* and *aes* (1,500 and 700 nodes, respectively) exhibit run times of the order of minutes for a constraint of 4-2 and of hours or cannot terminate for higher constraints.

Finally, one can wonder what the benefits are of identifying disconnected subgraphs. Of course, if only connected subgraphs are considered

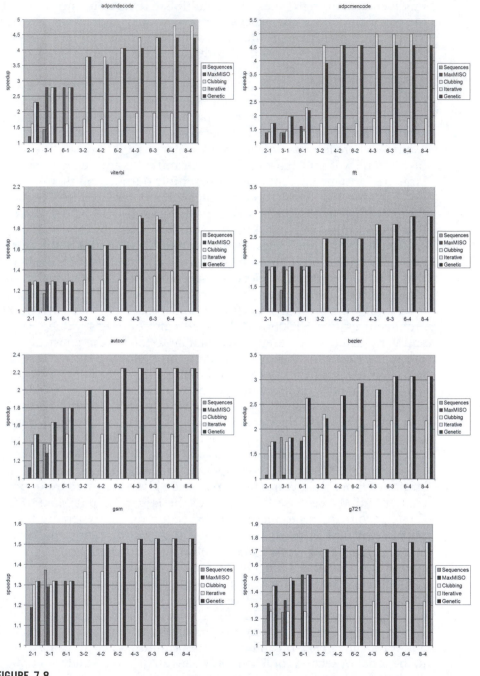

▪ **FIGURE 7.8**

Analysis of the performance of the algorithm described in this section, denoted with Iterative. Up to 16 ISEs were chosen for each bookmark.

	2-1	4-2	4-3	8-4
gsmdecode	0.002	0.001	0.001	0.001
g721encode	0.000	0.003	0.012	0.027
adpcmencode	0.001	0.038	0.837	10
adpcmdecode	0.000	0.025	0.467	4
viterbi	0.001	0.023	0.001	0.003
fft	0.000	0.004	0.027	0.138
autcor	0.000	0.000	0.001	0.001
bezier	0.000	0.004	0.033	0.131
md5	0.200	71	—	—
aes	0.093	77	8865	—

■ **FIGURE 7.9**

Run times of Iterative, in seconds, for input/output constraints of 2-1, 4-2, 4-3, and 8-4.

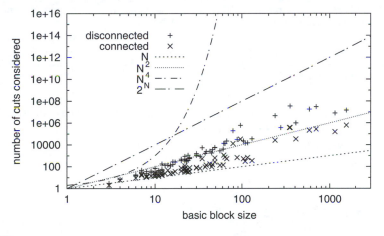

■ **FIGURE 7.10**

Number of cuts considered by Iterative when identifying potentially disconnected graph versus connected-only graph, for $N_{in} = 4$ and $N_{out} = 2$.

for ISE identification, the search space decreases considerably, and some authors have successfully and explicitly exploited such complexity reduction [3]. Figs. 7.10 and 7.11 show how the run time complexity decreases in practice when considering connected-only graphs versus possibly disconnected ones. Almost all state of the art proposals consider connected graphs only, but an algorithm able to identify large, disconnected graphs, such as Iterative, has the important capability of discovering additional forms of parallelism. In benchmarks where parallelism is abundant, such as *aes*, the difference in performance when considering disconnected subgraphs is considerable, as shown in Fig. 7.12. In particular, note that in the case of connected graphs only, performance fails to increase with the output constraint—i.e., it fails to take advantage of register port availability.

	2-2	4-2	6-2	8-2	2-3	4-3	6-3	8-3	8-4
disconnected	3.8e+6	3.0e+7	1.6e+8	6.4e+8	4.4e+8	2.1e+9	—	—	—
connected	4.0e+4	2.7e+5	8.9e+5	1.9e+6	4.8e+5	3.0e+5	9.6e+5	1.9e+6	1.9e+6

▪ FIGURE 7.11

Complexity difference (in terms of number of cuts considered for identifying the first ISE in the largest basic block of 700 nodes) for potentially disconnected and connected-only graphs, for benchmark *aes*, and algorithm Iterative.

	2-2	4-2	6-2	8-2	2-3	4-3	6-3	8-3	8-4
disconnected	2.17	2.80	2.93	3.14	2.30	3.14	3.28*	3.41*	3.88*
connected	2.02	2.47	2.76	2.76	2.02	2.47	2.76	2.76	2.76

▪ FIGURE 7.12

Speedup difference for potentially disconnected and connected-only graphs, for benchmark *aes*, 16 ISEs chosen, and algorithm Iterative. ('*' indicates a heuristic, based on genetic algorithms and not discussed here; it was used where Iterative could not terminate.)

7.3   HEURISTICS

The algorithm presented in the previous section is exact in solving Problem 1, and it has been experimentally shown to be able to handle basic blocks of the order of 1,000 nodes for I/O constraints of 4/2. In this section, we will see why and how one can go beyond this kind of rather extreme limitation.

7.3.1   Motivation

The core of the *aes* encryption algorithm is the "round transformation," operating on a 16-byte state and described in Fig. 7.13. The state is a two dimensional array of bytes consisting of four rows and four columns; the columns are often stored in four 32-bit registers and are inputs of the round transformation. First, a nonlinear byte substitution is applied on each of the state bytes by making table lookups from substitution tables stored in memory. Next, the rows of the state array are rotated over different offsets, and then a linear transformation called the Mix-Column transformation is applied to each column. Finally, an XOR with the round key is performed, and the output of a round transformation becomes the input of the next one. In the C code we have used, the core has been unrolled twice by the programmer, so that the size of the main basic block amounts to around 700 assembler-like operations.

The exact algorithm proposed in the previous section cannot handle such size for high (e.g., 8-4) I/O constraints. However, applications such as this one are of extreme interest for ISE, even for higher I/O constraints: specialized instructions consisting mostly of very simple bitwise calculations have great potential to improve performance significantly

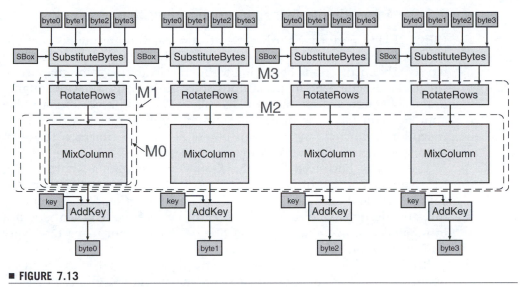

■ FIGURE 7.13

Motivational example from the *aes* encryption benchmark.

and have relatively modest complexity in hardware. The MixColumn block (see Fig. 7.13) represents the most computationally intensive part of the algorithm; it is free of memory access, and is a most interesting candidate for ISE. Moreover, it has a very low I/O requirement of one input and one output (graph M0). Availability of three further inputs would allow inclusion of a RotateRows block (graph M1), and of course all four MixColumns in parallel should be chosen when four input and four output ports are available (graph M2), and all MixColumns and RotateRows when sixteen inputs and four outputs are given (graph M3).

7.3.2 Types of Heuristic Algorithms

A heuristic that can automatically identify the previously mentioned ISEs is of immediate interest and obvious importance. We will only suggest here two recent heuristics, which represent well very different attempts to handle large graphs. The first one, ISEGEN [11], sticks to the formal problem definition of Section 7.2 and uses some well known partitioning heuristics (widely used in the hardware/software codesign community) to find a solution to the problem. The second one [12] is representative of a wide class of solutions that aim at the same result but do not necessarily pass through a full formulation of the problem: such heuristics have started developing very early, especially in the context of reconfigurable computing [13, 14]. The one we will present is distinctive in the way the heuristic is built to target relatively large clusters of operations much in the same way exact algorithms do.

7.3.3 A Partitioning-Based Heuristic Algorithm

ISEGEN [11] performs hardware-software partitioning at instruction-level granularity—i.e., it creates partitions of DFGs representing basic blocks. The instructions belonging to the hardware partition map to an ISE to be executed on a special functional unit, while those belonging to the software partition individually execute on the processor core. The approach considers the basic blocks in an application—based on their execution frequency and speedup potential—and performs a series of successive bipartitions into hardware and software.

The idea is borrowed from the *Kernighan-Lin* min-cut partitioning heuristic and consists of steering the *toggling* of nodes in the DFG between software (S) and hardware (H) based on a gain function M_{toggle} that captures the designer's objective. The effectiveness of the K-L heuristic lies in its ability to overcome many local maxima without using too many unnecessary moves.

ISEGEN is an iterative improvement algorithm that starts with all nodes in software and tries to toggle each unmarked node, n, in the graph either from S to H or from H to S in every iteration. Within each iteration of ISEGEN, the best cut found so far is retained, and the decision to toggle n with respect to the best cut is based on a gain function, $M_{toggle}()$. Note that the chosen cut at every iteration may be violating input/output constraints and convexity constraints: a cut is allowed to be illegal, giving it an opportunity to eventually grow into a valid cut. Experimentally, it was observed that five passes are enough for finding a satisfactory solution.

The gain function M_{toggle} is a linear weighted sum of the following five components that act as control parameters for the algorithm. The five components (1) estimate the speedup, (2) assign a penalty to input-output violations, (3) assign a penalty to nonconvex intermediate solutions, (4) favor large cuts, and (5) enable independent connected components. The weights for each component are determined experimentally, which appears to be the main drawback of the technique. ISEGEN performs on average 35% better compared to a previously published genetic algorithm heuristic [6], and is able to generate cuts with good reusability.

7.3.4 A Clustering Heuristic Algorithm

A heuristic recently proposed by Clark et al. [15], as an extension of a previous work by the same authors [12], also proposes ways to overcome the limits of exact algorithms, such as the one presented in Section 7.2. Candidates for ISEs are grown by considering each node in the DFG as a possible seed and adding nodes by traversing the DFG downward along edges.

An observation is made that only few edges have good potential, while many would not lead to a good quality candidate because, for example, they might not be on the critical path. A guide function is therefore needed to eliminate nonpromising edges. Effectively, eliminating edges from a DFG creates a partition of it, and since this technique only considers connected graphs as candidate for ISE (as all "clustering-based" algorithms), DFG partitioning causes a decrease of the size of the problem.

The guide function is used to determine which dataflow edges would likely result in nonperforming candidates and is conceptually similar to the gain function M_{toggle} described in the previous section. It uses three categories to rank edges: (1) criticality—whether an edge lies on a long path, (2) latency—whether an edge links nodes likely to be chained in the same cycle when executed in hardware, and (3) area—to what extent ISE area increases due to inclusion of the edge's target node. Candidates are also combined when equivalent and are generalized to maximize reusage across applications. Final ISE selection among generated candidates is based on two proposed methods, one greedy and one based on dynamic programming. The paper also proposes algorithms for compiling applications to given ISEs—i.e., discovering in a DFG subgraphs equivalent to preselected ISEs. This is an issue that is not discussed in any of the other methods described in this chapter.

Finally, an important remark should be made regarding algorithms discovering connected versus disconnected graphs as ISEs. Heuristic techniques are needed, because the exact algorithm presented in Section 7.2, which can also identify disconnected graphs as ISEs, presents a limit in the size of basic blocks it can handle: around 1,000 nodes for an I/O constraint of 4/2. It should be noted, however, that if only connected graphs are of interest, then the search space decreases considerably, and the exact algorithm can handle 1,000 nodes *even for very high I/O constraints* (e.g., 8/4). Therefore, it should be noted that ISE identification heuristics that can only discover connected graphs, such as the one presented in this section, should be considered useful only in extreme cases—i.e., for sizes and I/O constraints larger than the ones just mentioned.

7.4 STATE-HOLDING INSTRUCTION-SET EXTENSIONS

Many applications access small portions of memory multiple times in a frequently executed part of code; it has already been realized that such memory parts would benefit from being moved from main memory to a scratchpad closer to a core and without continuously polluting the cache. On the other hand, it would be even more beneficial to move them directly inside the computation core—i.e., *inside the (application-specific) functional units, which therefore mainly process data locally available inside them*. In this section we discuss a recently published contribution to ISE identification [7] that presents a technique for identifying not only the operations that are best moved to an *application-specific functional unit (AFU)* (as it was the case in Section 7.2), but also the associated data that are best stored locally in the AFUs.

7.4.1 Motivation

Consider a portion of the *fast Fourier transform (fft)* kernel shown in Fig. 7.14. The innermost loop is run $2^n/2^k = 2^{n-k}$ times. Therefore, for each k, there are $2^{k-1} \cdot 2^{n-k}$ or 2^{n-1} accesses to memory. For $n = 8$, k goes from 1 to 8 leading to $8 \cdot 127 = 1,024$ memory accesses for each array variable in the critical region. Since there are 6 memory reads and 4 memory writes corresponding to array variables RealBitRevData[] and ImagBitRevData[], overall there are 6,144 memory reads and 4,096 memory writes in the *fft* kernel, for $n = 8$.

ISE techniques that are not able to include memory-access operations, such as the one presented in the previous section, would identify instructions such as the butterfly, leaving memory-accesses to the processor core. However, if the *fft* kernel executes in an AFU with a small local memory with a storage space for 256 elements, all the costly 10,240 accesses to main memory can be redirected to the fast and energy-efficient AFU-resident local memory. Of course, a *direct memory access (DMA)* operation is required to copy the 256 elements into and out of

```
for(k = 1; k <= n; k++){
   n1 = 1<<k;
   n2 = n1>>1;
   ...
   for (j = 0; j < n2; j++){
      ...
      for (i = j; i < 2^n; i += n1){
         1 = i + n2;
         tReData = (WRe * ReBRData[1])  +  (WIm * ImBRData[1]);
         tImData = (WRe * ImBRData[1])  +  (WIm * ReBRData[1]);
         tReData = tReData >> SCALE_FACTOR;
         tImData = tImData >> SCALE_FACTOR;
         ReBRData[1] = ReBRData[i]  - tReData;
         ImBRData[1] = ImBRData[i]  - tImData;
         ReBRData[i] += tReData;
         ImBRData[i] += tImData;
      }
   }
}
```

■ **FIGURE 7.14**

The *fft* kernel.

the AFU memory at the beginning and end of the outer loop, but this requires far less accesses than it saves.

In general, the advantages of an AFU-resident memory are manifold: it lowers cache pollution, it increases the scope of ISE algorithms, it increases resulting performance, and it reduces energy consumption. This section will discuss a formal framework where ISEs taking advantage of AFU-resident memories can be automatically identified [7].

As a further example, Fig. 7.15 shows the code of the inner loop of the application depicted in Fig. 7.2—i.e., *adpcmdecode*. It can be seen that loop-carried dependences exist on some variables—e.g., index and step. If these variables could be held inside the AFU where an ISE is implemented, then they would not require I/O (i.e., register access), apart from at the entry and exit of the loop. Moreover, vectors stepsizeTable and indexTable are constant (this is not visible in Fig. 7.15), and therefore they could also be kept inside an AFU and eliminate the need for the correspondent loads. The whole loop can then fit in a 1-input 1-output ISE; the technique discussed in the following section can identify such a solution.

7.4.2 Local-Memory Identification Algorithm

Problem 1 is still the problem to solve: find a constrained subgraph S of the DFG of a basic block, maximizing the speedup achieved when

```
for (len=LEN; len>0; len--){
  ...
  index += indexTable[delta];
  if (index < 0) index = 0;
  if (index > 88) index = 88;
  sign = delta & 8;
  delta = delta & 7;
  vpdiff = step >> 3;
  if (delta & 4) vpdiff += step;
  if (delta & 2) vpdiff += step>>1;
  if (delta & 1) vpdiff += step>>2;
  if (sign)
    valpred -= vpdiff;
  else
    valpred += vpdiff;
  if (valpred > 32767)
    valpred = 32767;
  elseif(valpred < -32768)
    valpred = -32768;
    step = stepsizeTable[index];
  *outp++ = valpred;
}
```

■ **FIGURE 7.15**

The *adpcmdecode* kernel.

the cut is implemented as a custom instruction. Speedup is measured by means of a merit function M(S). However, in the work described in this section, a different definition of *forbidden nodes* applies: in fact, nodes representing memory access, previously forbidden, can now be chosen in an S. It is the merit function M(S) that needs to encapsulate the benefit or the disadvantage of including certain memory accesses into a cut.

When a cut includes some or all memory accesses to a certain vector, as shown in Fig. 7.16(a), it means that a state-holding ISE is taken into consideration. Architecturally, in the most general case, the vector in question needs to be transferred from the main memory to the AFU local memory by DMA before loop execution. As a result, all memory accesses to the vector are now essentially internal accesses to the local memory instead. Finding the most favorable code position for the transfer is crucial; at the end of the loop, again in the most favorable position and only if needed, the vector is copied back to the main memory.

The merit function M(S) per unit execution of a *critical basic block* (*CBB*)—that is, the basic block where the ISE is currently being identified—is expressed as follows:

$$M(S) = \lambda_{sw}(S) - \lambda_{hw}(S) - \lambda_{overhead}(S), \tag{7.1}$$

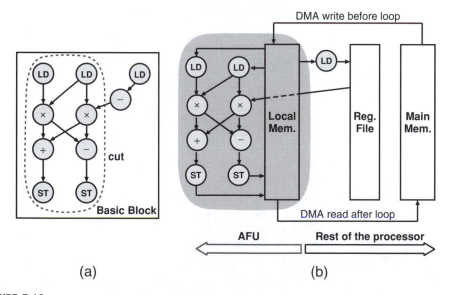

(a)   (b)

■ **FIGURE 7.16**

(a) A cut includes some memory-access operations to a given vector. (b) The corresponding AFU contains an internal memory; the vector is loaded in and out of such memory at an appropriate time; all memory instructions in this basic block access the AFU internal memory instead of main memory.

where $\lambda_{sw}(S)$ and $\lambda_{hw}(S)$ are the estimated software latency (when executed natively in software) and hardware latency (when executed on AFU with ISE) of the cut S, respectively, and $\lambda_{overhead}$ estimates the transfer cost. Considering a DMA latency of λ_{DMA}, and supposing that the DMA write and read operations required will be placed in basic blocks (*WBB* and *RBB*) with execution counts N_{WBB} and N_{RBB}, respectively (ideally much smaller than the execution count of *CBB*, N_{CBB}, where the ISE is identified), the transfer cost can be expressed by the following equation:

$$\lambda_{overhead} = \frac{N_{WBB} + N_{RBB}}{N_{CBB}} \cdot \lambda_{DMA}. \qquad (7.2)$$

Note that all the considerations are valid for inclusion of loop-carried scalar accesses—such as those to variables index and step mentioned in the *adpcmdecode* motivational example—and not only for vectors. However, in the case of scalar accesses, the transfer will be much cheaper, as it does not involve DMA setup and transfer overhead.

For a given CBB, the steps for generating ISEs that include architecturally visible storage are: (1) Find vectors and scalars accessed in CBB; (2) Search for the most profitable code positions for inserting memory transfers between the AFU and the main memory; and (3) Finally, execute the process of ISE generation using a slightly modified version of the algorithm presented in Section 7.2 and using the updated merit function (M (S) in Equation 7.1).

For details of each step, the reader is referred to the original paper [7]. With respect to step (2), which is the most complex, we only hint here that memory transfer operations to the AFU should be performed in a basic block with the least execution count; however, for correctness, they should be placed in a basic block that always reaches *CBB* and step (2) guarantees that no more writing of the vector being transferred to the AFU can happen between such basic block and *CBB*.

7.4.3 Results

The described memory-aware ISE generation algorithm was also implemented within the MachSUIF framework and was run on six benchmarks: *adpcmdecode* (ADPCM decoder), *adpcmencode* (ADPCM encoder), *fft* (fast Fourier transform), *fir* (FIR filter), *des* (data encryption standard), and *aes* (advanced encryption standard), taken from Mediabench, EEMBC, and cryptography standards. The cycle-accurate SimpleScalar simulator [16] for the ARM instruction set was used, and modified in following way. For vectors, a DMA connection between the local memory inside an AFU and the main memory was introduced by adding four new instructions to the instruction set: two for setting the source and destination addresses and two for setting the command registers to transfer data from main memory to AFU memory and vice versa. For handling scalars, two additional instructions were added to set and get local registers inside the AFU. Of course, we also added the application-specific ISEs identified by our ISE generation algorithm.

The hardware latency for each instruction was obtained by synthesizing the constituent arithmetic and logic operators on the UMC 0:18-µ*m* CMOS process using the Synopsys Design Compiler. The access latency of the internal memory (modeled as an SRAM) was estimated using the Artisan UMC 0.18-µ*m* CMOS process SRAM Generator. The default SimpleScalar architecture is equipped with 4 integer ALUs, 1 integer multiplier/divider, 4 floating-point adders, 1 floating-point multiplier/ divider, and a three-level memory hierarchy for both instruction and

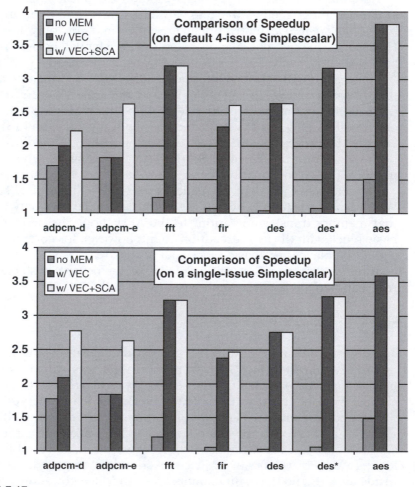

■ **FIGURE 7.17**

Comparison of speedup for I/O constraints of 4/2 obtained on a four-issue (default) and a single-issue SimpleScalar with the ARM instruction set.

data. The sizes of L1 and L2 data caches are 2 KB and 32 KB, respectively. The main memory has a latency of 18 cycles for the first byte and 2 cycles for subsequent bytes. The same latency is also used when transferring data between main memory and AFU by DMA.

The baseline case is pure software execution of all instructions. An I/O constraint to 4 inputs and 2 outputs was set, and *a single cut* was generated and added as an ISE to the SimpleScalar architecture. First, the cut did not allow memory inclusion ("no MEM"). Then, local memory with vector accesses was added ("w/VEC") and, subsequently, scalars accesses also ("w/VEC+SCA"). For these three cases, Fig. 7.17 shows a comparison of speedup on several applications obtained on the default SimpleScalar architecture (4-width out-of-order issue) as well as on the single-issue SimpleScalar architecture.

Observe that (1) the speedup is raised tangibly when state holding AFUs are considered (1.4 times on average for the case with no memory, to

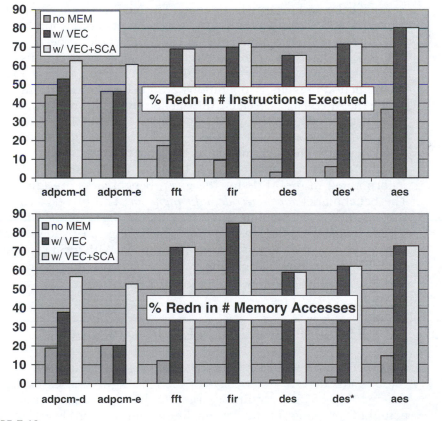

■ **FIGURE 7.18**

Percentage reduction in the number of instructions executed and percentage reduction in the number of memory accesses.

TABLE 7.1 ■ Summary of local memories selected for the different benchmarks

Benchmarks	Array Identifier	Size (bytes)
adpcmdecode	stepsizeTable	356
	indexTable	64
adpcmencode	stepsizeTable	356
	indexTable	64
fft	RealBitRevData	512
	ImagBitRevData	512
fir	inbuf16	528
	coeff16	128
des	des_SPtrans	2048
aes	Sbox	256

2.8 times for the "w/VEC+SCA" case, on the default architecture), and (2) the trend of speedups obtained on the two different configurations of the SimpleScalar architecture is the same. The label *des∗* indicates the results for *des* with 3 ISEs rather than with a single one (*des* is the only benchmark where a single cut was not enough to cover the whole kernel).

Fig. 7.18 shows the reduction in the number of instructions executed and in the number of memory accesses. Interestingly, there is an average 9% reduction in memory operations even before incorporating memory inside the AFU. This is because ISEs generally reduce register need (multiple instructions are collapsed into one) and therefore reduce spilling. With the incorporation of memory inside the AFU, the average reduction in memory instructions is a remarkable two thirds, hinting at a very tangible energy reduction. Table 7.1 shows that the sizes of the vectors incorporated in the AFUs for the given benchmarks are fairly limited. The benchmarks *adpcmdecode*, *adpcmencode*, and *fir* very clearly show the advantage of including scalars by exhibiting a marked increase in speedup due to the increased scope of ISE. Fig. 7.19 shows the kernel of *fft*, with the omission of address arithmetic. With local storage inside the AFU, a single cut covers the entire DFG, thus almost doubling the speedup obtained without memory in the AFU.

7.5 EXPLOITING PIPELINING TO RELAX I/O CONSTRAINTS

A further improvement on the state of the art—an orthogonal method to the one just described—consists of recognizing the necessity to exploit the speedup of ISEs requiring large I/O, but at the same time limiting actual I/O by using concurrent pipelining and register file access sequentialization [17]. In fact, a particularly expensive asset of the processor core is the number of ports to the register file that AFUs can use. While this number is typically kept small in available processors—indeed, some

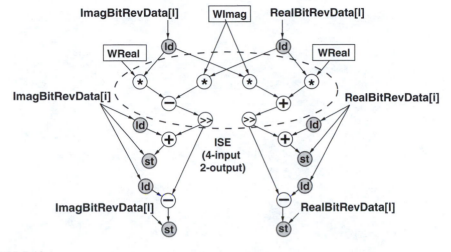

■ FIGURE 7.19

DFG of *fft*. The whole kernel is chosen when architecturally visible storage is allowed; only the cut in a dotted line is chosen otherwise.

only allow two read and one write ports—it is also true that input/output allowance significantly impacts speedup. A typical trend can be seen in Fig. 7.20, where the speedup for various combinations of I/O constraints is shown for an application implementing the *des* cryptography algorithm. This example shows that, on a typical embedded application, the I/O constraint impacts strongly on the potentiality of ISE: speedup goes from 1.7 for 2 read and 1 write ports, to 4.5 for 10 read and 5 write ports. Intuitively, if the I/O allowance increases, larger portions of the application can be mapped onto an AFU, and therefore a larger part can be accelerated. In this section, we describe a recently proposed method to relax I/O constraints while taking advantage of the necessity of pipelining most complex AFUs [17].

7.5.1 Motivation

As a motivational example, consider Fig. 7.21(a), representing the DAG of a basic block. Assume that each operation occupies the execution stage of the processor pipeline for one cycle when executed in software. In hardware, the delay in cycles (or fraction thereof) of each operator is shown inside each node. Under an I/O constraint of 2/1, the subgraph indicated with a dashed line on Fig. 7.21(a) is the best candidate for ISE. Its latency is one cycle (ceiling of the subgraph's critical path), while the time to execute the subgraph on the unextended processor is roughly 3 cycles (one per operation). Two cycles are therefore saved every time the ISE is used instead of executing the corresponding sequence of instructions. Under an I/O constraint of 3/3, on the other hand, the whole DAG

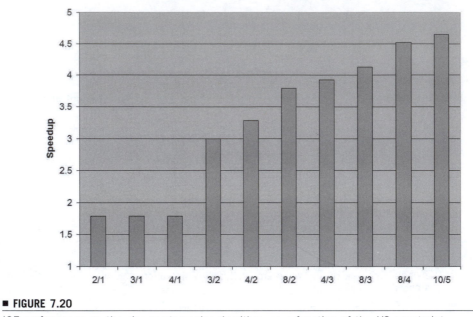

■ FIGURE 7.20

ISE performance on the *des* cryptography algorithm, as a function of the I/O constraint.

can be chosen as an AFU (its latency in hardware is 2 cycles, its software latency is approximately 6 cycles, and hence 4 cycles are saved at each invocation). Fig. 7.21(b) shows a possible way to pipeline the complete basic block into an AFU, but this is exclusively possible if the register file has 3 read and 3 write ports. If the I/O constraint is 2/1, a common solution is to implement the smaller subgraph instead and reduce significantly the potential speedup.

An ISE identification algorithm could instead search for candidates exceeding the constraint and then map them on the available I/O by serializing register port access. Fig. 7.21(c) shows a naïve way to implement serialization, which simply maintains the position of pipelines' registers as it was in Fig. 7.21(b) and adds registers at the beginning and at the end to account for serialized access. As indicated in the I/O access table, value A is read from the register file in a first cycle, then values B and C are read and execution starts. Finally, two cycles later, the results are written back in series into the register file, in the predefined (and naïve) order of F, E, and D. The schedule is legal, since only at most 2 read and/or 1 write happen simultaneously. Latency, calculated from the first read to the last write, is now 5 cycle: only 1 cycle is saved. However, a better schedule for the DAG can be constructed by changing the position of the original pipeline registers to allow register file access and computation to proceed in parallel. Fig. 7.21(d) shows the best legal schedule, resulting in a latency of 3 cycles and hence a gain of 3 cycles: searching for larger AFU candidates and then pipelining them in an efficient way in order to serialize register file access and to ensure

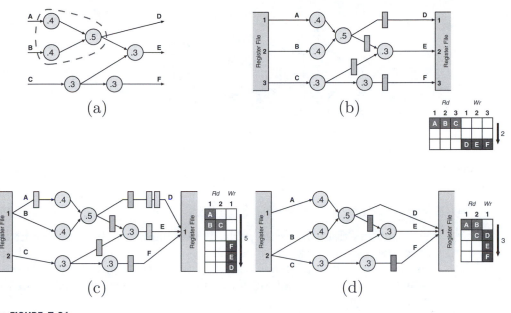

■ FIGURE 7.21

Motivational example: (a) The DAG of a basic block annotated with the delay in hardware of the various operators. (b) A possible connection of the pipelined datapath to a register file with 3 read ports and 3 write ports (latency = 2). (c) A naive modification of the datapath to read operands and write results back through 2 read ports and 1 write port, resulting in a latency of 5 cycles. (d) An optimal implementation for 2 read ports and 1 write port, resulting in a latency of 3 cycles. Rectangles on the DAG edges represent pipeline registers. All implementations are shown with their I/O schedule on the right.

I/O legality can be beneficial and, as will be seen in the experimental section, it can boost the performance of ISE identification.

This section describes an ISE identification algorithm that recognizes the possibility of serializing operand-reading and result-writing of AFUs that exceed the processor I/O constraints. It also presents an algorithm for input/output constrained scheduling that minimizes the resulting latency of the chosen AFUs by combining pipelining with multicycle register file access. Measurements of the obtained speedup show that the proposed algorithm finds high-performance schedules resulting in tangible improvement when compared to the single-cycle register file access case.

7.5.2   Reuse of the Basic Identification Algorithm

Problem 1 is again addressed here: find a constrained subgraph S of the DFG of a basic block, maximizing the speedup achieved when the cut is implemented as a custom instruction.

In the work described in this section, our goal differs from the basic formulation of Problem 1 because of two related reasons: (a) we allow the cut S to have more inputs than the read ports of the register file and/or

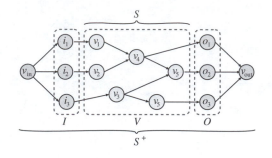

■ FIGURE 7.22

A sample augmented cut S^+.

more outputs than the write ports; if this happens, (b) we account for successive transfers of operands and results to and from the specialized functional unit in the latency of the special instruction. We take care of point (b), while we also introduce pipeline registers, if needed, in the datapath of the unit.

The way we solve the new single-cut identification problem consists of three steps: (1) We generate the best cuts for an application using any ISE identification algorithm (e.g., the algorithm of Section 7.2.3) for all possible combinations of input and output counts equal and above N_{in} and N_{out}, and below a reasonable upper bound—e.g., 10/5. (2) We add to the DFG of S both the registers required to pipeline the functional unit under a fixed timing constraint (the cycle time of the host processor) and the registers to temporarily store excess operands and results. In other words, we fit the actual number of inputs and outputs of S to the microarchitectural constraints. (3) We select the best ones among all cuts. Step (2) is the new problem that we formalize and solve in the present section.

7.5.3   Problem Statement: Pipelining

We call again $S(V, E)$ the DAG representing the dataflow of a potential special instruction to be implemented in hardware; the nodes V represent primitive operations, and the edges E represent data dependencies. Each graph S is associated to a graph

$$S^+(V \cup I \cup O \cup \{v_{in}, v_{out}\}, E \cup E^+)$$

which contains additional nodes I, O, v_{in}, and v_{out}, and edges E^+. The additional nodes I and O represent, respectively, input and output variables of the cut. The node v_{in} is called *source* and has edges to all nodes in I. Similarly, the node v_{out} is the *sink* and all nodes in O have an edge to it. The additional edges E^+ connect the source to the nodes I, the nodes I to V, V to O, and O to the sink. Fig. 7.22 shows an example of cut.

Each node $u \in V$ has associated a positive real weight, $\lambda(u)$; it represents the latency of the component implementing the corresponding operator. Nodes v_{in}, v_{out}, I and O have a null weight. Each edge $(u, v) \in E$ has an associated positive integer weight, $\rho(u, v)$; it represents the number of registers in series present between the adjacent operators. A null weight on an edge indicates a direct connection (i.e., a wire). Initially all edge weights are null (that is, the cut S is a purely combinatorial circuit).

Our goal is to modify the weights of the edges of S^+ in such a way as to have (1) the critical path (maximal latency between inputs and registers, registers and registers, and registers and outputs) below or equal to some desired value Λ, (2) the number of inputs (outputs) to be provided (received) at each cycle below or equal to N_{in} (N_{out}), and (3) a minimal number of pipeline stages, R. To express this formally, we introduce the sets W_i^{IN}, which contain all edges (v_{in}, u) whose weight $\rho(v_{in}, u)$ is equal to i. Similarly the sets W_i^{OUT} contain all edges (u, v_{out}) whose weight $\rho(u, v_{out})$ is equal to i. We write $|W_i^{IN}|$ to indicate the number of elements in the set W_i^{IN}. The problem we want to solve is the particular case of scheduling described next.

Problem 3 *Minimize R under the following constraints:*

1. **Pipelining.** *For all combinatorial paths between $u \in S^+$ and $v \in S^+$—that is, for all those paths such that $\sum_{\text{all edges } (s, t) \text{ on the path}} \rho(s, t) = 0$,*

$$\sum_{\text{all nodes } k \text{ on the path}} \lambda(k) \leq \Lambda. \tag{7.3}$$

2. **Legality.** *For all paths between v_{in} and v_{out},*

$$\sum_{\text{all edges } (u, v) \text{ on the path}} \rho(u, v) = R - 1. \tag{7.4}$$

3. **I/O schedulability.** $\forall i \geq 0$,

$$\left| W_i^{IN} \right| \leq N_{in} \text{ and } \left| W_i^{OUT} \right| \leq N_{out}. \tag{7.5}$$

The first list item ensures that the circuit can operate at the given cycle time, Λ. The second ensures a legal schedule—that is, a schedule that guarantees that the operands of any given instruction arrive together. The third item defines a schedule of communication to and from the functional unit that never exceeds the available register ports: for each edge (v_{in}, u), registers $\rho(v_{in}, u)$ do not represent physical registers, but the schedule used by the processor decoder to access the register file. Similarly, for each (u, v_{out}), $\rho(u, v_{out})$ indicates when results are to be written back. For this reason, registers on input edges (v_{in}, u) and on output edges (u, v_{out}) will be called *pseudoregisters* from now on; in all figures, they are shown with a lighter shade than physical registers. As an example, Fig. 7.23 shows the graph S^+ of the optimized implementation shown in Fig. 7.21(d) with the pseudoregisters that express the register file access schedule for reading and writing. Note that the graph satisfies the legality check expressed earlier: exactly two registers are present on any given path between v_{in} and v_{out}.

7.5.4   I/O Constrained Scheduling Algorithm

The algorithm proposed for solving Problem 1 first generates all possible pseudoregister configurations at the inputs, meaning that pseudoregisters are added on input edges (v_{in}, u) in all ways that satisfy the input schedulability constraint—i.e., $|W_i^{\text{IN}}| \leq N_{\text{in}}$. This is obtained by repeatedly applying the *n choose r* problem—or *r combinations of an n set*—with $r = N_{\text{in}}$ and $n = |I|$, to the set of input nodes I of S^+, until all input variables have been assigned a read-slot—i.e., until all input edges (v_{in}, u) have been assigned a weight $\rho(v_{\text{in}}, u)$. Considering only the r combinations ensures that no more than N_{in} input values are read at the same time. The number of *n choose r* combinations is $\binom{n}{r} = \frac{n!}{r!(n-r)!}$. By repeatedly applying *n choose r* until all inputs have been assigned, the number of total configurations becomes $\frac{n!}{(r!)^x(n-xr)!}$, with $x = \left\lceil \frac{n}{r} \right\rceil - 1$. Note that the complexity of this step is nonpolynomial *in the number of inputs of the graph*, which is a very limited quantity in practical cases (e.g., on the order of tens). Fig. 7.24 shows the possible configurations for the simple example of Fig. 7.21: I = A, B, C and the configurations, as defined earlier, are AB → C, AC → B, and BC → A. Note that this definition does not include, for example, A → BC. In fact, since we are scheduling for minimum latency, as many inputs as possible are read every time.

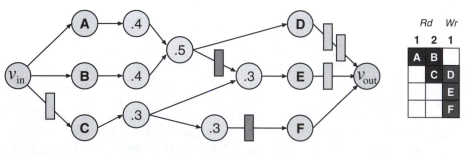

■ **FIGURE 7.23**

The graph S^+ of the optimized implementation shown in Fig. 7.21(d). All constraints of Problem 3 are verified, and the number of pipeline stages R is minimal.

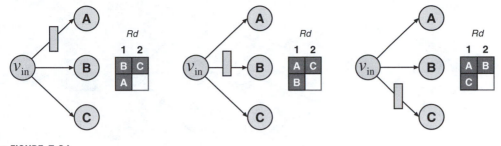

All possible input configurations for the motivational example, obtained by repeatedly applying an *n choose r* pass to the input nodes.

Then, for every input configuration, the algorithm proceeds in three steps:

1. A scheduling pass is applied to the graph, visiting nodes in topological order. The algorithm essentially computes an ASAP schedule, but it differs from a general ASAP version because it considers an initial pseudoregister configuration (details can be found in the original paper [17]). It is an adaptation of a retiming algorithm for DAGs [18], and its complexity is $O(|V| + |E|)$. Fig. 7.25(a) shows the result of applying the scheduling algorithm to one of the configurations.

2. The schedule is now feasible at the inputs but not necessarily at the outputs. Since only a limited number of results (N_{out}) can be written in any given cycle, at most N_{out} edges to output nodes must have 0 registers (i.e., a weight equal to 0), at most N_{out} edges must have a weight equal to 1, and so on. If this is not the case, a line of registers on all output edges is added until the previously mentioned condition is satisfied. Fig. 7.25(b) shows the result of this simple step.

3. Registers at the outputs are transformed into pseudoregisters—i.e., they are moved to the right of output nodes, on edges (u, v_{out})—as shown in Fig. 7.25(c). The schedule is now feasible at both inputs and outputs.

All schedules of minimum latency are the ones that solve Problem 3. Among them, a schedule requiring a minimum number of registers is then chosen. Fig. 7.25(d) shows the final schedule for another input configuration, which has the same latency but a larger number of registers (3 versus 2) than the one of Fig. 7.25(c).

7.5.5 Results

The experimental setup is very similar to the one described in Section 7.2. However, due to some added precision in the hardware and

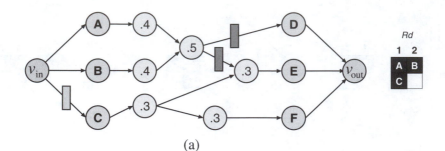

(a)

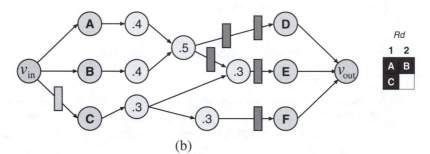

(b)

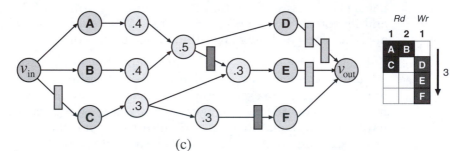

(c)

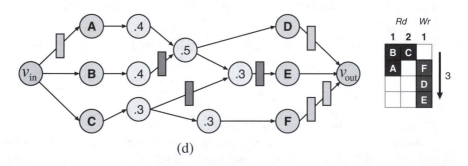

(d)

▪ **FIGURE 7.25**

Proposed algorithm: (a) A modified ASAP scheduling pass is applied to the graph, for the third initial configuration of Fig. 7.24. The schedule is feasible at the inputs but not at the outputs. (b) One line of registers is added at the outputs. (c) Three registers at the outputs are transformed into pseudoregisters, in order to satisfy the output constraint. (d) The final schedule for another input configuration—its latency is also equal to three, but three registers are needed; this configuration is therefore discarded.

software latencies, speedup estimations are slightly different. As described in Section 7.5.2, we proceed in two steps: (1) We run the ISE algorithm of Section 7.2.3 with any combination of constraints greater than or equal to the constraints allowed by the microarchitecture, up to 10/5. Throughout this section, we call the former constraints *temporary*, and the latter *actual*, or *microarchitectural*. (2) We schedule the identified AFUs using the algorithm described in Section 7.5.4, in order to generate an implementation that satisfies the actual constraints and to calculate the new latency.

In these experiments, we tried 10 different combinations of temporary I/O constraints, from 2/1 to 10/5, and the best 16 AFUs for each combination were identified. Then, all of the AFUs were mapped from the temporary constraint combinations onto actual, and smaller, microarchitectural constraints. Finally, the 16 I/O-constrained AFUs from the temporary combination that gave the best speedup were selected as final ones. Experiments were repeated for 10 different combinations of microarchitectural I/O constraint, as well.

Fig. 7.26 shows the speedup obtained for each of the benchmarks, for various I/O constraints. Two columns are present for each I/O combination. The first shows the speedup of the original algorithm of Section 7.2.3, where only AFUs with an I/O requirement no greater than the microarchitectural allowance are considered. The second shows the speedup obtained by using the selection strategy and the I/O constrained scheduling algorithm presented in this section, where all AFUs with I/O requirements greater than or equal to the actual microarchitectural are considered and then are rescheduled accordingly.

It can be observed that speedup is raised tangibly in most cases. The improvement for the 2/1 constraint is on average 32%, and in some cases is as much as 65%, showing that this method represents yet another advance on the state of the art on automatic identification and selection of ISE. Some further points that can be noted follow intuition: (1) the difference in speedup between first and second columns decreases as the I/O increases—there is less margin for improvement—and therefore the proposed technique is particularly beneficial for low microarchitectural constraints; (2) as I/O increases, the speedup values obtained by the strategy discussed in this section (second columns) tend toward the value of the 10/5 first column, without actually reaching it since, of course, register-file access serialization increases latency in general.

Fig. 7.27 shows an example of 8/2 cut that has been pipelined and whose inputs and outputs have been appropriately sequentialized to match an actual 2/1 constraint. The example has an overall latency of 5 cycles and contains only eight registers (and six of them are essential for correct pipelining). With the naïve solution illustrated in Fig. 7.21(c), 12 registers (one each for C and D, two each for E and F, and so forth.) would have been necessary to resynchronize sequentialized inputs (functionally replaced here by the two registers close to the top of the cut), and

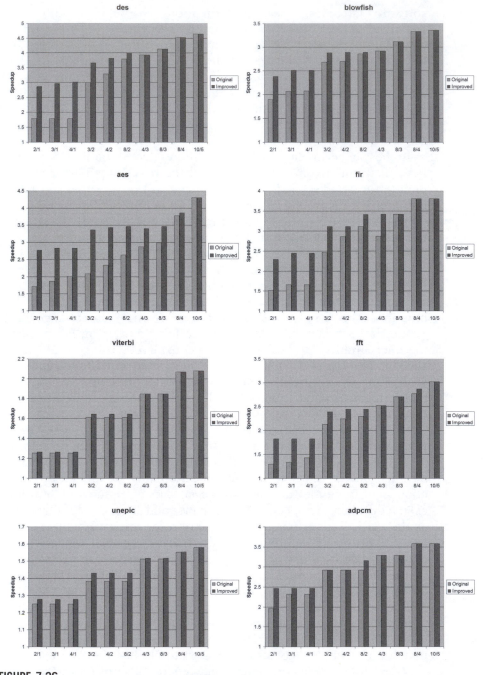

■ FIGURE 7.26

Analysis of the performance of the pipelining method discussed here.

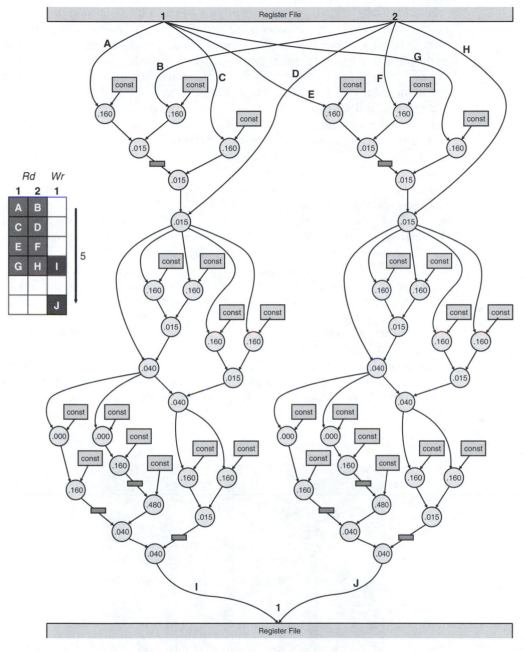

■ **FIGURE 7.27**

Sample pipelining for a 8/2 cut from the *aes* cryptography algorithm with an actual constraint of 2/1.
Compared to a naïve solution, this circuit saves 11 registers and shortens the latency by a few cycles.

one additional register would have been needed to delay one of the two outputs: our algorithm makes good use of the data independence of the two parts of the cut and reduces both hardware cost and latency. This example also suggests some ideas for further optimizations: if the symmetry of the cut had been identified, the right and left datapath could have been merged, and a single datapath could have been used successively for the two halves of the cut. This would have produced the exact same schedule at approximately the half hardware cost, but the issues involved in finding this solution go beyond the scope of this chapter.

Finally, Fig. 7.28 shows the speedup relative to all temporary I/O constraints, for an actual constraint of 2/1, for the *des* application. The best value, seen in columns 8/2 and 10/5, is the one reported in Fig. 7.26 for *des* (as second column of constraint 2/1). As it can be seen, the speedup does not always grow with the temporary constraints (contrast, for instance, 4/2 with 4/3). This is due to the fact that sometimes selection under certain high temporary constraints might include new input nodes of moderate benefit that later cannot be serialized without a large loss in speedup (too large as increase in latency). Note that the value for constraint 2/1 in this graph is the same as the value of the first column for 2/1 on Fig. 7.26 for *des*. Mapping AFUs, generated with a temporary constraint, onto an identical actual constraint is the same as using the original, nonimproved, ISE algorithm of Section 7.2.3.

As a final remark, note that the presented I/O constrained algorithm ran in fractions of seconds in all cases. In fact, the complexity of the I/O constrained scheduling algorithm proposed is exponential in the number of inputs of the input graph, which is a very limited number in all

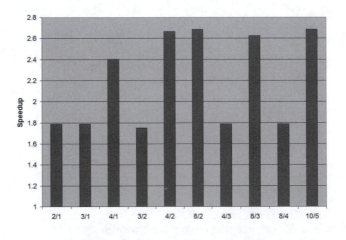

■ **FIGURE 7.28**

ISE performance on the *des* cryptography algorithm as a function of the temporary constraint for an actual constraint of 2/1.

practical cases. The cycle time was set, in these experiments, to the latency of a 32×32 multiply-accumulate. Of course, any other user-given cycle time can be adopted.

7.6 CONCLUSIONS AND FURTHER CHALLENGES

This chapter has shown that it is possible to automate the process of ISE identification to a nonnegligible extent, and in many cases results are close to those found by expert designers. This opens possibilities for a futuristic compiler that not only generates code for a given machine, but chooses a machine description and then compiles for it.

Despite the many and rapid advances, many questions remain open in the field of automatic ISE. Recurrence of complex patterns, especially recurrence of similar rather than identical patterns, is an important topic—especially if one considers that applications could be modified after the initial identification has been performed and the extension manufactured already; current techniques hardly address the issue. Also, code-transformation techniques specifically targeted to prepare the code for ISE identification are necessary and are being investigated. Finally, cases where automated and manual choices show tangible differences still exist, and some reasons why this happens are discussed in the next chapter.

CHAPTER 8

CHALLENGES TO AUTOMATIC CUSTOMIZATION

Nigel Topham

Architectural customization is a powerful technique for improving the performance and energy efficiency of embedded processors. In principle, by identifying application-specific hardware functionality, it is possible to obtain a close fit between what hardware provides and the computations performed by the target application [1]. In contrast, a general-purpose processor (GPP) will typically execute a sequence of generic instructions to implement the same functionality found in a single application-specific instruction. An obvious benefit of application-specific instructions is to simply reduce the number and overall size of executed instructions, allowing the same real-time performance constraints to be met by a device that either operates at a lower clock frequency or simply fetches fewer bits from an instruction cache [2]. In turn this may allow other energy-saving optimizations, such as a reduction in operating voltage or the use of slower and more energy-efficient transistors [3]. Another very important energy-saving implication of application-specific instructions is the beneficial impact of a closer match between hardware and application. When a fragment of a data-flow graph is implemented directly in hardware, the signal transitions that occur during function evaluation should dissipate close to the minimum theoretical energy for that particular function, assuming the hardware is efficiently synthesized. However, when the same function is implemented by a sequence of generic instructions, there will be an inevitable increase in signal transitions and energy dissipation due to the update of intermediate processor state, additional memory accesses, and the computation of unnecessary side effects. Previous work in the area of synthetic instruction encoding [4] has shown, for example, that energy-aware synthesis of entire instruction sets yields significant energy savings. The case for application-specific architecture customization is therefore compelling, particularly in embedded mobile ASIC- and FPGA-based systems, where high performance is required but energy efficiency is critical. Aside from the nontechnical issues of how to manage the deployment of customized architectures, the two primary technical considerations are how to *identify* and *implement* potential customizations.

In recent years tools and techniques have advanced to the point where some opportunities for customization can be identified automatically [5–7]. By their very nature, these customizations must be derived from an internal representation of the application. Customizations obtained in this way provide a more effective method of computing the functionality defined by an application usually expressed in C or C++. However, they are as yet unable to take a global view of the application and restructure the computation, the data organization, and the flow of information through the system.

In this chapter we focus on the types of customization that are hard to automate and yet provide some of the most substantial performance improvements and energy savings. The ideas in this chapter are explained with particular reference to the ARCompact™ architecture, a 32-bit extentable RISC architecture designed for energy and cost-sensitive embedded applications, produced by ARC International. We therefore begin by examining ARC's extension mechanisms and consider the microarchitectural challenges of stateful extension to an architecture that supports a virtual machine progamming model.

We illustrate the limitations of simple instruction set extensions through the example of motion-video entropy decoding. We show how a top-down analysis of system behavior leads to a radical repartitioning of hardware and software that is intuitive and yet hard to automate. Through this example we examine the types of analysis and transformation needed by the tightly integrated compilers and architecture configuration tools of the future. After summarizing some of the current limitations of fully automated customization, we point out the significant benefits from customization that are available today.

8.1 THE ARCOMPACT™ INSTRUCTION SET ARCHITECTURE

There are a number of commercial architectures offering the possibility of instruction set extension, each with its own particular features. In this section we present a brief overview of one of the most widely deployed architectures that supports user customization, notably the ARCompact instruction set from ARC International [8]. As its name suggests, the ARCompact instruction set is optimized for compactness as well as defining an extendable architecture. The architecture evolved through a process of careful program analysis, aimed at identifying the most space-efficient ways to represent frequently occuring instructions while at the same time being able to compile performance-sensitive code using the fewest possible instructions.

The primary compactness feature is the use of a variable-length instruction format supporting freely intermixed 16-bit and 32-bit instructions.

Both 16-bit and 32-bit instruction formats may also contain an optional embedded 32-bit literal operand. Hence, instructions can be 16, 32, 48, or 64 bits in length. The most significant 5 bits of every instruction contain a common field that defines one of 32 *major opcodes*. The overall instruction format is shown in Table 8.1.

The first eight major opcodes encode 32-bit instructions, whereas the remaining major opcodes define 16-bit instructions. A summary of all major opcodes is shown in Table 8.2.

TABLE 8.1 ■ ARCompact™ 32-bit and 16-bit formats

5	27
format	opcode and 3 operands

5	11
format	opcode and 2 operands

TABLE 8.2 ■ ARCompact™ major formats, showing space set aside for extension instructions in both 16-bit and 32-bit formats

(a) Base-Case ARCompact Instruction Groupings

Major Opcode	Bits	Instruction Groups	Notes
0	32	Bcc	Conditional branch
1	32	BLcc, BRcc	Branch & link, compare & branch
2	32	LD	Load instructions
3	32	ST	Store instructions
4	32	*op* a,b,c	32-bit generic format
8,15, 23	16	*op*_S b,b,c	16-bit generic format
12	16	LD_S, ADD_S	Short load/add reg-reg
13	16	ADD_S, SUB_S, shift_S	Register plus immediate
14	16	MOV_S, CMP_S, ADD_S	One full-range register operand
16–19	16	LD_S	Load instructions
20–22	16	ST_S	Store instructions
24	16	*op*_S %sp	Load/store/add with %sp
25	16	*op*_S %gp	Load/add with %gp
26	16	LD_S %pc	%pc-relative load
27	16	MOV_S	Move short immediate to register
28	16	ADD_S, CMP_S	Add/compare with short immediate
29	16	BRcc_S	Branch on compare ==, !=
30	16	Bcc_S	Conditional branch
31	16	BLcc_S	Conditional branch & link

(b) Extension Instruction Groupings

5–7	32	*op* a,b,c	Extension instructions
9–11	16	*op*_S b,b,c	Extension instructions

In general, the 32-bit formats provide three register operands, or two register operands and a 6-bit literal. Predicated execution is also possible in these formats, using a 6-bit field to select one of 64 Boolean conditions on which to enable the instruction. The 16-bit formats generally provide two distinct register operands, one of which is used as both source and destination. Again, one of the operands can be replaced by a short literal. Whereas the 32-bit format is able to address up to 64 general-purpose registers, the 16-bit format addresses a noncontiguous subset of 16 registers, which have been chosen to provide a mix of callee-saved and caller-saved registers.

Six out of the thirty-two major opcode formats are set aside for extension instructions; three of them are 32-bit, and three of them are 16-bit formats. System designers using processor cores based on the ARCompact™ instruction set are free to define custom instruction set extensions and encode them in these formats.

On paper the ARCompact™ instruction encoding is moderately complex. This complexity allows both a compact encoding and extension regions for both 16-bit and 32-bit instructions. In practice, the complexity of encoding has little impact on real implementations of the instruction set, which are able to operate at comparable clock frequencies to similar synthesized processors while having significantly lower gate counts.

The ARCompact™ programming model defines 32 default general purpose registers (GPRS) r0 – r31 and a further 28 extension registers r32 – r59. Registers r60 – r63 have special meanings [8]. There is also an extended space of 2^{32} *auxiliary registers*. These cannot be used directly as operands but instead are loaded or stored, to or from GPRS, using lr and sr instructions. A number of default auxiliary registers is defined in a base-case configuration of an ARC core. These include status registers, exception registers, default I/O registers, and so on. Again, the system designer is free to define custom auxiliary registers as necessary. Typically these are used to provide I/O registers for interfacing custom logic blocks where an instruction-level interface is not appropriate or desirable.

Although the ARCompact™ architecture is a customizable architecture, it is also designed to be efficient in its base-case configuration for supporting a wide range of embedded applications. These range from the deeply embedded systems with a single application running without an operating system to Linux-based systems with virtual memory support [9] and, increasingly, multi-processor systems. Throughout this wide range, the architecture has to remain efficient in its support for architecture extensions. A summary of the default base-case instructions provided by the ARCompact™ architecture is presented in Table 8.3. In addition to providing the necessary support for all high-level language operators, there are are also a number of bit-manipulation operators that are less obvious for a compiler to generate but which are extremely useful in many embedded applications.

TABLE 8.3 ■ ARCompact™ base-case instruction set, showing partial equivalence between instructions coded in 32-bit formats and 16-bit formats

Mnemonic	32-Bit Instructions Operation	Mnemonic	16-Bit Instructions Operation
ld	Load from memory	ld_s	Load from memory
st	Store to memory	ld_s	Store to memory
		push_s	Store on stack
		pop_s	Load from stack
lr	Load from aux. reg.		
sr	Store to aux. reg.		
mov	Copy value to register	mov_s	Copy value to register
add	Add	add_s	Add
adc	Add with carry		
sub	Subtract	sub_s	Subtract
subc	Subtract with borrow	neg_s	Negate
rsub	Reverse subtract		
add$n$	Add with left shift by n	add$n$_s	Add with left shift by n
sub$n$	Subtract with left shift by n		
sub1	Subtract with left shift by 1		Subtract with left shift by 1
ext	Unsigned extension	ext_s	Unsigned extension
sex	Signed extension	sex_s	Signed extension
tst	Test		
cmp	Compare		
rcmp	Reverse compare		
min	Minimum of 2 values		
max	Maximum of 2 values		
abs	Absolute value	abs_s	Absolute value
not	Logical bit inversion	not_s	Logical bit inversion
and	Logical AND	and_s	Logical AND
or	Logical OR	or_s	Logical OR
xor	Logical XOR	xor_s	Logical XOR
bic	Bit-wise inverted AND		
asl	Arithmetic shift left	asl_s	Arithmetic shift left
lsr	Logical shift right	lsr_s	Logical shift right
asr	Arithmetic shift right	asr_s	Arithmetic shift right
ror	Rotate right		
rrc	Rotate right through carry		
bset	Set specified bit to 1	bset_s	Set specified bit to 1
bclr	Set specified bit to 0	bclr_s	Set specified bit to 0
bmsk	Logical AND with bit mask	bmsk_s	Logical AND with bit mask
btst	Clear specified bit	btst_s	Clear specified bit
bxor	Toggle specified bit		
bcc	Branch if condition true	bcc_s	Branch if condition true
brcc	Compare and branch	brcc_s	Compare and branch
blcc	Branch and link conditionally	bl_s	Branch and link
bbit$n$	Branch if bit is n (0,1)		
jcc	Jump if condition true	j_s	Jump
jlcc	Jump and link conditionally	jl_s	Jump and link
lp	loop with no branch overhead		
flag	Write to status register		
sleep	Place CPU in sleep mode		
		nop_s	No operation

8.1.1 Mechanisms for Architecture Extension

The ARCompact™ architecture can be extended in five orthogonal ways:

1. New instructions, encoded in appropriate groups

2. Additional core registers

3. Additional auxiliary registers

4. Application-specific condition codes

5. Arbitrarily sophisticated IP blocks, interfaced via registers defined in the auxiliary address map and read or written using `lr` and `sr` instructions.

High-end implementations of the ARCompact™ architecture support virtual memory and page-based address translation. Hence, memory referencing instructions that raise a page fault must be restartable. It also means that instructions in the shadow of faulting memory reference instructions must also be restartable. This requirement for precise exceptions carries over into architecture customizations and means, for example, that extension instructions must not modify processor state before they reach the *commit point* of the pipeline. The extension interface in the ARC 700 family of processors is designed with this in mind and ensures that extensions to processor state are protected from premature update. With this facility, applications that use customizations can be run within a fully featured multitasking operating system such as Linux.

Through experience with the kinds of extensions that are widely used in the field, a number of predefined and preverified extension options have been identified. These include a set of extended arithmetic instructions that perform saturation, a collection of instructions to assist floating-point libraries, a set of digital signal processing (DSP) extensions aimed primarily at audio signal processing, and most recently a major set of extensions for video processing [10].

8.1.2 ARCompact Implementations

The ARCompact™ instruction set is realized as a product in the form of two distinct synthesizable processor core families, the ARC 600™ and ARC 700™, both of which can be customized by licensees. The key implementation characteristics of the ARC 600™ and ARC 700™ are given in Table 8.4. All measurements are taken from cores that are synthesized for the TSMC 0.13-μm LVOD process [11] using standard cell libraries from Virage Logic. Operating frequencies are quoted for worst-case conditions and process variation, and power consumption is quoted under typical conditions and process variation. Dhrystone MIPS

TABLE 8.4 ■ Implementation characteristics for ARC 600™ and ARC 700™ customizable processor cores in their base-case configurations

Design Characteristic	ARC 600™	ARC 700™	Units
Pipeline length	5	7	stages
Logic area	0.31	1.23	mm²
Performance	1.30	1.58	Dhrystone 2.1 MIPS/MHz
Operating frequency	400	533	MHz
Power consumption	0.06	0.16	mW/MHz

measurements are taken under conditions that conform to Dhrystone benchmarking rules [12].

The ARC 600™ is smaller than the ARC 700™, due mostly to its shorter pipeline. As a result, only the compare-and-branch instructions require any form of branch prediction to minimize branch cost. For this, the ARC 600™ employs a simple, low-cost static branch prediction method based on the *backwards taken forwards not taken*, or BTFN, rule [13]. The ARC 700™ has a deeper pipeline and therefore implements a dynamic branch prediction method for all control transfers, based on a modified *gshare* scheme [14].

8.2 MICROARCHITECTURE CHALLENGES

When a simple processor architecture is extended with new instructions, there are relatively few microarchitecture challenges to overcome. It is usually possible to incorporate the functionality of relatively simple instructions within the execute stage of a typical RISC pipeline. With more complex functions it becomes necessary to pipeline them over several stages, or implement them as blocking multicycle operations, perhaps with out-of-order completion. In a high-performance processor, with sophisticated data forwarding between pipeline stages to minimize instruction latencies, such multicycle operations can lead to additional complication in the detection of stall conditions. However, in most cases these can be fully automated, and from an engineering perspective they present few problems.

When the architectural state of a more complex processor architecture is extended through the introduction of new registers, the microarchitectural challenges become more significant. When we talk of a more complex architecture we mean that, for example, the processor may have a precise exception model to allow it to support a TLB-based memory management unit and page-based virtual memory.

There are three important issues raised by the introduction of a new processor state:

1. Running a multitasking OS in conjunction with stateful extensions touches many aspects of the system design, particularly in relation to state save and restore functionality.

2. In a pipeline that supports restartable precise exceptions, extension state cannot be speculatively updated unless it can also be rolled back when an exception occurs.

3. The greater the scope of an extension, and the greater the autonomy of the logic associated with the extension state, the more decoupled it can become from the normal flow of baseline instructions. This presents both opportunities for increased parallelism and challenges for automation.

The introduction of new state may therefore require extensions to the operating system. This may simply require an extension to the process control block to allow the extended state to be saved. It may require new device drivers and interrupt handlers, if the processor extensions introduce new types of exceptions or interrupts.

In a processor with precise exceptions, any state extensions must obey the normal rules for speculative update. For example, if an extension contains a state machine, the state must not be updated in the shadow of a memory referencing instruction that may generate an exception. If such an update is needed for performance reasons, then it must be performed speculatively and the previous state must be restored if a state-updating instruction is killed before it commits. For some extensions it makes sense to delay the state update operation until the end of the pipeline (i.e., at the write-back stage). However, in other cases this may introduce a stall if a subsequent extension operation wishes to use the most recently computed extension state. As with base-case instructions, a range of pipeline design options must be considered for each extension if an extended core is to retain the efficiency of the base-case core.

When processor extensions become autonomous, there are issues of how best to integrate and program the new functionality. Consider, for example, a block motion estimation (BME) engine included as an autonomous IP block to enhance video encoding within a customized media processor. In this case a great deal of overlap can be tolerated between the BME engine and the CPU. This type of extension effectively creates a *heterogeneous multiprocessor*. While the performance benefits may be high, the autonomy of the extension logic seriously complicates the programming model. At present this kind of extension cannot be fully automated and remains a challenge to research.

8.3   CASE STUDY—ENTROPY DECODING

In the previous section we observed that automating the creation of some of the most powerful architecture extensions can be highly complex—possibly so complex as to require manual intervention. Whereas the skilled design engineer can absorb the entirety of a system and propose software refactoring and system design choices, identifying these automatically is a major challenge at the present time. An instructive way to explore this challenge further is to consider a case study in which extensive, though manual, architecture optimization yields a significant benefit.

Video decoding is an important application area of embedded systems that also presents a heavy computational load when implemented by an embedded processor rather than by dedicated hardware [15]. There is an increasing reliance on software-based video decoders, as they offer the flexibility of being able to cope easily with evolving video coding standards and to support a wide range of coding standards using the same hardware. Fig. 8.1 illustrates the structure of a typical MPEG-2 video decoder.

The primary purpose of video coding is to compress a sequence of pictures. Each picture comprises a mosaic of square blocks, typically 8×8 pixels. Each picture is thus represented as a sequence of coded blocks.

The coding of blocks typically employs a range of techniques to eliminate spatial redundancy within individual pictures and temporal redundancy between successive pictures. Some blocks are predicted from blocks nearby in the same picture. Others are predicted from blocks that are nearby but in previous or successive pictures. Predicted blocks may not be an exact match, in which case the residual difference between the prediction and the actual block is coded in the frequency domain using

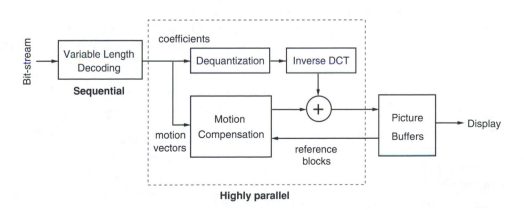

■ **FIGURE 8.1**

Structure of a typical video decoder.

block transforms such as the two-dimensional discrete cosine transform (DCT) [16]. This is followed by quantization of the DCT coefficients, which reduces the number of bits required to code the coefficients at the cost of reduced image quality. The process of coding blocks in this way results in a stream of symbols describing both the coding decisions and the coded picture contents. To further compress the stream of coded picture symbols, some form of *entropy coding* is normally employed [17]. This relies on the observation that symbols appear with differing frequencies in typical video sequences and the relative frequency of many common symbols are *a priori* predictable. For this reason most video coding standards use a set of variable-length codes, often based on Huffman coding [18], to compress coded picture symbols.

The structure of a typical variable-length symbol decoder (VLD) is shown in Fig. 8.2. Here we see a small input buffer containing the next 32–64 bits to be decoded from a video stream. This is fed with data from a stream buffer, usually several kilobytes in size. The symbols in the input buffer are of unknown size until they have been decoded. Hence, we cannot know the size or position of symbol i until all symbols $j < i$ have been decoded. This constraint imposes a strict sequentiality on the decoding of variable-length symbols. This contrasts sharply with the highly parallel nature of the subsequent pixel computations involved in motion compensation, dequantization, and inverse DCT, in which SIMD and VLIW techniques can be deployed to significant effect [10, 19, 20]. The sequential nature of VLD computation represents a potential bottleneck in most video decoders. Perhaps not surprisingly, therefore, research interest often focuses instead on the more glamorous aspects of video decoding that can be parallelized. For example, an early study in [21] demonstrated high levels of potential ILP in media applications. Similarly, data parallel operations but have also been shown to improve energy efficiency as well as enhance performance [22].

However, Amdahl's Law [23] also tells us that we cannot ignore the sequential parts of a video decoder, such as the VLD. Rather than trying to exploit parallelism within the VLD itself, we must minimize the number of clock cycles needed to decode each symbol. Where possible,

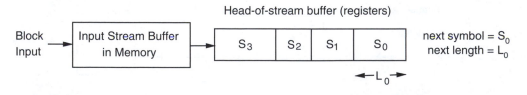

■ FIGURE 8.2

Symbol extraction in variable-length decoding.

we should also try to overlap VLD computations with parallelized block computations.

8.3.1 Customizing VLD Extensions

In order to identify useful VLD customizations automatically, one must work from an existing baseline source code implementation. There are many existing software video decoder implementations in widespread use. Some of them, for example `ffmpeg` and `mplayer`, are in the public domain. The ideal process of customization would take a VLD implementation from such a baseline software implementation and automatically derive a set of architecture extensions to minimize the time required to decode each variable-length symbol. So what problems might be encountered by architecture customization tools in their attempt to generate VLD extensions? One way to address this question is to explore manually the kinds of customizations that one might expect to see coming from an automated process.

Fisher et al. in Chapter 3 highlight the difficulty posed by not having source code available early enough in a typical product lifecycle. However, when application code is available in advance, it will often be *legacy code* derived from previous projects that has been optimized for best performance on a previous architecture. For example, if one examines one of the many public domain applications for video decoding or encoding of any of the MPEG or H.26x standards, one will find code that has been heavily optimized for one (or perhaps several) specific architectures, usually with SIMD media instructions. So we immediately see an important instance of a more generic problem: that automated customizations must, of necessity, be derived from source code that is already heavily optimized for a specific architecture. Naturally, the source code must not require any manual modification prior to the customization process, as it could not then be considered genuinely automatic. There is no reason to believe that the majority of real applications of architecture customization are any different, and we have to accept that automated customization tools must be capable of *undoing* the manual optimizations that have been previously applied. The process of returning the source code to a canonical state is likely to be more difficult than the subsequent optimization and customization.

Perhaps the simplest case of undoing a manual optimization is the case where function evaluation has been replaced during software optimization by a lookup table. This kind of optimization occurs frequently in signal processing and video coding where the domain and range of values is often small enough permit unary functions to be implemented as a table lookup. Consider, for example, the process of converting an 11-bit signed integer value to an 8-bit unsigned integer with saturation of unrepresentable values. A possible C implementation of this is shown in Fig. 8.3.

```
static inline unsigned char clip8 (int x)
{
   if (x > 255) return 255;
   if (x < 0)   return 0;
   return (unsigned char)x;
}
```

▪ **FIGURE 8.3**

Typical example of a function that might be replaced by a table lookup in manually optimized code.

The code of Fig. 8.3 will translate to a sequence of several instructions even when the compiler is able to inline the function appropriately. Conversely, by creating a lookup table with 4,096 entries, one can replace the call to clip8 with a single array reference using X as the array index. Any reasonable tool for generating automated instruction set extensions should be capable of generating a clip8 instruction with an efficient hardware implementation. However, if clip8 has already been replaced by a table lookup, few tools (if any) will be able to customize the application using clip8. It is common practice to initialize lookup tables at run time, unless the contents are trivial or the table is very short. It is usually easier and less error-prone to write a table-initialization function than it is to express them as a constant array initializer. Analysis tools will therefore perceive this functionality as a memory reference to a nonconstant array and will most likely search elsewhere for easier customizations.

There is another, more complex, example of replacing function evaluation with table lookup that lies at the heart of variable-length symbol decoding. One way to decode variable-length symbols using a Huffman code is to examine each bit of the input in sequence, traversing a binary Huffman tree based on the value of each bit until a leaf-node representing a coded symbol has been found. Unfortunately this method is rather slow, requiring several instructions per input bit. A common optimization is to again use a table lookup approach. This is complicated somewhat by the variable length of each sequence of input bits. However, this is easily overcome by trading additional table space for a simple algorithm. The basic idea is to first determine the maximum number of bits to be decoded in one step. If this is n bits, then a table with 2^n entries is required. To decode the coded symbol c that has $m < n$ bits, one places the decoded output at all table entries with indices i such that $c.2^{(n-m)} \leq i < (c + 1).2^{(n-m)}$. This ensures that a table lookup of n bits, in which the most significant m bits are the coded symbol c, will find an entry which decodes symbol c. Short symbols have many identical entries, whereas long symbols have few. There is an added twist in cases where coded symbol sizes can be larger than n. To avoid excessive table sizes, it is necessary to construct a hierarchical table in which short symbols are decoded with a single probe but which require multiple probes to decode a long symbol.

Using table lookup for VLD operations is a particularly effective optimization for a CPU with a fixed instruction set. A sequence of many instructions can be replaced with a small number of table accesses. When a symbol has been located in the table, its decoded value and coded symbol length are extracted. The symbol length is used to determine how many bits to discard from the input buffer, and the symbol value is used in later stages of video decoding.

Given the sophisticated structure of software VLD tables, they are invariably constructed during the initialization phase of the video decoder, possibly from more compressed representations of the required Huffman codes. As with the earlier table lookup example, it is infeasible to determine at compile time that VLD tables are *dynamically read only*. If this were possible, then one could use such information to replace the table lookup with custom logic implementing the same function.

It is surprisingly easy to perform a semi-automated replacement of table-based Huffman decoding with customized architectural extensions. The first step is to construct a generator that is able to generate logic from an abstract specification of a Huffman code. Consider the `ffmpeg` Huffman code shown in Fig. 8.4. This table is used to decode MPEG B-type macroblock parameters. This has just 11 table entries and is therefore possible to show in its entirety. Most other MPEG tables are very much larger and are impractical to use as illustrative examples.

In this case study a simple tool was constructed to create a direct logic implementation of the VLD tables for the MPEG-1, MPEG-2, and MPEG-4 video standards. The generator uses the source representation of the original VLD tables to produce Verilog HDL for the logical equivalent of a table lookup. The logic is contained within the `vld_logic` module shown in Fig. 8.5. This is a stateless implementation that takes

```
static uint16_t table_mb_btype[11][2] = {
    { 3, 5 }, // MB_INTRA
    { 2, 3 }, // MB_BACK
    { 3, 3 }, // MB_BACK      |MB_PAT
    { 2, 4 }, // MB_FOR
    { 3, 4 }, // MB_FOR       |MB_PAT
    { 2, 2 }, // MB_FOR       |MB_BACK
    { 3, 2 }, // MB_FOR       |MB_BACK    |MB_PAT
    { 1, 6 }, // MB_QUANT     |MB_INTRA
    { 2, 6 }, // MB_QUANT     |MB_BACK    |MB_PAT
    { 3, 6 }, // MB_QUANT     |MB_FOR     |MB_PAT
    { 2, 5 }, // MB_QUANT     |MB_FOR     |MB_BACK   |MB_PAT
};
```

■ **FIGURE 8.4**

`ffmpeg` table for MPEG macroblock B type code.

```
module vld_logic (vld_table, bitstream, symbol, length)
   input   [5:0] vld_table;
   input   [25:0] bitstream;

   output [15:0] symbol;
   output [4:0] length;

   reg    [15:0] sym;
   reg    [4:0] len;

   task table_entry;
      input   [4:0] e_len;
      input [15:0] e_sym;
   begin
      sym = e_sym;
      len = e_len;
   end
   endtask

   always @( vld_table or bitstream )
   begin
      sym = 16'd0;
      len = 5'd0;

      casez(vld_table)
      `include"table1.v"; // logic for MPEG   macroblock B-type code
      `include"table2.v"; // logic for MPEG   macroblock P-type code
                           // add further tables as required
      endcase
   end

   assign symbol  = sym;
   assign length  = len;
endmodule
```

▪ **FIGURE 8.5**

Verilog module for variable-length symbol decoding.

a 26-bit `bitstream` vector and a 6-bit `vld_table` number as inputs, and produces a 16-bit `symbol` and 5-bit `length` value.

A doubly nested `casez` construct provides a suitable structure for defining the mapping from table number and input bit pattern to symbol value and symbol length.

The generated Verilog for the Huffman table of Fig. 8.4 is illustrated in Fig. 8.6. In this example the `table_entry` task is simply a notational convenience that enables a more compact representation of the table in Verilog source code. Each `table_entry` statement defines the symbol length and symbol value matching the input pattern given by the case label. The use of ? tokens in the input pattern allows variable length patterns to be defined in a compact and efficiently synthesized way.

```
6'd14: // MPEG macroblock B-type table
  casez (bitstream [25:20])
  6'b00011?:table_entry (5, 0);
  6'b010???:table_entry (3, 1);
  6'b011???:table_entry (3, 2);
  6'b0010??:table_entry (4, 3);
  6'b0011??:table_entry (4, 4);
  6'b10????:table_entry (2, 5);
  6'b11????:table_entry (2, 6);
  6'b000001:table_entry (6, 7);
  6'b000010:table_entry (6, 8);
  6'b000011:table_entry (6, 9);
  6'b00010?:table_entry (5, 10);
  endcase
```

■ **FIGURE 8.6**

Generated Verilog fragment for MPEG macroblock B-type code.

So far we have derived a block of combinational logic from the Huffman table structures present in the original C source code of a video decoder. Although the derivation of logic from C has been automated, the realization that this would be a beneficial transformation was entirely manual. And, at the time of writing, the reintegration of the VLD functionality back into the software video decoder is also a manual task. However, the process of optimization does not stop here: we need to determine how we can best use this customized functionality and how to encapsulate it within an API.

Let us now relate the example of VLD table optimization to the four primary mechanisms of architecture extension highlighted earlier. The first and simplest way to incorporate VLD table logic is via instruction-set extensions. The most useful extension instruction would be an instruction that takes a coded bit pattern from the head of the input bit-stream and a Huffman table identifier as its two operands. This instruction would interrogate the selected table and return the decoded symbol. This is the instruction called vld_sym in Table 8.5. There also needs to be a related instruction, shown as vld_len, that returns the length of the next symbol in the coded bit-stream.

A simple extension of the processor state with a special-purpose register containing the length of the symbol decoded by the vld_sym instruction would allow both the symbol value and length to be computed simultaneously. If the input bits are held in a conventional

TABLE 8.5 ■ Candidate ISA extensions for VLD operations

Instruction	Operands	Functionality
vld_sym	(table, bits)	Decode the next symbol
vld_len	(table, bits)	Compute length of next symbol
bvld	label	Branch to label if VLD buffer needs refill

processor register, they must be shifted by the symbol length before the next symbol can be decoded. If the special-purpose symbol length register can be used as the distance operand to a shift instruction, then no actual instructions are required to determine the length of each symbol. However, the manipulation of the bit-stream buffer is moderately complex. It requires that a register is shifted by an arbitrary amount and the vacated space is filled with the same variable number of bits from the bit-stream input buffer in memory.

Most experienced hardware engineers at this point will notice that the head of the input bit-stream buffer could be implemented directly in hardware, surrounded by sufficient logic to perform all bit-stream operations. As the bit-stream buffer provides inputs directly to the `vld_logic` module, and is itself later modified by shifting out the number of bits given by the `length` output from the `vld_logic` module, an integrated design combining both buffer and logic will eliminate the need for distinct instructions to access the `vld_logic` function and shift the bit-stream buffer. However, we still need to detect when the buffer register is empty or nearly so. With a small number of instructions we can read a status register, test its value, and branch conditionally on the outcome. Occasionally the program would then branch to a section of code to refill the buffer from memory. However, a more efficient customization would be to invent a new branch condition, and hence a new conditional branch instruction, to sense the state of the buffer directly. In the ARCompact™ instruction set, half of all branch conditions are available to be used in this way. The result of this customization would be the `bvld` instruction illustrated in Table 8.5.

Implementing the bit-stream buffer directly in hardware has major implications for the rest of the system. For example, there are occasions when a video decoder needs to extract bits directly from the bit-stream and use them verbatim. These elements of video decoder functionality are not necessarily located in the same compilation module as the elements of functionality from which the bit-stream buffer logic and the Huffman decode logic is synthesized. The introduction of new state into the processor, and the association of application data with that state, not only changes the way in which data is manipulated, but also changes the data representation. The main challenge to automation in this case is that a number of separately compiled modules share a common view of certain data, and those data are the subject of a customization that will alter their representation. This is an example of the kind of refactoring [24] that is needed in order to successfully deoptimize source code before automated customizations are attempted. Although manual refactoring can be performed, the immaturity of refactoring tools for preprocessed C code presents a major challenge to automated architecture customization where global data transformations are required.

There are a number of reasons why replacing table lookup with customized logic would be beneficial. The most obvious reason is that it eliminates memory references. In principle, by eliminating references to VLD tables, pressure on the data cache will be eased. We can gauge the significance of this effect from Fig. 8.7, which shows the number of distinct data cache blocks required by VLD table references during the decoding of an MPEG-2 video sequence. The parameters of the MPEG-2 test sequence are summarized in Table 8.6.

In Fig. 8.7 we see 250 to 350 distinct 32-byte cache blocks referenced by VLD table accesses in each half-frame interval. This corresponds, on average, to approximately 30% of the blocks in a 32-KB data cache being touched repeatedly by the VLD table accesses alone.

The VLD table data must coexist alongside other cached data, which in practice means the repeated accesses to VLD table entries, spanning 30% of the available cache locations, are likely to collide and evict other cached data. In effect, to eliminate the negative impact of VLD table

TABLE 8.6 ■ MPEG-2 test sequence parameters

Parameter	Value
Test sequence	Trailer for *Bend It Like Beckham*
Bit rate	2 Mbps
Resolution	720 × 576, 25 fps

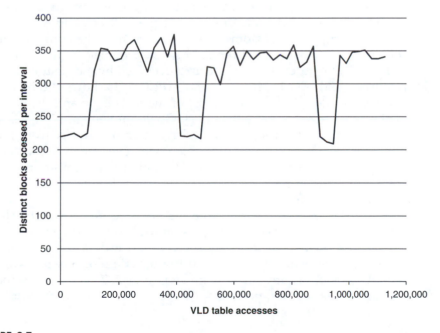

■ **FIGURE 8.7**

Memory footprint of VLD table accesses measured over a succession of half-frame intervals.

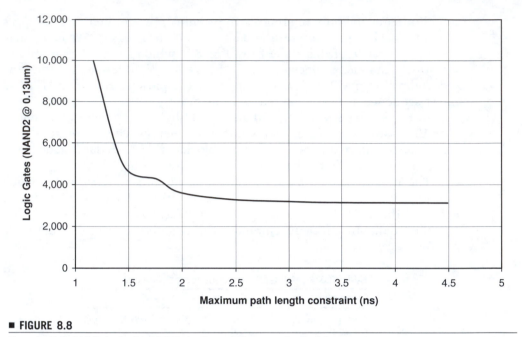

FIGURE 8.8

Complexity of VLD logic for MPEG-1, 2, and 4 versus logic delay.

accesses on cache performance, additional cache capacity at least equal
to the footprint of the VLD table in cache will be required. The extra
die area occupied by custom VLD extensions must be compared against
the die area of that part of the data cache occupied by the VLD tables.

Fig. 8.8 illustrates the total die area occupied by the custom VLD logic
required by MPEG-1, MPEG-2, and MPEG-4. It shows the relationship
between the delay (in nanoseconds) of the VLD logic and the gate count
of the synthesized logic network. The logic implementation follows the
scheme described earlier and was derived automatically from the VLD
tables present in the `ffmpeg` source code. The generated Verilog was
synthesized using Synopsis Physical Compiler using minimum physical
constraints, targeting a 0.13-μm standard cell library under worst-case
commercial conditions.

An approximate measure of the die area required to cache the VLD
table accesses in the original noncustomized code is shown in Table 8.7.

The RAM cells of a four-way set-associative data cache with 32-KB
capacity, implemented in the same 0.13-μm technology used to synthesize
the VLD logic, occupy an area greater than 227 Kgates.[1] As the VLD
table accesses will require approximately 30% of the capacity of such

---

[1] In this chapter, *gate count* is computed as die area divided by the area of a 2-input
NAND gate with unit drive strength.

TABLE 8.7 ■ VLD cache area estimation, based on 30% occupancy of a 4-way set-associative 32Kbyte data cache with 32-byte blocks

Memory component	Area (gate-equivalents)	Number	Total	30%
2048 × 32 data RAM	49,500	4	198,000	59,400
256 × 20 tag RAM	7,300	4	29,200	8,760

a data cache, an area equivalent to at least 68,000 gates will be occupied permanently by cached VLD accesses. In comparison, we see from Fig. 8.8 that the area of the customized VLD logic varies from 3,000 to 10,000 gates depending on logic delay constraints. From the perspective of die area, this customization clearly has a positive benefit. One would also expect to see much improved data cache performance when VLD table accesses are removed from the software video decoder and replaced with custom instructions to access the VLD logic directly. Hence, for the same original performance one would expect to be able to reduce the cache capacity by approximately one third. Not only is this optimization difficult to automate, but the die area cost function used to evaluate its effectiveness must include cache memory as well as CPU logic area.

The replacement of tables and memory references with custom logic will also affect energy consumption. In most embedded systems, energy efficiency is as important as performance and die area. Customizations must be beneficial in all three areas, or at the very least offer a good tradeoff between additional die area and improved performance and energy consumption. If the decision on whether or not to apply each customization is fully automated, then some way of evaluating the likely impact on energy consumption of each customization is needed. In general, this is a hard problem, since it requires a model of energy consumption before and after customization, as well as a way of applying program profile information to quantify actual energy consumption under typical conditions.

8.4   LIMITATIONS OF AUTOMATED EXTENSION

Elsewhere in this book we have seen how instruction-set extensions can be generated automatically using profile-guided analysis. The automation of instruction-set extensions takes much of the manual effort out of customizing an architecture for a particular class of applications. We may think of it as a natural extension of profile-guided compiler optimization. However, we saw from our case study that simple instruction-set extensions are not the only form of extension, nor even the most effective.

From our earlier examination of the manual process, we can highlight a number of limitations of the automated process:

■ Extensions are typically identified through *local* analysis of data-flow graphs and, at present, global optimization is still an open research topic.

■ The fusion of program fragments into extension instructions is limited by the number of live-in and live-out values that can be encoded in one instruction (usually 2-in and 1-out).

■ Automatic identification of extensions may result in large numbers of extensions, each with relatively low impact. However, taken together they may reduce the total instruction count significantly. This has the potential to bloat the CPU core, as the logic for each additional instruction is added. Manual intervention is required to establish the set of included instructions.

■ Until the gate-level representation of a customized core is laid out, we do not know whether each additional instruction actually speeds up or slows down the application. For example, we may decide that adding 50 new instructions minimizes the number of instructions executed, but later we may discover that adding the logic for those 50 instructions also extends the actual clock period and leads to a slower overall solution.

■ Another growing problem is the increasing disparity between the logical and physical domains, and the fact that area, timing, and power predictions from logic synthesis are inaccurate at very deep submicron technologies. Making accurate energy and area trade-offs clearly requires far more information about the post-synthesis impact of customizations than is usually available to a compiler and architecture customization tool. Simple linear models of die area, timing, and power consumption break down below 130-nm design rules, due to the increasing impact of wire delay and wire load. Automating the design tradeoffs at 90 nm and below will require physical models that take account of the *layout* of gates rather than simply the *number* of gates.

■ Automated extensions, by implication, have no domain knowledge and cannot make high-level restructuring decisions. Refactoring the code may have a big impact on the effectiveness of automated extensions. Dealing with the interaction between code refactoring and automated ISA extensions is a highly creative process and beyond the capabilities of current tools.

- The initial push toward automated extensions has focused on optimizing the mapping of *operators* into hardware—for obvious good reasons. But there are other areas that need to be considered too. In particular:

  1. Translating data access into logic, for sparse read-only data or functions that have been optimized at source code into a lookup table.

  2. Transforming the way in which data is stored and manipulated, so that item (1) can be more effectively applied. Note: VLD lookup tables can be modified in this way prior to recoding the multilevel table lookup into VLD decode logic.

  3. Innovating in the way data is moved between memory and processor. For example, this may include application-specific DMA devices or cache prefetch capabilities.

8.5 THE BENEFITS OF ARCHITECTURE EXTENSION

The limitations of fully automated architecture customization that we have outlined in this chapter serve as challenging research goals for those of us who continue to explore this fascinating area. In the meantime, there remain strong commercial and technical reasons why customizable and configurable architectures have an important role to play in the design of embedded systems. Four of the most important benefits of architecture customization that apply today are as follows:

1. Architecture customization is an important enabling technology for hardware-software codesign.

2. The ability to customize a processor provides a performance buffer zone and leads to greater confidence in the choice of embedded CPU.

3. The rapid and reliable construction of domain-specific platform IP relies on the existence of customizable and configurable processor architectures.

4. Architecture configuration provides a powerful tool through which sophisticated IP users can optimize and differentiate their products.

In the following four subsections, we elaborate on these four benefits.

8.5.1 Customization Enables Codesign

In Chapter 3, Fisher et al. explain how the product lifecycle can pose difficulties for automated architecture customization. For example, the late availability of software in a stable form may not allow time for hardware design-space exploration. Although this may be a problem in some cases, there are other cases where a new product is created from an existing software base, with the goals of improved performance or lower power consumption. Often in such cases, the performance-critical kernels may be quite stable, whereas other aspects of functionality and user-interface may remain fluid until quite late in the product development cycle. The ability to customize a CPU enables the process of hardware-software codesign for the performance-critical kernels, provided customization does not undermine the stability of the baseline CPU functionality and the tool chain.

A clean separation between baseline and extended functionality in the CPU core and the tool chain is a notable feature of all successful customizable processors. Keeping a clean separation encourages stability in operating system and library code, all of which can be prebuilt to target baseline configurations while continuing to function correctly within an extended CPU core. This argues for a baseline CPU with a programming model that is *extended* rather than *redesigned* for each customized core. For example, an ARC core is extended in a clean, well defined way by the introduction of additional instructions, registers, Boolean conditions, and so on. In comparison, two instances of a customizable VLIW processor may issue different numbers of instructions per cycle, and therefore require separately compiled operating system binaries and libraries.

8.5.2 Customization Offers Performance Headroom

Choosing a processor IP core is one of the most important decisions for engineering managers of system-on-chip design projects. Such decisions are based on many factors, including cost, performance, power consumption, software tool-chain quality, and the availability of third-party software. Once IP decisions have been made, it is extremely costly and time consuming to change course and choose a different processor core. If the performance estimates on which the processor choice was made turn out to have been overly optimistic, the whole design process can be placed at risk. However, if the processor is customizable, there will usually be a way of customizing the processor to gain additional performance and hence meet the performance goals. Even if the customization takes place late in the design process, the overall cost to the project will be lower than if the design is restarted with a different processor core (where the same uncertainty over application performance may still exist). For this reason, enlightened engineering managers sometimes choose a customizable processor core even though they have no initial expectation of actually exploiting the customizability. So, customizability can serve as a safety net, simply to be there if needed.

8.5.3 Customization Enables Platform IP

In Chapter 3, Fisher et al. highlight a number of important commercial difficulties with the concept of "one CPU per application." For example, the cost of verification can be as much as 60% of the engineering cost for new system-on-chip devices. And verification costs are growing as system-on-chip devices become increasingly complex. However, there are other ways to exploit customization. One particularly appealing option is for customizable IP providers to perform the customization in-house to create preverified, preoptimized processor extensions, each targeting a specific application domain. By combining domain-specific knowledge with a deep understanding of the processor design, IP providers can remove much of the engineering risk that might otherwise impede the acceptance of a customized processor. It also reduces the design and verification effort expended by IP providers in developing highly optimized standard solutions to a range of popular applications.

8.5.4 Customization Enables Differentiation

The ability to differentiate a system-on-chip device from competing devices provides an important advantage in the market place. Other contributions in this book have shown, in a variety of ways, how customization can yield combined hardware-software solutions that are smaller, faster, and more energy efficient than fixed solutions. Although the engineering of customized solutions requires additional effort, the payback in terms of market advantage is often sufficiently valuable to make it worthwhile. The balance between risk and benefit will be different in each case. New incumbents in growing markets are often willing to deploy customized architectures as a means of differentiating their product from similar products offered by other vendors. The customizations will be completely hidden of course from the end user, but the user will hopefully perceive an enhanced and more attractive product.

8.6 CONCLUSIONS

In this chapter we have investigated some of the issues raised by processor customizations that go beyond the introduction of new instructions. We saw how the ARCompact™ instruction set provides a range of mechanisms for architecture customization, including but not limited to the creation of new instructions. Other mechanisms are also available: the extension of register files, the creation of new branch conditions and branch instructions, and extensions to the set of auxiliary registers. With such state extension, it is possible to construct interfaces to autonomous processing engines, offering the possibility of heterogeneous parallelism.

We saw in the case of architecture extensions for VLD that significant performance gains are achieved by defining a complete API for VLD operations rather than by inventing a few extra instructions. Extension instructions are usually pure functions, operating on just one or two operands. This limits their applicability to cases in which an application-specific sequence of simple operations can be fused into a compound function.

The next step beyond instruction-set extensions is to create a new processor state that can be referenced implicitly by custom instructions. The customization process must map key elements of application data to the new processor state—a process that in general will require a refactoring of the original source code. Of course, there will be cases where this is not required, but many interesting and commercially important cases will involve legacy software where the key elements of application data are global and perhaps referenced by pointers passed as arguments through a contorted call graph.

As an application-specific processor state is created, the set of functions needed to maintain it also increases. In most processors, there must be a method of saving and restoring all processor states, including the application-specific processor state, when a context switch occurs. And when all of the operations that must take place on the new application-specific processor state are brought together, one has then effectively created a new class of object with its own API. If the customization process is entirely automated, the API may never be documented and externalized in the ways we normally expect for a manually generated API. And yet, it would still exist, somewhere inside the architecture customization and compiler toolset.

At this point one may reasonably ask where the boundary lies between *optimized translation*, from one form of program into another, and the *synthetic design* of a target system? We believe the boundary is crossed when an automated architecture customization tool *learns from its experience* and applies that learning over time. This requires a certain *persistence* in the knowledge gained through customization. Indeed, if such systems were able to learn, their knowledge could be as valuable a commodity as the tools themselves.

CHAPTER 9

COPROCESSOR GENERATION FROM EXECUTABLE CODE

Richard Taylor and David Stewart

Previous chapters have discussed various techniques for the configuration of embedded processors. Both manual specification of instruction extensions using ADLs (Chapter 4) and largely automated techniques for instruction discovery (Chapters 6 and 7) have been covered. Customized instructions require customized tool chains in order for them to be easily targeted, and this has been discussed in Chapters 5 and 8. This chapter covers a somewhat different approach to the problem that is motivated by the desire for an efficient design flow in the context of multiprocessor SoCs. In this approach, the ASIP is treated as an autonomously executing coprocessor, acting as a slave to some other embedded control processor in the system. The coprocessor's data paths and instruction set are derived automatically via profile-driven analysis, and the software tool flow is greatly simplified by the use of static binary translation techniques to program the coprocessor using the native instruction set of the controlling processor.

9.1 INTRODUCTION

As we look to the future, we witness two opposing forces that will have a profound impact. On the one hand, increasing software content is inexorably driving up the data processing requirements of SoC platforms, while on the other hand, we see embedded processor clock frequencies being held in check as we struggle to stay within stringent power envelopes. Given this backdrop, we must look to parallelism, at both the microarchitectural and system level, to sustain a winning design approach. It seems unlikely that this gap will be bridged with complex out-of-order superscalar processors, as has been the trajectory in the desktop processor market [1, 2]. Their area and power inefficiency cannot be justified in platforms where predictable execution patterns can be leveraged to achieve greater efficiency.

We will see the continued rise of heterogeneous multiprocessor SoCs as the solution of choice. This heterogeneity is a result of the need to bring not just increased processing performance but *application-specific* processing to the platform. The use of ASIPs allows us to reduce the number of instructions that need be executed to solve a given problem and thus supply a more power-efficient solution without significantly impinging on programmability, on the condition that the custom instructions are judiciously selected [3]. However, in addition to ASIPs, we will still require the thread-level parallelism enabled by multiprocessor systems. It seems unlikely that the SoC of the future will be architected around a uniprocessor, even if it is highly application-specific and configurable.

The key challenge we face is not simply how to design a single ASIP, but how to map our software functionality onto the rich tapestry of a multiprocessor, multi-instruction set environment. We must deal with the problem as an intrinsically system-level one. Ignoring or sidelining system-level bus traffic and memory architecture issues while concentrating on the minutiae of custom instructions will not yield a practically applicable design flow.

We also have to accept the typical state of software developed within embedded systems. The rise of multicore systems in the embedded space will ultimately furnish software engineers with a greater awareness of multithreading. However, the reality of embedded software is that multithreading is not typically integrated into the application as a foundation model of computation but as an afterthought used to adapt an application to a particular embedded environment.

As ASIPs become easily accessible, more functionality will be retained in the software domain, but we must acknowledge that not everything will evolve as quickly as we would like. We want to map from our standard software to offload functionality onto an autonomous hardware block in order to free up the main processor for other tasks. We don't want to have to make radical changes to our main processor, such as moving to a new architecture or vendor, as this is the heart of a system in which there is huge investment in legacy software, design, and expertise. While we retain a relatively simple threading model, the concept of the central control processor is not going to disappear.

Thus the pragmatic solution that fits within the constraints of embedded systems is one that offloads functionality from the main processor directly onto a loosely coupled programmable accelerator, or coprocessor.

9.2   USER LEVEL FLOW

CriticalBlue's Cascade solution allows software functionality implemented on an existing main CPU to be migrated onto an automatically optimized and generated loosely coupled coprocessor. This is realized as an automated design flow that provides a bridge from a standard embedded software implementation onto a soft core coprocessor described in RTL. Multiple coprocessors may be attached to the main CPU, thus architecting a multiprocessor system from a software-orientated abstraction level.

The generated coprocessors are not fixed-function hardware accelerators, but programmable ASIPs. Their combination of functional units, internal memory architecture, and instruction set are derived automatically by analysis of the application software; as such, they represent a powerful instrument for system partitioning. Migration of functionality onto coprocessors does not impose the loss of flexibility implied by the existing manual processes of developing hardwired RTL accelerator blocks.

The primary usage of such coprocessors is for offloading computationally intensive algorithms from the main processor, freeing up processing resources for other purposes. Clearly there will be a limited set of blocks within a typical SoC where a custom hardware implementation remains the only viable option, perhaps due to physical interfacing requirements or extreme processing needs beyond that practically achievable with programmable IP. It should be noted that the Cascade flow includes the capability for a customer IP to be utilized within the generated coprocessor RTL, thus widening the applicability of the approach.

Fundamentally, the advantage provided by this design trajectory is about exploitation of parallelism. The parallel execution of the main processor alongside coprocessor(s) provides macro-level parallelism while the microarchitecture of the coprocessors themselves aggressively exploit instruction level parallelism (ILP). In this sense the goals of the flow are identical to those of the manual process of offloading functionality onto a hardwired block.

Since this process is heavily automated, the user can rapidly explore many different possible architectures. Being a fully programmable machine, a coprocessor can be targeted with additional functionality beyond computationally intensive kernels with little additional area overhead. This allows design exploration unencumbered by typical concerns of RTL development time and overreliance on inflexible hardwired implementations.

One unique aspect of the Cascade solution is that it implements directly from an executable binary targeted at the main processor, rather than from the C implementation of an algorithm. A static binary translation is performed from the host instructions into a very different application-specific instruction set that is created for each coprocessor. We have found that this flow provides a significant usability enhancement that eases the adoption of this technology within typical SoC projects. Since existing software development tools and practices can still be used, the existing environments and knowledge can be leveraged to full effect. Moreover, the offloaded functionality remains in a readily usable form for use on the main CPU, allowing greater flexibility in its deployment for different product variants.

Fig. 9.1 provides a flow diagram of the steps in using Cascade from a user perspective. The initial steps for targeting the flow are identical to those of a fully manual approach. The application software is developed and may then be profiled using a function-level profiler or more sophisticated hardware trace-level data capture. This highlights which particular software functions, or indeed function call hierarchies, are responsible for a significant proportion of the consumed processor cycles. In most instances this will simply serve to confirm the expectations that the designer had earlier in the design phase. If possible, the engineer will take the path of least resistance and optimize the code further, but there are always cases when the performance requirement is too high, even if code optimizations are taken right down to the assembler coding level.

The first step in using Cascade is to read in the executable binary. From this, Cascade is able to identify the individual software functions and present this information to the user. Guided by their profiling results, the user can select individual functions or the roots of function call hierarchies that are to be offloaded. Cascade is then able to produce an instrumented C model of the instruction streams within those functions. These C models are integrated into an instruction set simulator (ISS), so that when the offloaded functions are executed, detailed instruction-level profiling and memory access trace information is automatically captured. Using this approach, no manual partitioning of the original application is required, allowing non-offloaded code to be used to stimulate the functionality to be offloaded and thus leveraging the existing unmodified software test suite. It is highly advantageous to stimulate the offloaded software functions with test data that is representative of the usage in the final application. Data captured during this step is used to drive the downstream adaptation of the coprocessor architecture for the application. However, additional test data sets may be used later for providing enhanced functional test coverage.

The trace files generated by this step may then be read into Cascade for further analysis. The memory trace information is used to explore the possible internal cache memory architectures for the coprocessor. Good data cache memory architecture design space exploration is a keystone for efficient ASIP generation. We must provide sufficient bandwidth to exploit application parallelism while maintaining high cache hit rates that minimize bus traffic. As such, this is a system-level optimization where the interaction between the main processor and coprocessor is modeled. By driving the architecture exploration and performance estimation from actual memory traces, Cascade is able to construct application-specific cache architectures.

Once a cache architecture has been selected, a series of candidate architectures may be automatically generated for the coprocessor. Cascade employs sophisticated synthesis and code-mapping algorithms (covered in more depth in later sections) to generate these architectures. Every candidate coprocessor architecture is capable of executing the

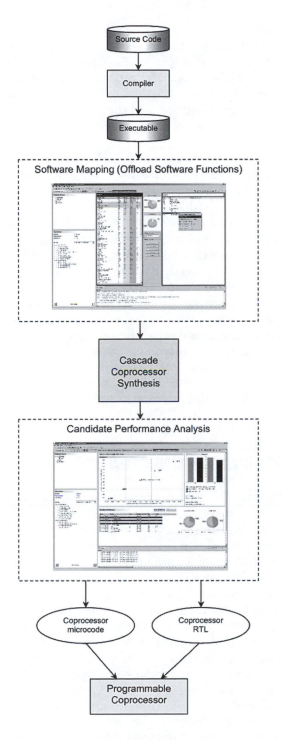

■ **FIGURE 9.1**

User level cascade flow.

required software functions and represents different tradeoffs in terms of area and throughput. Cascade is typically able to generate hundreds of candidates in just a few hours, providing an efficient tool for rapid design space exploration. The process is designed to be interactive and indeed iterative. Detailed performance and area breakdown may be obtained for any candidate, and the user may choose to modify the original application code or the set of offloaded functions in response to the findings at this stage.

Based on the user's area and performance budget, a particular candidate architecture is selected for continuation through the final stages of the flow. The user may generate either a Verilog or VHDL top level for the coprocessor that instantiates all of the required library blocks and interfaces to the main processor. This generated coprocessor IP is designed to operate as a slave device on the system bus. Cascade is also able to utilize the stimulus and responses captured at the software level to generate an RTL testbench for the coprocessor. This validates that the coprocessor is able to execute in a functionally identical manner to the original software implementation, without the requirement for any manual testbench or verification environment development.

It is important to note that the coprocessor IP is fully programmable and is controlled by a microcode stream also generated by Cascade. This is held in main memory and loaded by the coprocessor upon demand. Modifications to the coprocessor operation may be accommodated via changes to the original offloaded functions at the main processor tool chain level.

9.3   INTEGRATION WITH EMBEDDED SOFTWARE

At design time Cascade minimizes disruption to the standard software flow by using executable code as its input. This philosophy must be extended to the runtime environment in order to maintain a simplified system-level integration.

Despite radical instruction set differences, the coprocessor maintains compatibility to the main processor at the application binary interface (ABI) level. Thus parameters may be passed into and out of the offloaded functions in an identical manner to the original code, allowing the transition from main processor to coprocessor execution to be achieved in a rapid and seamless manner by transferring register state. Moreover, the coprocessor operates within the same address space of the main processor, allowing pointers to be exchanged between the systems with only minimal restrictions.

Cascade automatically modifies the offloaded functions to redirect execution to a coprocessor control driver. Once the driver is entered, the register state is transferred to the coprocessor. Depending upon the application, larger blocks of data will need to be transferred between the

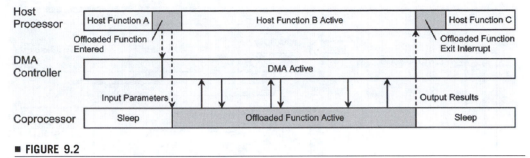

Main processor and coprocessor interaction.

main processor and coprocessor. A combination of user input and memory trace analysis is used to discover the relevant data blocks. The transfer may be achieved by block copies performed by the main processor or by the use of DMA transfers. If DMA is being used, and the user has asserted that coprocessor parallel operation is legal, then the called software function may return immediately, allowing the main processor to continue with other activities while the coprocessor is active. When the coprocessor activity is complete, an interrupt or periodic polling informs the main processor that any results may be read. Fig. 9.2 illustrates this, showing the execution timelines for the host processor, DMA, and coprocessor for a particular function offload.

Clearly cache coherency is a key issue here, especially since many embedded SoC environments do not support the snooping cache coherency protocols typical of higher end systems. The coprocessor itself flushes its cache contents using an optimized flushing mechanism after each offload, since it only needs to hold highly transient data. Transfers performed by the main processor itself are coherent with its caches, and transfers via DMA must also provide coherency.

The overhead of performing an offload is at least tens, usually hundreds, of clock cycles. Thus for effective amortization of the benefits, it is necessary for the offload to represent a relatively coarse grain block of functionality that would typically take thousands of main processor clock cycles. This suits the loosely coupled, autonomously executing, coprocessor execution model. Since the coprocessor is microcode programmable and targeted directly from application-level software, a suitable offload point in the software call hierarchy can be easily utilized.

9.4 COPROCESSOR ARCHITECTURE

Each coprocessor is automatically designed to fit the requirements of a particular application. However, there are some constituent components that are present in all coprocessor designs, as illustrated in Fig. 9.3.

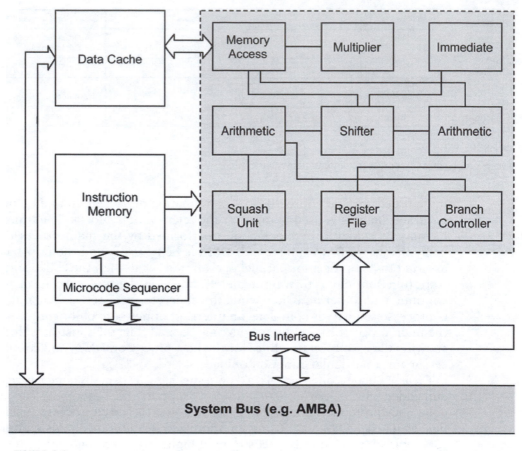

FIGURE 9.3

Coprocessor architecture.

- *Functional units*: A number of interconnected hardware units that are responsible for executing software operations. The number of these units and how they are interconnected is highly configurable.

- *Data cache*: One or more memory units used to form an application-specific cache architecture.

- *Instruction memory*: Storage for the microcode used to control the coprocessor. At its simplest, this is a memory for holding all of the microcode, initialized after reset. In more sophisticated coprocessors, this is a fully fledged instruction cache that loads microcode on demand in response to the dynamic execution pattern.

- *Microcode sequencer*: Control logic for sequencing the microcode to the functional units and dealing with branches and exceptions.

- *Bus interface*: Interface between the core coprocessor logic and the external environment using an industry-standard bus interface.

The functional units are extracted from a library that is supplied as part of Cascade. The individual units implement the basic operations of computation (such as arithmetic, shifts, multiplication) that are present within the instruction set of the main processor. There is not a direct correspondence between these components and the instruction set; rather, they are a set of operations that are sufficient, via various compositions, to cover the instruction set. As such, they represent a subRISC decomposition into the elemental operations of any instruction set and are fairly portable across a wide range of main processors. Additionally, library blocks are provided for access to the application-specific data cache and for performing branching and predication operations that are more closely associated with the coprocessor's own microarchitecture.

One powerful aspect of this approach is that the component library may be augmented with the customer's own hardware IP blocks. Provided that the blocks conform to some basic interfacing conventions, they may be targeted directly via explicit software-level invocations. This provides a powerful means to seamlessly intermix automatically generated microarchitectures with these user "custom units."

The functional unit array is parameterized in multiple dimensions. Individual coprocessor candidates vary both the multiplicity and the types of units, driven by automated analysis of the application code. Moreover, the connectivity networks between these units are also heavily customized. Such attention to connectivity as well as execution logic is an essential strategy in the deep submicron (DSM) era, where interconnect issues have a tendency to dominate.

In a typical RISC microarchitecture, the key component is the centralized register file [4]. This is usually a highly ported memory component providing operand read and result bandwidth for the instruction stream. As the level of instruction parallelism is scaled, the bandwidth pressure on the register file grows, requiring a many-ported implementation that is power-hungry and can quickly become a timing bottleneck [5]. The basic latency of feeding data between instructions via the register file can be of the order of two cycles. This is unacceptable since simple dependent instructions must be executed on successive cycles to obtain reasonable throughput. Thus implementations also have complex bypass forwarding networks so that results may be fed from one instruction to the next without such latency.

The coprocessor architecture also includes a centralized register file with which certain functional units are able to communicate, although communication may be made directly between units without transfers through this register file. Indeed, for heavily optimized computationally intensive code, this is typically the case. Distributed register storage is provided so that data may be held in the functional unit array without the need to communicate it to the main register file. All data transfers and the management of distributed register resources are in direct control of the microcode and are thus ultimately choreographed by Cascade. This significantly increases the complexity of the code-generation process but yields significant benefits in reducing hardware complexity and area, and provides a more scalable solution. It represents a radical departure from the typical configurable RISC approach, where new functional units (i.e., instruction extensions at the software level) may be added but their performance is constrained by the operand and result communication fabric.

Apart from some basic connectivity motivated by the requirements of programmability, most of the data paths in a given coprocessor are created as a direct response to the application data flow requirements. In effect, this provides a highly customized bypass network between units. Although the microcode controls the functional units in a traditional manner by issuing operation codes, the microarchitecture has many similarities to a transport triggered architecture (TTA) [6]. In such architectures, the bypass network is compiler controlled and a great emphasis is placed on connectivity resources.

The coprocessor is, in essence, a horizontally microcoded custom machine. Individual bit fields from the instruction memory are used to directly control individual functional units and their input and output connectivity. To help mitigate against the code expansion typical of such a representation, support is provided to allow certain functional units to share control bits. Moreover, the control fields may be presented on different execution cycles from their usage in order to improve scheduling freedom.

The high level of configurability in the approach allows a very parallel coprocessor architecture to be generated, circumventing many of the typical restrictions on scalability. This parallelism is directly architected to the requirements of a particular application, thus avoiding the overhead of hardware resources irrelevant to the application. Of course, a coprocessor's actual parallelism can only ever be as good as our ability to reliably extract it from real software applications. It is on this topic that we shall focus next.

9.5   ILP EXTRACTION CHALLENGES

Extraction of parallelism from C code is known to be a difficult problem. It remains a constant challenge to compiler optimizer developers and to the implementers of dynamic issue processors that seek to extract parallelism at run time.

Fortunately, several factors make the task easier in the context of acceleration in embedded systems:

- The type of computationally intensive code that we wish to accelerate typically has abundant parallelism. It is often expressed in kernel loops representing strong hot spots in the application.

- Since embedded systems are designed to implement a relatively fixed set of computationally intensive tasks, there is a high level of predictability in their operation—much more so than in general-purpose computing. Thus by gathering extensive profiling information, we are able to gain a greater insight into the behavior of the application.

- We are able to perform a much more extensive analysis of the profile and subsequent generation of potential solutions. We are not operating in the heavily interactive mode of a compiler, and therefore longer tool analysis periods are more acceptable.

ILP extraction at an executable code level is an even more difficult problem than similar extraction at the C code level. However, the design flow advantages of working at this level outweigh these additional complexities, which may be largely hidden from end users. The information required for ILP extraction is generally still available at the executable code level—it is just more obfuscated.

The first problem we face is that of code discovery. Many binary formats rather freely intermix read-only data and instruction code; therefore, we must perform a control flow analysis to discover which blocks are reachable from our known function entry points. This is achieved through standard control flow analysis techniques, augmented with some additional heuristics for code discovery through indirect branches. Once this is complete, we can perform a data flow analysis to track the liveness of registers throughout the program.

A further area of particular importance and analytic difficulty is alias analysis. In order to perform efficient code motions and optimizations later in the design flow, we must try to determine whether a given store operation is potentially aliased with other store or load operations. We do this through a combination of detailed data flow analysis and empirical information gathered from profiling the offloaded functions. This allows us to recreate most of the information that was lost during the compilation process. Moreover, we can use our empirical information (in conjunction with user assertions) to improve the quality of our alias analysis beyond that which was available even to a C compiler. The somewhat cavalier approach that C/C++ has to pointer semantics makes the process of reliable alias analysis notoriously difficult even with the luxury of full source code visibility.

The compiler will have made instruction selection choices to map from the original source code operations to those supported in the main processor instruction set. In modern RISC architectures, with a small core instruction set and good orthogonality between operation and register assignment, the instruction selection is relatively trivial. Thus, little relevant information is lost at the executable level.

Despite operating from an executable input rather than source code, the early stages in the Cascade flow lean heavily on techniques developed in the field of VLIW compilation technology [7]. After control and data flow analysis, the input code is partitioned into a number of regions, sometimes referred to as hyperblocks. These are groups of one or more basic blocks with the property of having a single entry point but any number of exit points. By combining multiple basic blocks into a single unit for optimization, we are able to perform advanced global optimizations and scheduling steps, and maximize our scope for ILP extraction. Construction of regions is driven by the detailed profiling information so that they are constructed to represent typical traces through the instruction stream.

Standard region-enlargement techniques are applied to further expand average region sizes and thus improve the scope for ILP, but with careful management to avoid excessive code expansion. Tail duplication is used to restructure control flow via code duplication so that control flow edges that would normally break regions can be redirected to allow greater region growth. Complete loop unrolling is a powerful technique that can be used for loops with a smaller number of iterations and fixed bounds. We determine the loop bounds by automatic analysis of the code at the boundary level and unroll key loops.

Within individual regions, there may be conditional basic blocks and conditional instructions. We use predication techniques built into the coprocessor microarchitecture to allow us to freely mix operations with different conditionality. This allows us to make code motions within an individual region (subject to data dependencies) and thus make efficient use of the computational resources within the architecture. Multiway branching is also supported from a single region.

The predication mechanisms are further leveraged to permit partial loop unrolling. This allows some unrolling of loops, even when the loop bounds are variable or data dependent. This allows operations from multiple iterations to be executed in parallel, greatly increasing ILP and thus the efficiency of the coprocessor architecture.

9.6 INTERNAL TOOL FLOW

This section provides an overview of the key internal steps within the Cascade flow. These are implemented in a highly integrated tool framework, and the end user is heavily insulated from most of these internal steps. However, Cascade provides various analysis views that allow more advanced users (and indeed the tool developers!) very detailed information about the results of each of these steps. The flow may be viewed as a series of data transforms from the original instruction stream, targeted at the main processor, to a parallel instruction stream for a custom-generated coprocessor. At each step we are able to perform full verification that a given application dataset is executed correctly. The internal steps are illustrated in Fig. 9.4.

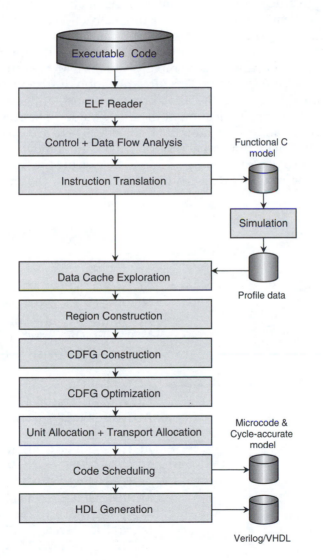

■ **FIGURE 9.4**

Cascade internal flow.

The first step in the process is the reading of the executable binary, typically in ELF, and then the control and data flow analysis steps previously discussed. The instruction stream is then translated into a lower level form that we term virtual machine operations (VMOPs). A VMOP is a basic operation that may be directly executed using our functional unit hardware library. VMOPs are very basic arithmetic and data move style operations that are formed from a decomposition of the original RISC instructions (much like that performed by a RISC processor hardware instruction decoder runtime). As such, these VMOPs are largely independent of the structure of the instruction set we supported at the input. Thus we are able to support different instructions by plugging in different instructions to VMOP translators into the Cascade framework.

Fig. 9.5 provides an example of a short ARM instruction sequence converted into VMOPs. The ARM assembly instructions are shown on the left side, performing a load from memory, an assignment of a constant, then a multiplication of these two values. The resulting sequential VMOP stream is shown on the right side. There are two primary classes of VMOP operation, an activation (ACT) and a move (MOV). An activation causes a particular operation to be executed, whereas a move transports the resulting data from one activation to the next. An activation

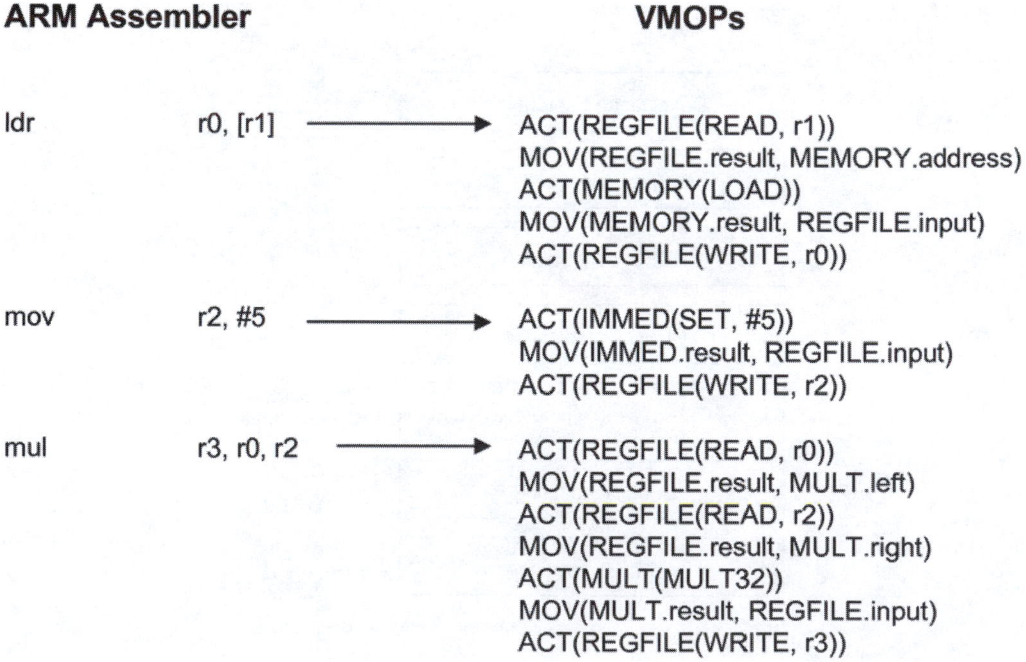

ARM Assembler   **VMOPs**

ldr   r0, [r1]   ⟶   ACT(REGFILE(READ, r1))
                               MOV(REGFILE.result, MEMORY.address)
                               ACT(MEMORY(LOAD))
                               MOV(MEMORY.result, REGFILE.input)
                               ACT(REGFILE(WRITE, r0))

mov   r2, #5   ⟶   ACT(IMMED(SET, #5))
                               MOV(IMMED.result, REGFILE.input)
                               ACT(REGFILE(WRITE, r2))

mul   r3, r0, r2 ⟶   ACT(REGFILE(READ, r0))
                               MOV(REGFILE.result, MULT.left)
                               ACT(REGFILE(READ, r2))
                               MOV(REGFILE.result, MULT.right)
                               ACT(MULT(MULT32))
                               MOV(MULT.result, REGFILE.input)
                               ACT(REGFILE(WRITE, r3))

■ **FIGURE 9.5**

ARM instruction to VMOP translation example.

specifies the type of functional unit required to perform the operation, the particular type of operation to be performed on that functional unit, and any other attributes that are required, such as a literal value. For instance, ACT(REGFILE(READ, r1)) indicates that a read of r1 should be performed on the register file. In the VMOP stream, accesses to the architectural register file, an aspect described in more detail later. For example, MOV(REGFILE.result, MEMORY.address) moves the result from the result of the register file to the address input of the memory unit for use by a subsequent activation. In this initial representation, it is an implicit assumption that there is only one instance of each unit type. During the subsequent synthesis and code mapping phases, multiple functional unit type instances may be created.

The VMOP stream provides a simple representation of the functionality of the original code sequence whose form is independent of the original instruction set. From the VMOP stream we are able to generate an equivalent C model. This can be plugged back into an ISS environment and executed instead of the original instruction stream. This C code includes heavy instrumentation, which is used to gather detailed information about control and data flows and memory access patterns for the offloaded application code.

Once this information has been gathered, the regions are identified. As discussed previously, these are selected to provide a framework for downstream steps that are able to operate in the constrained scope of a region but can still perform optimizations across multiple basic blocks. Since we have detailed profiling information, we are also able to determine how important a contribution each region makes to the total execution time of the application, given the representative test data used to drive the profiling step. This region-weighting information is used extensively downstream to judiciously budget tool runtime and architecture resource additions toward regions in proportion to their potential impact on overall application performance. Transforms applied to individual regions always maintain the same live-in and live-out data flow state, so that optimizations may be applied independently on regions with an assurance that the recomposed regions will remain consistent.

Once we have identified the regions, we are able to construct a control and data flow graph (CDFG) for each region. The CDFG is built from a traversal of the VMOP stream associated with each region, along with the control flow subgraph that the region covers. The CDFG embodies all the information we have extracted regarding the data and control flow dependencies between operations. Each CDFG node represents an individual activation VMOP, and move VMOPs are transformed into data edges. The CDFG represents the dependencies between operations and thus elucidates the parallelism that is potentially available.

Fig. 9.6 shows the example in the form of a CDFG, showing the direct correspondence between nodes with activations and edges with moves. Alias information extracted from analysis performed on the VMOP

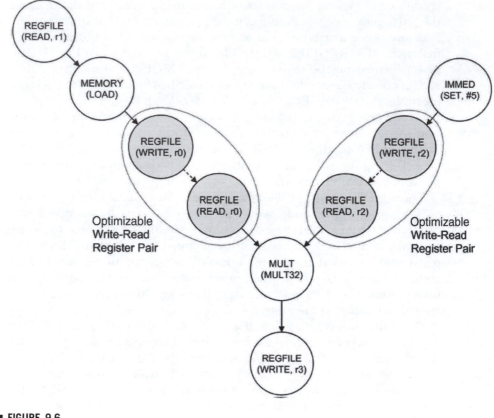

▪ **FIGURE 9.6**

Example CDFG constructed from VMOPs.

stream is embodied by alias edges. For instance, the edge between the write of r0 and the subsequent read of r0 shows that the activations alias the same register. Thus the read will always consume the data value written by the write, and therefore the read must always be performed after the write. These alias relationships must be maintained correctly even when there is more complex control flow, including conditionally executed activations.

The CDFG is then optimized using a term-rewriting engine to produce a functionally equivalent **CDFG** that contains fewer operations and a reduced number of critical path dependencies. This is illustrated in Fig. 9.6, where aliased pairs of register writes and reads may be optimized from the CDFG to produce direct data flow between activations. In the subsequently generated code schedule, this corresponds to performing a direct data transport between two operations rather than using the indirection of a register file access. These optimizations are extended to much more complex cases involving known idioms of groups of operations, aliased memory accesses, and multiple reads of a single register or memory write. Conditional activations may also be transformed into speculatively executed operations with multiplexer hardware units to steer data from the appropriate source depending upon a dynamic condition at runtime.

The next step is to bind individual CDFG operations and the data transports to particular hardware resources in the coprocessor architecture. This is a complex optimization algorithm that is partitioned into unit binding (unit allocation) and connection binding (transport allocation) steps. These are heavily interrelated problems, and the actual implementation performs these steps in an iterative fashion, with information being passed between phases. When a new coprocessor architecture is being created (as opposed to an existing coprocessor being reprogrammed), a capability is provided to create new unit and connectivity resources in the architecture in response to opportunities to exploit parallelism. These mechanisms are discussed in more detail in the following sections. A key point is that there is a strong association between the mechanisms used for synthesis and those used for code mapping: they are in fact different modes of the same fundamental algorithms. It is essential that any generated coprocessor architecture can be efficiently exploited by the code-mapping technology, and this is neatly achieved by the unification of the algorithms.

Once the binding steps have been completed, a final scheduling step can be performed on the CDFG that allows each set of operations in a region to be allocated to a particular time step. This schedule is then rendered in microcode form and is used to dynamically control the coprocessor architecture at run time.

The final hardware architecture is held in an XML representation readable by the Cascade tool. It may be rendered into either Verilog or VHDL as the first step in the hardware realization process. Alternatively, an equivalent C or SystemC model may be generated that can be integrated into a system-level modeling framework.

9.7 CODE MAPPING APPROACH

Cascade employs sophisticated unit and transport allocation algorithms utilizing a number of proprietary techniques to map arbitrary code sequences onto a custom architecture. A general high-level overview of these algorithms is provided here.

All allocation is performed in the context of regions, and earlier analysis allows us to accurately estimate how the schedule length of an individual region will influence the overall application performance. As the allocation steps proceed, trial schedules of partially allocated regions are repeatedly generated, allowing feedback on how allocation decisions are influencing the best case schedule length. The algorithms are very computationally intensive, but our knowledge about application hotspots at the region level allows us to concentrate effort on regions with the most potential benefit. Moreover, since the use of Cascade is out with the day-to-day software development flow, we are not as restricted in terms of tool runtimes as a typical optimizing compiler would be.

The unit allocation must map each of the operations in the CDFG onto a particular functional unit in the architecture that has the capability to perform the required operation. The asymmetric nature of the connectivity within the architecture means that this is not a simple task. Different instantiations of the same unit type will have different connectivity to other units in the architecture. A naïve allocation that seeks only to maximize parallelism by allocating operations that can proceed in parallel to different functional units will result in a very poor transport allocation, ultimately leading to an equally poor schedule. The unit allocator must take account of both opportunities for parallelism and the spatial allocation of operations to achieve a good connectivity fit. This requires a spatial analysis of each operation and the neighboring CDFG operations with which data is communicated (both operand feed and data output). The extent to which an allocation compromises the best possible connectivity is related to how much that might influence the potential for parallelism and thus the achievable schedule length for the region. Since the allocation of neighboring nodes in the CDFG could impact subsequent allocations, the order in which CDFG allocations are performed is also highly influential. To maximize the efficiency, the allocations are completed in a heuristically guided "wave front" order through the CDFG that is not directly related to a time ordering of the schedule.

Fig. 9.7 presents a highly simplified example of a CDFG subgraph and its mapping onto a part of the architecture, driven by connectivity constraints. The CDFG segment consists of two addition operations, a shift, and a multiplication. This must be mapped onto an architecture containing two arithmetic units, a multiplier, and a shifter. The bindings for the multiplier and shifter are trivial, as there is only a single instance of these functional unit types. The binding for the additions takes account of the connectivity provided by each of the functional unit instances. In this case, the first addition is bound to the arithmetic unit shown on the top left, as it has a direct connection to the multiplier, which is the consumer of the resulting value. The other addition is bound to the other arithmetic unit, as it has connectivity with the shifter. As discussed later, these connections between units are themselves derived from the dataflow during the coprocessor synthesis. Thus common recurring patterns of dataflow are typically driven to the same sets of bindings during the code-mapping process.

After unit allocation, a transport allocation must be completed for the region to allocate the connectivity resources that will be used for transporting data items in the schedule. As discussed earlier, the Cascade

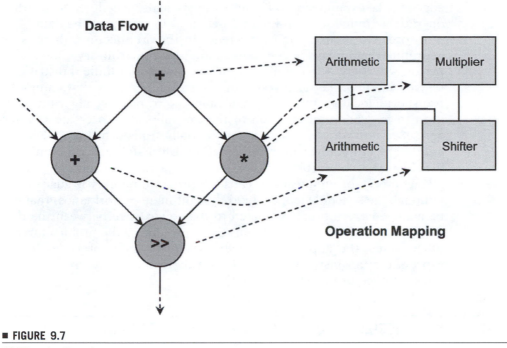

Data Flow

Operation Mapping

■ FIGURE 9.7

CDFG unit allocation.

approach treats connectivity as a first-class attribute of the architecture, and all data transport is under direct microcode control. At this stage, each data edge in the CDFG will be between two specific unit instances. The transport allocator must bind each such data edge to a particular connection in the architecture.

The unit allocation is also performed without direct reference to time steps. Although two units might communicate data and there might even be a suitable connection available, there is no guarantee that the consuming operation can be scheduled right away. It might have some other CDFG dependency that results in a delay of many clock cycles before the data can be consumed. Depending on various factors, including the number of cycles before consumption, the data may have to be transmitted back to the register file for transient storage. However, overreliance on this as the only architectural storage mechanism recasts the register file as a centralized bottleneck. Instead, distributed register resources may be employed to hold data for a number of cycles before forwarding to the appropriate consumer. Thus the transport allocation problem is a combined one of both storage and connectivity transport.

If a data transport needs to be made between units that are not directly connected, the architecture includes mechanisms for individual functional units to forward data items without operating upon them. The transport allocator includes path-finding algorithms to find a route through the architecture to the required consumer. The initial data edge may be expanded into a number of statically scheduled forwarding operations (much like a miniature network) to transport the data.

In order to restrict the connectivity in the architecture, there is no guarantee that any given communication between arbitrary units can be performed. Moreover, the lifetime of data items and limited distributed register storage may also result in a set of allocations that cannot be transport allocated. In these circumstances, proprietary algorithms are used to feed back information to the unit allocator to create a different set of unit allocations and to potentially transform the CDFG into a different, but equivalent, representation. Ultimately, after a number of iterative steps any arbitrary CDFG may be a mapped onto any generated coprocessor architecture. Thus full programmability is guaranteed, although the schedule lengths (and thus ultimate performance) that can be achieved are intimately related to the congruence between the data flows in the CDFG and the connectivity present in the architecture. In other words, the application-specific nature of the coprocessor is not expressed at an instruction level but in the connectivity present in the microarchitecture itself.

9.8   SYNTHESIZING COPROCESSOR ARCHITECTURES

The synthesis process is controlled by a subset of the regions in the application. These are the most performance-critical regions, whose cumulative execution time represents some predefined proportion of the total. By concentrating on these regions, a more rapid exploration of the design space is possible with only minimal impact on the quality of architectures that are generated.

As previously discussed the same core algorithms are used for both mapping a CDFG onto an architecture and for generating an architecture from a CDFG. The key difference is that in synthesis mode, additional options are available to create a new unit or add connectivity resources. All mapping decisions are made by calculating a cost metric formed from the increased area cost of the mapping minus the weighted performance benefit of that mapping. The weighting is determined by the global importance of the region being processed so that resources are more readily added for important regions. Typically the decision that minimizes the cost metric is selected. For code mapping, the area cost term is always zero, since no new resources may added, and thus the decisions are always performance driven. When performing synthesis, an improved performance may be attainable through the addition of a new resource in the architecture. The performance benefit of the resource is thus weighed against the additional area.

The impact is controlled by a parameter that describes the relationship between area and performance. Essentially, it determines how much area the synthesis algorithms are willing to add in order to obtain a given level of overall performance improvement. This parameter (along with various other control parameters) is swept across suitable ranges in order to explore the design space and provide a population of different candidate coprocessor architectures.

Different possible microcode layouts are explored as part of the synthesis process. These provide different tradeoffs in terms of instruction memory depth (and thus area) and the increased level of operation serialization mandated by reduced width.

9.9 A REAL-WORLD EXAMPLE

This section places the described design flow in the context of a real-world example. CriticalBlue was asked by a customer to benchmark our approach using an open source error correction protocol (BCH) [8]. It is capable of correcting a fixed number of data errors within each data packet and is typically deployed in wireless protocols. BCH is a computationally intensive algorithm that would benefit from implementation in a custom coprocessor to either free up cycles on the main ARM9 processor or to reduce total system power by lowering implementation frequency. An example system is shown in Fig. 9.8.

The application is coded in a fairly portable C style but uses typical software constructs of complex control flow structures, arbitrary pointer dereferencing, and variable strides through memory. It would be very difficult, for example, to target this code at a behavioral synthesis flow without significant adaptation.

The application itself is composed of a number of loop kernels, which together represent a significant proportion of the total run time. These

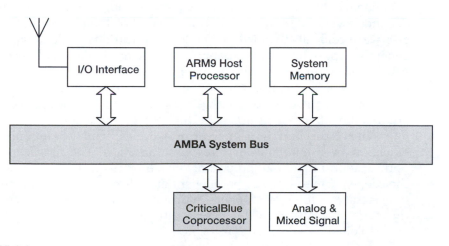

■ **FIGURE 9.8**

Simplified example system architecture.

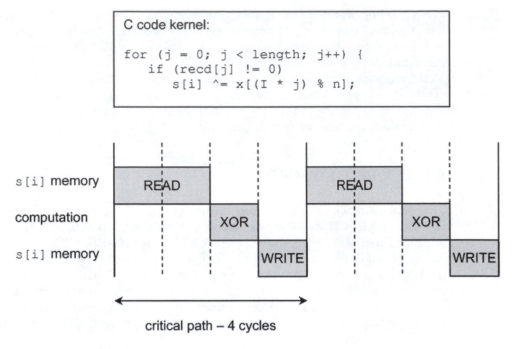

```
C code kernel:

for (j = 0; j < length; j++) {
    if (recd[j] != 0)
        s[i] ^= x[(I * j) % n];
```

■ **FIGURE 9.9**

Example loop kernel and critical path timing.

are complex loops with variable bounds, internal conditional statements, and pointer dereferences. One typical loop kernel is shown in Fig. 9.9, along with a timeline representing the minimum theoretical initiation interval of 4 cycles. This is driven by the presence of a potential data dependency through memory. A small code change could eliminate this dependency and permit an initiation interval of 1 cycle, but in this exercise we specifically avoided any code modifications to demonstrate the general-purpose nature of our approach. The original compiler-generated ARM9 code requires an average of 12 cycles per iteration because of the full set of operations that need to be performed for each iteration.

The goal of the Cascade synthesis process is to produce a customized architecture that is able to make full use of the available parallelism and achieve the initiation interval of 4 cycles, while maintaining area efficiency by minimizing new resource addition. The first step in the optimization process is the partial unrolling of the loop. This expands the scope for extraction of parallelism and allows operations from multiple iterations to be overlapped. Additional unit resources are added on the basis of analysis, in this case multiple arithmetic units, and these are exploited to allow overlap of non-data-dependent operations from different loop iterations in the generated schedule. Fig. 9.10 shows a representation of the schedule achieved with the optimized architecture,

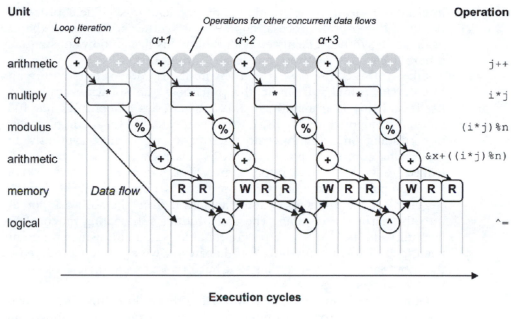

Execution cycles

■ **FIGURE 9.10**

Illustration of the achieved operation schedule.

showing part of the data flow for four unrolled iterations of the loop. Each of the unit types are shown, along with data flow between those units. Certain units (e.g., the memory) have interleaved operations scheduled upon them.

In effect, a custom pipeline has been generated for efficient execution of the loop. The same resources are reused (through the use of a different controlling microcode sequence) for the other loop kernels that are not identical but do have some similarity in their operation mix and required data flows.

An overall speedup of 4.78 times was achieved relative to the ARM9 executable that was used as the original input to Cascade. The coprocessor required a total of 85K gates of logic plus 96 Kb of memory for instruction and data memories. These results fully met the system requirements and were achieved in just a few hours after compiling the application onto the ARM9—a dramatic speedup compared to any manual RTL implementation of this relatively complex algorithm. The results were also achieved without the need to study and understand the source code itself.

9.10   SUMMARY

We have covered the key concepts of the Cascade design flow, provided an impression of it from a user perspective, and detailed how the various internal steps lead to the creation of an application-specific coprocessor. We term this overall process "coprocessor synthesis." We think the use of the term synthesis neatly embodies how this design flow substantially differs from other approaches for generating ASIPs.

Like synthesis, the flow is fundamentally top down in its nature and intent. The primary input to the process is the software application itself, and this is transformed, through a series of complex but fully automated algorithmic steps, into an equivalent software implementation running on a highly customized coprocessor. Like synthesis, the underlying coprocessor architecture is relatively free form, being controlled by synthesis rules rather than a fixed conformity to some predefined template architecture. Our approach extracts some of the dataflow "essence" of the application and transfers it from the software domain into the hardware domain. The end product still retains much of the flexibility of a fully software programmable system, and since we do not need to transform everything into hardware, we are unencumbered by the input language restrictions that have typically characterized behavioral synthesis approaches.

Our top-down approach of synthesizing the architecture directly from the application strongly differentiates it from traditional configurable processor design. This approach also seeks to bring the architecture into alignment with the application but achieves it via the indirection of instruction set design, an abstraction that is not wholly relevant either to the application or microarchitectural efficiency perspectives. Such an approach requires that the ASIP be built upon a fixed-base platform that is extended with customized instructions.

By enabling the architecture to be constructed automatically from the application, there is no need to expose the indirection of the instruction set to the user. The embedded software engineer can focus on the efficient coding of the application at the C level, and the hardware engineer carfocus on the implementation of RTL in the downstream flow. Moreover, by including custom units in an architecture, the RTL engineer can contribute directly to the efficiency of a coprocessor without having to become deeply embroiled within the control aspects of an algorithm that can be more freely expressed via software abstraction.

We believe that there is a growing class of users with design challenges to which our approach of directly architecting an application-specific coprocessor from application software provides a rapid and efficient solution. As the complexity of multicore heterogeneous SoCs grows, and embedded software increasingly drives system functionality, we believe that such a synthesis-oriented approach will be mandatory.

CHAPTER 10

DATAPATH SYNTHESIS

Philip Brisk and Majid Sarrafzadeh

As stated in Section 2.3.6 of Chapter 2, configurable processors require applications that are large enough to necessitate a complex base architecture. Portions of the application are identified as complex custom instructions using a technique such as the one described in Chapter 7. This chapter discusses the process of synthesizing a datapath that is capable of implementing the functionality of all of the custom instructions. The primary concern of this chapter is to avoid area bloat, the third major limitation of automated extension listed in Section 8.4 of Chapter 8. Techniques for resource sharing during datapath synthesis are described in detail. This work is an extension of a paper that was first published at DAC 2004 [1].

10.1 INTRODUCTION

Customizable processors will inevitably be limited by *Amdahl's Law*: the speedup attainable from any piece of custom hardware is limited by its utilization, i.e., the relative frequency with which the custom instruction is used during program execution. The silicon that is allocated to a datapath that implements the custom instructions increases the overall cost of the system in terms of both area and power consumption. Resource sharing during synthesis can increase the utilization of silicon resources allocated to the design. Moreover, resource sharing allows for a greater number of custom instructions to be synthesized within a fixed area.

Without resource sharing, custom instruction set extensions will never scale beyond a handful of limited operations. As the amount of silicon dedicated to custom instructions increases, the physical distance between custom resources and the base processor ALU will also increase. The interconnect structures to support such designs will increase both cost and design complexity. The next generation of compilers for extensible processor customization must employ aggressive resource sharing in order to produce area-efficient and competitive designs.

10.2 CUSTOM INSTRUCTION SELECTION

Compilers that fail to judiciously allocate silicon resources will produce designs that perform poorly and fail economically. A compiler targeting an extensible processor will generate a large number of candidate custom instructions. Most candidate generation techniques, including those in Chapter 7, are based on semiexhaustive enumeration of subgraphs of the compiler's intermediate representation of the program.

Custom instruction selection is a problem that must be solved in conjunction with candidate enumeration. Let $Candidates = \{I_1, I_2, \ldots, I_n\}$ be a set of candidate custom instructions. Each instruction $I_j \in Candidates$ has a performance gain estimate p_j and an estimated area cost a_j. These values can be computed by synthesizing a custom datapath for I_j and examining the result. Custom instruction selection arises when the area of the custom instructions in Candidates exceeds the area budget of the design, denoted $A_{\max}$. $A_{\max}$ limits the amount of on-chip functionality that can be synthesized. The problem is defined formally as follows:

Problem: *Custom Instruction Selection. Given a set of custom instructions Candidates and a fixed area constraint $A_{\max}$, select a subset Selected $\subseteq$ Candidates such that the overall performance increase of the instructions in Selected is maximized and the total area of synthesizing a datapath for Selected does not exceed $A_{\max}$.*

A typical approach to solving this problem is to formulate it as an *integer-linear program* (*ILP*) and use a commercially available tool to solve it. A generic ILP formulation is shown next. It should be noted that many variations of the instruction selection problem exist, and this ILP may not solve each variation appropriately or exactly; nonetheless, this approach can easily be extended to encompass any variation of the problem.

Define a set of binary variables $x_1, x_2, \ldots, x_n$, where $x_i = I$ if instruction I_i is selected for inclusion in the design; otherwise, $x_i = 0$. The ILP can now be formulated as follows:

$$Maximize: \qquad \sum_{i=1}^{n} x_i p_i \tag{10.1}$$

$$Subject\ To: \qquad \sum_{i=1}^{n} x_i a_i \leq A_{\max} \tag{10.2}$$

$$x_i \in \{0,\ 1\} \qquad 1 \leq i \leq n \tag{10.3}$$

To date, the majority of existing solutions to the aforementioned custom instruction problem have employed either greedy heuristics or used an ILP formulation. One exception is that of Clark et al. [2], who used a combination of identify operation insertion [3] and near-isomorphic graph identification to share resources. The algorithm was not described in detail, and an example was given only for paths, not general *directed acyclic graphs (DAGs)*.

The primary drawback of this problem formulation is that resource sharing is not considered. Area estimates are assumed to be additive. The cost of synthesizing a set of instructions is estimated as the sum of the costs of synthesizing each instruction individually. This formulation may discard many custom instructions that could be included in the design if resource sharing was used during synthesis.

As an example, consider Figs. 10.1–10.3. Fig. 10.1 shows a set of resources and their associated area costs. These resources will be used in examples throughout the chapter. Fig. 10.2 shows two DAGs, G_1 and G_2, representing candidate custom instructions I_1 and I_2. The arrows beginning with dark dots represent inputs, which are assumed to have negligible area; outputs are also assumed to have negligible area. The areas of each custom instruction are estimated to be the sum of the areas of the resources in each DAG. Then $a_1 = 17$ and $a_2 = 25$. Now, let $A_{max} = 30$. Then either G_1 or G_2, but not both, could be chosen for synthesis.

Fig. 10.3 shows an alternative solution to the problem instance in Fig. 10.2. Fig 10.3(a) shows a common supergraph G' of G_1 and G_2. Fig. 10.3(b) shows a datapath synthesized from G' that implements the functionality of both G_1 and G_2. The area of the datapath in Fig. 10.3, once again based on a sum of its component resources, is 28. The capacity of the datapath in Fig. 10.3(b) is within the allocated design area, whereas separate synthesis of both G_1 and G_2 is not. Unfortunately, the custom instruction selection problem as formulated earlier would not even consider Fig. 10.3 as a potential solution. Fig. 10.3 exemplifies the

■ FIGURE 10.1

Set of resources with area costs.

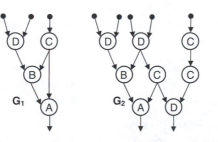

■ FIGURE 10.2

DAGs G_1 and G_2 representing custom instructions I_1 and I_2; $a_1 = 17$ and $a_2 = 25$.

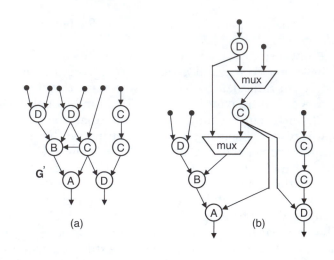

(a) G', an acyclic common supergraph of G_1 and G_2; (b) a datapath synthesized from G' that uses multiplexers to implement the functionality of G_1 and G_2.

folly of relying on additive area estimates during custom instruction selection. The remainder of this paper describes how to construct an acyclic supergraph of a set of DAGs and how to then synthesize the supergraph.

In Fig. 10.2, A, B, and D are binary operators and C is unary. In G_3, one instance of resource B has three inputs, and one instance of C has two. Multiplexers are inserted into the datapath in Fig. 10.3(b) to rectify this situation. Techniques for multiplexer insertion during the synthesis of the common supergraph are discussed as well.

10.3   THEORETICAL PRELIMINARIES

10.3.1   The Minimum Area-Cost Acyclic Common Supergraph Problem

Here, we formally introduce the *minimum area-cost acyclic common supergraph problem*. Bunke et al. [4] have proven a variant of this problem NP-complete.

Problem: *Minimum Area-Cost Acyclic Common Supergraph.* Given a set of DAGs $G = \{G_1, G_2, \ldots, G_k\}$, construct a DAG G' of minimal area cost such that G_i, $1 \leq i \leq k$, is isomorphic to some subgraph of G'.

Graphs $G_1 = (V_1, E_1)$ and $G_2 = (V_2, E_2)$ are *isomorphic* to one another if there is a one-to-one and onto mapping $f : V_1 \rightarrow V_2$ such that $(u_1, v_1) \in E_1 \Leftrightarrow (f(u_1), f(v_1)) \in E_2$. Common supergraph problems are similar in principle to the NP-complete subgraph isomorphism problem [5].

10.3.2 Subsequence and Substring Matching Techniques

Let $X = \langle x_1, x_2, \ldots, x_m \rangle$ and $Y = \langle y_1, y_2, \ldots, y_n \rangle$ be character strings. $Z = \langle z_1, z_2, \ldots, z_k \rangle$ is a *subsequence* of X if there is an increasing sequence $\langle i_1, i_2, \ldots, i_k \rangle$ of indices such that:

$$x_{i_j} = z_j \qquad j = 1, 2, \ldots, k \qquad (10.4)$$

For example, if $X = ABC$, the subsequences of X are A, B, C, AB, AC, BC, and ABC.

If Z is a subsequence of X and Y, then Z is a *common subsequence*. Determining the *longest common subsequence* (*LCSeq*) of a pair of sequences of lengths $m \leq n$ has a solution with time complexity $O(mn/logm)$ [6]. For example, If $X = ABCBDA$ and $Y = BDCABA$, the LCSeq of X and Y is $BCBA$. A *substring* is defined to be a *contiguous subsequence*. The solution to the problem of determining the *longest common substring* (*LCStr*) of two strings has an $O(m + n)$ time complexity [7]. X and Y in the preceding example have two LCStrs: AB and BD.

If a path is derived from a DAG, then each operation in the path represents a computation that must eventually be bound to a resource. The area of the necessary resource can therefore be associated with each operation in the path. To maximize area reduction by resource sharing along a set of paths, we use the *maximum-area common subsequence* (*MACSeq*) or *substring* (*MACStr*) of a set of sequences, as opposed to the LCSeq or LCStr. MACSeq and MACStr favor shorter sequences of high-area components (e.g., multipliers) rather than longer sequences of low-area components (e.g., logical operators), which could be found by LCSeq or LCStr. The algorithms for computing the LCSeq and LCStr are, in fact, degenerate cases of the MACSeq and MACStr, where all operations are assumed to have an area cost of 1.

Fig. 10.4 illustrates the preceding concepts.

P₁: ACDDAAACBCADDD
P₂: CDABCBCADDDAAA

LCStr: ACDDAAA**CBCADDD**
(AREA = 22)   CDAB**CBCADDD**AAA

MACStr: AC**DDAAA**CBCADDD
(AREA = 26)   CDABCBCAD**DDAAA**

LCSeq: A**CD**DAA**ACBCADDD**
(AREA = 34)   **CDABCBCADDD**AAA

MACSeq: **AC**DD**AAA**CBCADDD
(AREA = 43)   CD**A**BCBC**A**DDD**AAA**

■ **FIGURE 10.4**

Substring and subsequence matching example.

10.4   MINIMUM AREA-COST ACYCLIC COMMON SUPERGRAPH HEURISTIC

10.4.1   Path-Based Resource Sharing

The heuristic proposed for common supergraph construction first decomposes each DAG into paths, from *sources* (vertices of in-degree 0) to *sinks* (vertices of out-degree 0); this process is called *path-based resource sharing (PBRS)*. The sets of input-to-output paths for DAGs G_1 and G_2 in Fig. 10.1 are $\{DBA, CBA, CA\}$ and $\{DBD, DBD, DCD, DCD, CCD\}$, respectively.

Let $G = (V, E)$ be a DAG and P be the set of source-to-sink paths in G. In the worst case, $|P| = O(2^{|V|})$; it is possible to compute $|P|$ (but not necessarily P) in polynomial time using a topological sort of G. Empirically, we have observed that $|P| < 3|V|$ for the vast majority of DAGs in the Mediabench [8] application suite; however, $|P|$ can become exponential in $|V|$ if a DAG represents the unrolled body of a loop that has a large number of loop-carried dependencies. To accommodate larger DAGs, enumeration of all paths in P could be replaced by a pruning heuristic that limits $|P|$ to a reasonable size.

10.4.2   Example

An example is given to illustrate the behavior of a heuristic that solves the minimum area-cost acyclic common supergraph problem. The heuristic itself is presented in detailed pseudocode in the next section. Without loss of generality, a MACSeq implementation is used.

Fig. 10.5(a) shows three DAGs, G_1, G_2, and G_3, that are decomposed into sets of paths P_1, P_2, and P_3, respectively. The MACSeq of every pair of paths in P_i, P_j, $i \neq j$, is computed. *DBA* is identified as the area-maximizing MACSeq of P_1, P_2, and P_3. *DBA* occurs only in P_1 and P_2, so G_1 and G_2 will be merged together initially. G', the common supergraph of G_1 and G_2 produced by merging the vertices in the MACSeq, is shown in Fig. 10.5(b). These vertices are marked *SHARED*; all others are marked 1 or 2 according to the DAG of origin. This initial merging is called the *global phase*, because it determines which DAGs will be merged together each step. The *local phase*, which is described next, performs further resource sharing among the selected DAGs.

Let $G_1{}'$ and $G_2{}'$ be the respective subgraphs of G' induced by vertices marked 1 and 2, respectively. As the local phase begins, $G_1{}'$ and $G_2{}'$ are decomposed into sets of paths $P_1{}'$ and $P_2{}'$ whose MACSeq is C. The vertex labeled C in $G_2{}'$ may be merged with any of three vertices labeled C in $G_1{}'$. In this case, the pair of vertices of type C having a common successor, the vertex of type A, are merged together. Merging two vertices with a common successor or predecessor shares interconnect resources. No further resource sharing is possible during the local phase, since no

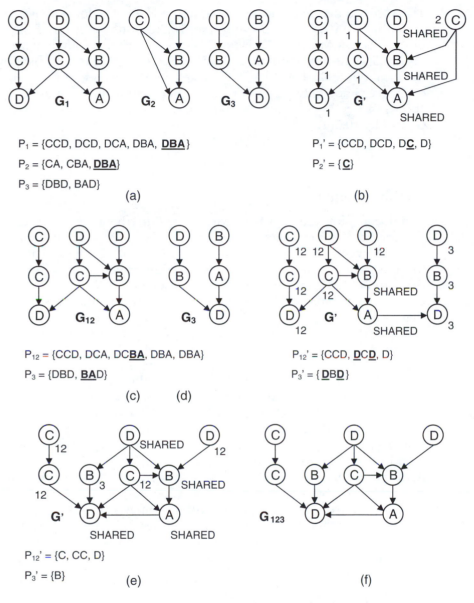

■ FIGURE 10.5

Example illustrating the minimum area-cost acyclic common supergraph heuristic.

vertices marked with integer value 2 remain. G_1 and G_2 are removed from the list of DAGs and are replaced with G', which has been renamed G_{12} in Fig. 10.5(c).

In Fig. 10.5(c), the set of DAGs now contains G_{12} and G_3. The global phase decomposes these DAGs into sets of paths P_{12} and P_3, respectively.

The MACSeq that maximizes are reduction is *BA*. The local phase begins with a new DAG, G', as shown in Fig. 10.5(d); one iteration of the local phase yields a MACSeq *DD*, and G' is shown after resource sharing in Fig. 10.5(e). At this point, no further resource sharing is possible during the local phase. In Fig. 10.5(f), $G' = G_{123}$ replaces G_{12} and G_3 in the list of DAGs. Since only G_{123} remains, the algorithm terminates. G_{123}, as shown in Fig. 10.5(f), is a common acyclic supergraph of G_1, G_2, and G_3 in Fig. 10.5(a).

The heuristic procedure used to generate G_{123} does not guarantee optimality.

10.4.3 Pseudocode

Pseudocode for the heuristic exemplified in the previous section is provided and summarized. A call graph, shown in Fig. 10.6, provides an overview. Pseudocode is given in Figs. 10.7–10.12. The starting point for the heuristic is a function called *Construct_Common_Supergraph* (Fig. 10.7). The input is a set of DAGs $G = \{G_1, G_2, \ldots, G_n\}$. When *Construct_Common_Supergraph* terminates, G will contain a single DAG that is a common supergraph of the set of DAGs originally in G.

The first step is to decompose each DAG $G_i \in G$ into a set of input-to-output paths, P_i, as described in Section 10.4.1. The set $P = \{P_1, P_2, \ldots, P_n\}$ stores the set of paths corresponding to each DAG. Lines 1–5 of *Construct_Common_Supergraph* (Fig. 10.7) build P.

Line 6 consists of a call to function *Max_MACSeq/MACStr* (Fig. 10.8), which computes the MACSeq/MACStr S_{xy} of each pair of paths $p_x \in P_i$, $p_y \in P_j$, $1 \le x \le |P_i|$, $1 \le y \le |P_j|$, $1 \le i \ne j \le n$. *Max_MACSeq/MACStr* returns a subsequence/substring $S_{\max}$, whose area, $Area(S_{\max})$ is maximal. The loop in lines 7–17 of Fig. 10.7 performs resource sharing. During each iteration, a subset of DAGs $G_{\max} \subseteq G$ is identified (line 8). A common supergraph G' of $G_{\max}$ is constructed (lines 9–12). The DAGs in $G_{\max}$ are removed from G and replaced with G' (lines 13–15).

The outer while loop in Fig. 10.7 iterates until either one DAG is left in G or no further resource sharing is possible. Resource sharing is divided into two phases, global (lines 8–9) and local (lines 10–12).

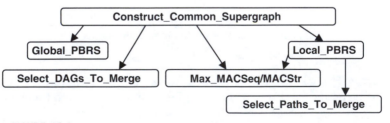

■ **FIGURE 10.6**

Program call graph for the minimum area-cost acyclic common supergraph heuristic.

```
          Algorithm:              Construct_Common_Supergraph( G )
          Input:                  G : Set of DAGs
          Local Variables:        i : Integer
                                  G_i ∈ G, G' : DAG
                                  P : Set of sets of paths
                                  P_i ∈ P, P_max, P' : Set of paths
                                  G_max : Set of DAGs
                                  S_max : Subsequence/Substring
```

1. $P \leftarrow \phi$
2. For i ← 1 to |G|
3.   Decompose G_i into set of paths P_i
4. $P \leftarrow P \cup P_i$
5. EndFor
6. $S_{max} \leftarrow$ Max_MACSeq/MACStr(P, ϕ)
7. While |G| > 1 and $S_{max} \neq \phi$

        /***** Global Phase *****/
8.   (G_{max}, P_{max}) ← Select_DAGs_To_Merge(G, P, S_{max})
9.   G'← Global_PBRS(G_{max}, P_{max}, S_{max})

        /***** Local Phase *****/
10.   While further resource sharing in G' is possible
11.   Local_PBRS(G_{max}, G')
12.   EndWhile
        /***** End Local Phase *****/

13.   Decompose G' into a set of paths P'
14. $G \leftarrow (G - G_{max}) \cup G'$
15. $P \leftarrow (P - P_{max}) \cup P'$
16. $S_{max} \leftarrow$ Max_MACSeq/MACStr(P, ϕ)
17. EndWhile
18. Return G

■ FIGURE 10.7

Minimum area-cost acyclic common supergraph construction heuristic.

The global phase consists of a call to *Select_DAGs_to_Merge* (Fig. 10.9) and *Global_PBRS* (Fig. 10.10).

The local phase consists of repeated calls to *Local_PBRS* (Fig. 10.11). The global phase determines which DAGs should be merged together at each step; it also merges the selected DAGs along a MACSeq/MACStr that is common to at least one path in each. The result of the global phase is G'—a common supergraph of G_{max} but not necessarily of minimal cost. The local phase is responsible for refining G' such that the area cost is reduced while maintaining the invariant that G' remains an acyclic common supergraph of G_{max}.

Line 8 of Fig. 10.7 calls *Select_DAGs_To_Merge* (Fig. 10.9), which returns a pair (G_{max}, P_{max}), where G_{max} is a set of DAGs as described earlier, and P_{max} is set of paths, one for each DAG in G_{max}, having S_{max} as a subsequence or substring. Line 9 calls *Global_PBRS* (Fig. 10.10), which constructs an initial common subgraph, G' of G_{max}. The loop in lines 10–12 repeatedly calls *Local_PBRS* (Fig. 10.10) to share additional

```
Subroutine:              Max_MACSeq/MACStr( P, G' )
Input:                   P : Set of sets of paths
                         G' : DAG
Output:                  Subsequence/Substring
Local Variables:         ΔA_max, i, j : Integer
                         S_xy, S_max : Subsequence/Substring
                         P_i, P_j ∈ P : Set of paths
                         p_x ∈ P_i , p_y ∈ P_j : Paths

/***** Identify the Maximal MACSeq/MAXStr *****/
1.       ΔA_max ← 0
2.       S_max ← φ
3.       For i = 1 to |P|
4.           For j = i+1 to |P|
5.               For x = 1 to |P_i|
6.                   For y = 1 to |P_j|
7.                       S_xy ← MACSeq/MACStr of p_x ∈ P_i and p_y ∈ P_j
8.                       If merging p_x with p_y via S_xy doesn't add
                            a cycle to G'
9.                           If Area(S_xy) > ΔA_max
10.                              S_max = S_xy
11.                              ΔA_max ← Area(S_xy)
12.                          EndIf
13.                      EndIf
14.                  EndFor
15.              EndFor
16.          EndFor
17.      EndFor
18.      Return S_max
```

▪ **FIGURE 10.8**

Subroutine used to identify paths having the maximal MACSeq or MACStr.

resources within G'. Lines 13–15 remove the DAGs in G_{max} from G and replaces them with G'; line 16 recomputes S_{max}.

Construct_Common_Supergraph terminates when G contains one DAG or no further resource sharing is possible. *Global_PBRS* and *Local_PBRS* are described in detail next. G_{max}, P_{max}, and S_{max} are the input parameters to *Global_PBRS* (Fig. 10.10). The output is a DAG, $G' = (V', E')$, an acyclic common supergraph (not necessarily of minimal area-cost) of G_{max}. G' is initialized (line 1) to be the DAG representation of S_{max} as a path; i.e., $S_{max} = (V_{max}, E_{max})$, $V_{max} = \{s_i | 1 \leq i \leq |S_{max}|\}$, and $E_{max} = \{(s_i, s_{i+1}) | 1 \leq i \leq |S_{max}| - 1\}$. Fig. 10.10 omits this detail.

DAG $G_i = (V_i, E_i) \in G_{max}$ is processed in the loop spanning lines 3–18 of Fig. 10.10. Let $S_i \subseteq V_i$ be the vertices corresponding to S_{max} in G'. Let $G_i' = (V_i', E_i')$, where $V_i' = V_i - S_i$, and $E_i' = \{(u, v) | u, v \in V_i'\}$. G_i' is copied and added to G'. Edges are also added to connect G_i' to S_{max} in G'. For each edge $e \in \{(u, v) | u \in S_i, v \in V_i'\}$, edge (m_i, v) is added to E', where m_i is the character in S_{max} matching u. Similarly, edge (u, n_i) is added to E' for every edge $e \in \{(u, v) | u \in V_i', v \in S_i\}$, where n_i is the character in S_{max} matching v, ensuring that G' is an acyclic supergraph of G_i.

All vertices in V_{max} are marked *SHARED*; all vertices in V_i' are marked with value i. *Global_PBRS* terminates once each DAG in G_{max} is

```
Subroutine:           Select_DAGs_To_Merge( G, P, Smax )
Input:                G : Set of DAGs
                      P : Set of sets of paths
                      Smax : Subsequence/Substring
Output:               (Set of DAGs, Set of Paths)
Local Variables:      i, x : Integer
                      found_match ← Boolean
                      Gmax : Set of DAGs
                      Pi ∈ P, Pmax : Set of Paths
                      Px ∈ Pi : Path
                      Gi ∈ G : DAG

1.     For i ← 1 to |G|
2.         j ← 1
3.         found_match ← FALSE
4.         Gmax ← ϕ
5.         Pmax ← ϕ
6.         x ← 0
7.         While x < |Pi| and found_match = FALSE
8.             If Smax is subsequence/substring of path px ∈ Pi
9.                 Gmax ← Gmax ∪ {Gi}
10.                Pmax ← Pmax ∪ {px}
11.                found_match ← TRUE
12.            EndIf
13.            x ← x + 1
14.        EndWhile
15.    EndFor
16.    Return (Gmax, Pmax)
```

■ **FIGURE 10.9**

Subroutine used during the global phase to select which DAGs to merge and along which paths to initially share vertices.

processed. To pursue additional resource sharing, *Local_PBRS* (Fig. 10.11) is applied following *Global_PBRS*.

Vertex merging, if unregulated, may introduce cycles to G'. Since G' must also be a DAG, this cannot be permitted. This constraint is enforced by line 8 of *Max_MACSeq/MACStr* (Fig. 10.8). This constraint is only necessary during *Local_PBRS*. The calls to *Max_MACSeq/MACStr* from *Construct_Common_Supergraph* (Fig. 10.7) pass a null parameter in place of G'; consequently, the constraint is not enforced in this context.

The purpose of labeling vertices should also be clarified. At each step during the local phase, the vertices marked *SHARED* are no longer in play for merging. Merging two vertices in G' that have different labels reduces the number of vertices while preserving G''s status as a common supergraph. A formal proof of this exists but has been omitted to save space. Likewise, if two vertices labeled i were merged together, then G_i' would no longer be a subgraph of G'.

Local_PBRS decomposes each DAG $G_i' \subset G'$ into a set of input-to-output paths P_i' (lines 3–8), where $P' = \{P_1', P_2', \ldots, P_k'\}$. All vertices in all paths in P_i' are assigned label i. Line 9 calls *Max_MACSeq/MACStr*

Subroutine: Global_PBRS(G_{max}, P_{max}, S_{max})
Input: G_{max}: Set of DAGs
                               P_{max}: Set of paths
                               S_{max}: Subsequence/Substring
Output: DAG (A Common Supergraph of G_{max})
Local Variables: i : Integer
                               G', $G_i \in G_{max}$: DAG
                               V', V_i, S_i : Set of Vertices
                               E', E_i : Set of Edges
                               $p_i \in P_{max}$: Path
                               u, v, m_i, n_i : Vertex
                               e : Edge

```
/* INVARIANT: Smax is a MACSeq/MACStr of each path pi ∈ Pmax */
 1.      G'=(V', E') ← {Smax| represented as a DAG}
 2.      Mark all vertices in V' as SHARED
 3.      For i ← 1 to |Pmax|
 4.          Let Vi be the set of vertices in DAG Gi ∈ Gmax
 5.          Let Ei be the set of edges in DAG Gi ∈ Gmax
 6.          Let Si be the set of vertices matching Smax as a subsequence/ substring in
             path pi ∈ Pmax
 7.          Assign integer label i to each vertex in Vi' = Vi-Si
 8.          V'← V' ∪ {Vi'}
 9.          For each edge e =(u, v) ∈ Ei
10.              If u ∈ Vi' and v ∈ Vi'
11.                  E'← E' ∪ {e}
12.              ElseIf u ∈ Si and v ∈ Vi'
13.                  E'← E' ∪ {(mi, v)| u corresponds to mi in Smax}
14.              ElseIf u ∈ Vi' and v ∈ Si
15.                  E'← E' ∪ {(u, ni)| v corresponds to ni in Smax}
16.              EndIf
17.          EndFor
18.  EndFor
19.  Return G'
```

▪ **FIGURE 10.10**

Applying PBRS to the DAGs along the paths selected by *Select_DAGs_to_Merge*.

(Fig. 10.8) to produce S_{max}', the MACSeq/MACStr of maximum weight that occurs in at least two paths, $p_x' \in P_i'$ and $p_y' \in P_j'$, $i \neq j$. Line 10 calls *Select_Paths_to_Merge* (Fig. 10.12), which identifies a set of paths—no more than one for each set of paths P_i'—that have S_{max}' as a subsequence or substring; this set is called P_{max}'. Line 11 of *Local_PBRS* (Fig. 10.10) merges the vertices corresponding to the same character in S_{max}', reducing the total number of operations comprising G'. This, in turn, reduces the cost of synthesizing G'.

Resource sharing along the paths in P_{max}' is achieved via vertex merging. Here, we describe how to merge two vertices, v_i and v_j, into a single vertex, v_{ij}; merging more than two vertices is handled similarly. Initially, v_i and v_j and all incident edges are removed from the DAG. For every

Subroutine:	Local_PBRS(G_{max}, G')
Input:	G_{max}: Set of DAGs
	G': DAG (Common supergraph of G_{max})
Output:	No Output /* Modifies G' */
Local Variables:	P' : Set of Sets Paths
	$P_i' \in P'$, P_{max}' : Set of Paths
	V_i' : Set of DAGs
	G_i' : DAG (Subgraph of G')
	S_{max}' : Subsequence/Substring
	p_{max}' : Path

```
1.     P' ← φ
2.     Pmax' ← φ
3.     For i = 1 to |Gmax|
4.         Let Vi' be the set of vertices of G' containing only vertices with label i (no
           shared vertices)
5.         Let Gi' be subgraph of G' induced by Vi'
6.         Decompose Gi' into a set of paths Pi':
7.             P' ← P' ∪ {Pi'}
8.     EndFor
9.     Smax' ← Max_MACSeq/MACStr( P', G' )
10.    Pmax' ← Select_Paths_To_Merge( P', Smax' )
11.    Merge all of the corresponding vertices along each path in Pmax' into a single
       path, pmax'
12.    Mark all vertices in pmax' as SHARED
13.    Return
```

■ **FIGURE 10.11**

The local phase, which comprises the core of the resource-sharing heuristic.

Subroutine:	Select_Paths_To_Merge(P', S_{max}')
Input:	P' : Set of sets of paths
	S_{max}' : Subsequence/Substring
Output:	Set of Paths
Local Variables:	P_{max}', $P_i' \in P_{max}'$: Set of Paths
	i, j : Integer
	found_match : Boolean
	$p_{ij}' \in P_i'$: Path

```
1.     Pmax' ← φ
2.     For i ← 1 to |P'|
3.         j ← 1
4.         found_match ← FALSE
5.         While j < |Pi'| and found_match = FALSE
6.             If Smax is subsequence/substring of path pij' ∈ Pi'
7.                 Pmax' ← Pmax' ∪ {pij'}
8.                 found_match ← TRUE
9.             EndIf
10.        EndWhile
11.    EndFor
12.    Return Pmax'
```

■ **FIGURE 10.12**

Assists *Local_PBRS* (Fig. 10.11) by determining which paths should be merged during each iteration.

edge (u, v_i) or (u, v_j), an edge (u, v_{ij}) is added to the DAG; likewise for every edge (v_i, v) or (v_j, v), an edge (v_{ij}, v) is added as well. In other words, the incoming and outgoing neighbors of v_{ij} are the unions of the respective incoming and outgoing neighbors of v_i and v_j.

The vertices that have been merged are marked *SHARED* in line 12 of *Local_PBRS*, thereby eliminating them from further consideration during the local phase. Eventually, either all vertices will be marked as *SHARED*, or the result of *Max_MACSeq/MACStr* will be the empty string. This completes the local phase of *Construct_Common_Supergraph* (Fig. 10.7). The global and local phases then repeat until no further resource sharing is possible.

10.5   MULTIPLEXER INSERTION

This section describes the insertion of multiplexers into the acyclic common supergraph representation of a set of DFGs—for example, the transformation from Fig. 10.3(a) into Fig. 10.3(b). In the worst case, multiplexers must be inserted on all inputs for all vertices in the acyclic common supergraph. Multiplexer insertion is trivial for unary and binary noncommutative operators. For binary commutative operators, multiplexer insertion with the goal of balancing the number of inputs connected to the left and right inputs is NP-complete, per operator.

10.5.1   Unary and Binary Noncommutative Operators

In Fig. 10.13(a), a unary operator U, assumed to be part of a supergraph, has three inputs. One multiplexer is inserted; its one output is connected to the input of U, and the three inputs to U are redirected to be inputs to the multiplexer. If the unary operator has only one input, then the multiplexer is obviously unnecessary. Now, let ● be a binary noncommutative operator, i.e., $a \bullet b \neq b \bullet a$. Each input to ● must be labeled as to whether it is the left or right operand, as illustrated in Fig. 10.13(b). The left and right inputs are treated as implicit unary operators.

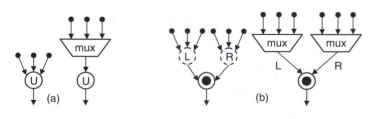

■ **FIGURE 10.13**

Multiplexer insertion for unary (a) and binary noncommutative (b) operators.

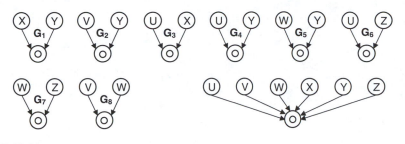

FIGURE 10.14

Eight DAGs and common acyclic supergraph sharing a binary commutative operator O.

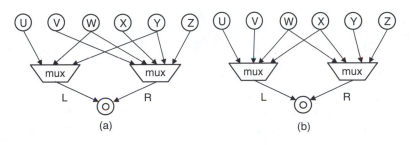

FIGURE 10.15

Two solutions for multiplexer insertion for the supergraph in Fig. 10.14.

10.5.2 Binary Commutative Operators

A binary commutative operator ∘ exhibits the property that $a \circ b = b \circ a$, for all inputs a and b. The problem for binary commutative operators is illustrated by an example shown in Fig. 10.14. Eight DFGs are shown, along with an acyclic common supergraph. All of the DFGs share a binary commutative operator ∘, that is represented by a vertex, also labeled ∘, in the common supergraph. Consider DFG G_1 in Fig. 10.14, where ∘ has two predecessors labeled X and Y. Since ∘ is commutative, X and Y must be connected to multiplexers on the left and right inputs respectively, or vice versa. If they are both connected to the left input, but not the right, then it would be impossible to compute X∘Y. The trivial solution would be to connect X and Y to both multiplexers; however, this should be avoided in order to minimize, area, delay, and interconnect.

The number of inputs connected to the left and right multiplexer should be approximately equal, if possible, since the larger of the two multiplexers will become the critical path through operator ∘. $S(k) = \lceil \log_2 k \rceil$ is the number of selection bits for a multiplexer with k inputs. The area and delay of a multiplexer are monotonically increasing functions of $S(k)$.

Fig. 10.15 shows two datapaths with multiplexers inserted corresponding to the common supergraph shown in Fig. 10.14. The datapath

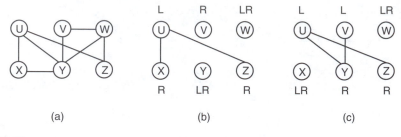

Graph G_o (a), corresponding to Fig. 10.14; induced bipartite subgraphs (b) and (c) corresponding to the solutions in Fig. 10.15(a) and (b), respectively.

in Fig. 10.15(a) has two multiplexers with 3 and 5 inputs, respectively; the datapath in Fig. 10.15(b) has two multiplexers, both with 4 inputs. $S(3) = S(4) = 2$, and $S(5) = 3$. Therefore, the area and critical delay through the datapath in Fig. 10.15(a) will both be greater than the datapath in Fig. 10.15(b). Fig. 10.15(b) is thus the preferable solution.

Balancing the assignment of inputs to the two multiplexers can be modeled as an instance of the *maximum-cost induced bipartite subgraph problem*, which is NP-complete [5]; a formal proof has been omitted. An undirected unweighted graph $G_o = (V_o, E_o)$ is constructed to represent the input constraints on binary commutative operator ∘. V_o is the set of predecessors of ∘ in the acyclic common supergraph.

An edge (u_o, v_o) is added to E_o if both u_o and v_o are predecessors of ∘ in at least one of the original DFGs. G_o for Fig. 10.14, is shown in Fig. 10.16(a). DFG G_1 from Fig. 10.14 contributes edge (X, Y) to G_o; G_2 contributes (V, Y); G_3 contributes (U, X); and so forth. Edge (u_o, v_o) indicates that u_o and v_o must be connected to the left and right multiplexers, or vice versa. Without loss of generality, if u_o is connected to both multiplexers, then v_o can trivially be assigned to either.

A solution to the multiplexer insertion problem for ∘ is a partition of V_o into three sets (V_L, V_R, V_{LR}): those connected to the left multiplexer (V_L), those connected to the right multiplexer (V_R), and those connected to both multiplexers (V_{LR}). For every pair of vertices $u_L, v_L \in V_L$, there can be no edge (u_L, v_L) in E_o since an edge in E_o indicates that u_L and v_L must be connected to separate multiplexers; the same holds for V_R. The subgraph of G_o induced by $V_L \cup V_R$ is therefore bipartite. Fig. 10.16(b) and (c) respectively show the bipartite subgraphs of G_o in Fig. 10.16(a) that correspond to the datapaths in Fig. 10.15(a) and (b).

The total number of inputs to the left/right multiplexers are $In_L = |V_L| + |V_{LR}|$ and $In_R = |V_R| + |V_{LR}|$. Since V_{LR} contributes to In_L and In_R, and all vertices not in V_{LR} must belong to either V_L or V_R, a good strategy is to maximize $V_L \cup V_R$. This is effectively the same as solving the maximum-cost induced bipartite subgraph problem on G_o.

To minimize the aggregate area of both multiplexers, $S(In_L) + S(In_R)$ should be minimized rather than $V_L \cup V_R$. To minimize the delay through ∘, $\max\{S(In_L), S(In_R)\}$ could be minimized instead. To compute a bipartite subgraph, we use a simple linear-time breadth-first search heuristic [9]. We have opted for speed rather than solution quality because this problem must be solved repeatedly for each binary commutative operator in the supergraph.

10.6 DATAPATH SYNTHESIS

Once the common supergraph has been constructed and multiplexers have been inserted, the next step is to synthesize the design. Two different approaches for synthesis are described here. One approach is to synthesize a pipelined datapath directly from the supergraph with multiplexers. The alternative is to adapt techniques from high-level synthesis, using the common supergraph to represent resource-binding decisions. The advantage of the pipelined approach is high throughput; the high-level synthesis approach, on the other hand, will consume less area.

10.6.1 Pipelined Datapath Synthesis

The process of synthesizing a pipelined datapath from a common supergraph with multiplexers is straightforward. One resource is allocated to the datapath for each vertex in the supergraph. Pipeline registers are placed on the output of each resource as well. For each resource with a multiplexer on its input, there are two possibilities: (1) chain the multiplexer together with its resource, possibly increasing cycle time; and (2) place a register on the output of the multiplexer, thereby increasing the number of cycles required to execute the instruction. Option (1) is generally preferable if the multiplexers are small; if the multiplexers are large, option (2) may be preferable. Additionally, multiplexers may need to be inserted for the output(s) of the datapath.

Pipelining a datapath may increase the latency of an individual custom instruction, since all instructions will take the same number of cycles to go through the pipeline; however, a new custom instruction can be issued to the pipeline every clock cycle. The performance benefit arises due to increased overall throughput. To illustrate, Fig. 10.17 shows two DFGs, G_1 and G_2, a supergraph with multiplexers, G', and a pipelined datapath, D. The latencies of the DFGs are 2 and 3 cycles, respectively (assuming a 1-cycle latency per operator); the latency of the pipelined datapath is 4 cycles.

10.6.2 High-Level Synthesis

High-level synthesis for a custom datapath for a set of DFGs and a common supergraph is quite similar to the synthesis process for a

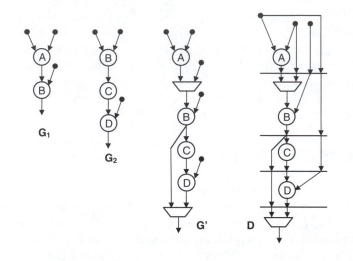

■ **FIGURE 10.17**

Illustration of pipelined datapath synthesis.

single DFG. First, a set of computational resources is allocated for the set of DFGs. The multiplexers that have been inserted into the datapath can be treated as part of the resource(s) to which they are respectively connected.

Second, each DFG is scheduled on the set of resources. The process of binding DFG operations to resources is derived from the common supergraph representation. Suppose that vertices v_1 and v_2 in DFGs G_1 and G_2 both map to common supergraph vertex v'. Then v_1 and v_2 will be bound to the same resource r' during the scheduling stage. This binding modification can be implemented independently of the heuristic used to perform scheduling.

The register allocation stage of synthesis should also be modified to enable resource sharing. The traditional left edge algorithm [10] should be applied to each scheduled DFG. Since the DFGs share a common set of resources, the only necessary modification is to ensure that no redundant registers are inserted as registers allocated for subsequent DFGs. Finally, separate control sequences must be generated for each DFG. The result is a multioperational datapath that implements the functionality of the custom instructions represented by the set of DFGs.

10.7 EXPERIMENTAL RESULTS

To generate a set of custom instructions, we integrated an algorithm developed by Kastner et al. [11] into the Machine SUIF compiler framework. We selected 11 files from the MediaBench application suite [8].

TABLE 10.1 ■ Summary of custom instructions generated

Exp	Benchmark	File/Function Compiled	Num Custom Instrs	Largest Instr (Ops)	Avg Num Ops/Instr
1	Mesa	blend.c	6	18	5.5
2	Pgp	Idea.c	14	8	3.2
3	Rasta	mul_mdmd_md.c	5	6	3
4	Rasta	Lqsolve.c	7	4	3
5	Epic	collapse_pyr	21	9	4.4
6	Jpeg	jpeg_fdct_ifast	5	17	7
7	Jpeg	jpeg_idct_4x4	8	12	5.9
8	Jpeg	jpeg_idct_2x2	7	5	3.1
9	Mpeg2	idctcol	9	30	7.2
10	Mpeg2	idctrow	4	37	20
11	Rasta	FR4TR	10	25	7.5

TABLE 10.2 ■ Component library

Component	Area (Slices)	Latency (cycles)	Delay (ns)	Frequency (MHz)
adder	17	1	10.669	93.729
subtractor	17	1	10.669	93.729
multiplier	621	4	10.705	93.414
shifter	8	1	6.713	148.965
and	8	1	4.309	232.072
or	8	1	4.309	232.072
xor	8	1	4.309	232.072
not	8	1	3.961	252.461
2-input mux	16	1	6.968	143.513
4-input mux	32	1	9.457	105.742

Table 10.1 summarizes the custom instructions generated for each application. The number of custom instructions generated per benchmark ranged from 4 to 21; the largest custom instruction generated per application ranged from 4 to 37 operations; and the average number of operations per custom instruction for each benchmark ranged from 3 to 20.

Next, we generated a VHDL component library for a *Xilinx VirtexE-1000* FPGA using *Xilinx Coregen*. We converted the VHDL files to *edif* netlists using *Synplicity Synplify Pro 7.0*. Each netlist was placed and routed using *Xilinx Design Manager*. Table 10.2 summarizes the component library.

We tested both MACSeq and MACStr implementations of the common supergraph construction heuristic in Section 10.4.3. In virtually

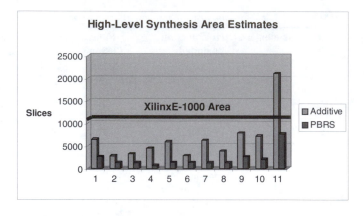

Results for high-level synthesis.

all cases, the MACSeq was, coincidentally, a MACStr. Results are only reported for the MACSeq implementation.

To synthesize the custom instructions, the two approaches described in Section 10.6 were used. The first approach synthesized the common supergraph (following multiplexer insertion) using a high-level synthesis and floorplanning system developed by Bazargan et al. [12] and Memik et al. [13] with modifications as discussed in Section 10.6.2. The second synthesizes a pipelined datapath as discussed in Section 10.6.1. The first approach will lead to an area-efficient design, whereas the second is designed for raw performance. These experiments demonstrate the importance of resource sharing during custom instruction set synthesis and datapath generation. Area results for the 11 benchmarks that we tested are presented in terms of slices.

Fig. 10.18 presents results using high-level synthesis. The columns labeled *Additive* show the area that would result from synthesizing each custom instruction individually; the columns labeled *PBRS* show the area of synthesizing the set of DFGs using the common supergraph to perform binding, as discussed in Section 10.6.2. The *Xilinx VirtexE-1000* FPGA has a total area of 12,288 slices; this value is shown as a horizontal bar.

PBRS reduced the area of the allocated computational resources for high-level synthesis by as much as 85% (benchmark 4) relative to additive area estimates. On average, the area reduction was 67%. In the case of benchmark 11, the additive estimates exceed the capacity of the FPGA; PBRS, on the other hand, fit well within the design. Clearly, sharing resources during custom instruction set synthesis can reduce the overall area of the design; conversely, resource sharing increases the number of custom instructions that can be synthesized within some

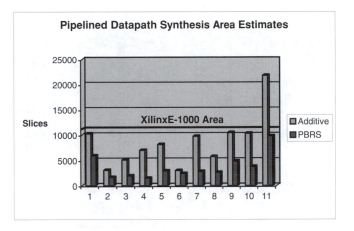

■ **FIGURE 10.19**

Results for pipelined datapath synthesis.

fixed area constraint. Resources are not shared between DAGs that are synthesized separately in the additive experiment. An overview of resource sharing during high-level synthesis can be found in Di Micheli's textbook [14].

Fig. 10.19 shows the results for pipelined datapath synthesis. For each benchmark, the area reported in Fig. 10.19 is larger than that reported in Fig. 10.18. Pipeline registers were inserted into the design during pipelined datapath synthesis; this is a larger number of registers than will be inserted during high-level synthesis. Second, the resource sharing attributable to high-level synthesis has not occurred. We also note that the results in Fig. 10.18 include floorplanning, while those in Fig. 10.19 do not. For pipelined datapath synthesis, PBRS reduced area by as much as 79% (benchmark 4) relative to additive area estimates; on average, the area reduction was 57%. Benchmark 6 yielded the smallest area reductions in both experiments. Of the five instructions generated, only two contained multiplication operations (in quantities of one and three, respectively). After resource sharing, three multipliers were required for pipelined synthesis and two for high-level synthesis. The area of these multipliers dominated the other elements in the datapath.

During pipelined synthesis, the introduction of multiplexers into the critical path may decrease the operating clock frequency of the system. This issue is especially prevalent when high clock rates are desired for aggressive pipelining. Some clock frequency degradation may be tolerable due to multiplexer insertion, but too much becomes infeasible. When multiplexers are introduced, we are faced with two options: store their output into a register, thus increasing the latency of every customized instruction by 1 cycle per multiplexer, or, alternatively, chain the multiplexer with its following operation, reducing clock frequency. To avoid

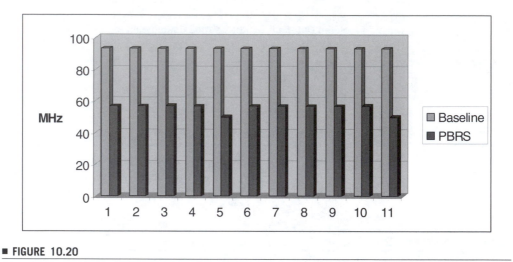

▪ **FIGURE 10.20**

Clock frequency degrading during pipelined synthesis due to multiplexer insertion.

the area increase that would result from inserting additional registers, we opted for a lower operating frequency. In only one case, a multiplexer required three selection bits; all other multiplexers required one or two selection bits, the vast majority of which required one.

Fig. 10.20 summarizes the clock degradation observed during pipelined datapath synthesis. The baseline frequency is the highest frequency supported by the multiplier, the limiting component in the library in Table 10.3. The PBRS frequency is the frequency at which the pipeline can run once multiplexers have been inserted into the datapath. The PBRS frequency ranged from 49.598 MHz (Exp. 5) to 56.699 MHz (Exp. 3 and 10). The variation observed in the final frequency column depends on the longest latency component that was chained with a multiplexer and the delay of the multiplexer. For example, if a multiplier (delay 10.705 ns) is chained with a 2-input multiplexer (delay 6.968 ns), the resulting design will run at 56.583 MHz.

The area and clock frequency estimates presented here were for the datapath only, and did not include a control unit, register file, memory, or other potential constraining factors in the design. One cannot assume that the frequencies in Table 10.2 represent the clock frequency of the final design. The baseline frequency is overly optimistic, because it is constrained solely by the library component with the longest critical path delay—the multiplier. Additionally, retiming during the latter stages of logic synthesis could significantly improve the frequencies reported in Fig. 10.20; this issue is beyond the scope of this research. It should also be noted that multipliers had a significantly larger area than all other components. None of the benchmarks we tested required division or modulus operations, which are likely to consume even larger amounts of chip area than multipliers. Consequently, designs that include these operations will see even greater reductions in area due to resource sharing than the results reported here.

10.8   CONCLUSION

As the experimental results in Section 10.7 have demonstrated, failure to consider resource sharing when synthesizing custom instruction set extensions can severely overestimate the total area cost. Without resource-sharing techniques such as the one described here, customization options for extensible processors will be limited both by area constraints and the overhead of communicating data from the processor core to the more remote regions of the custom hardware.

A heuristic procedure has been presented that performs resource sharing during the synthesis of a set of custom instruction set extensions. This heuristic does not directly solve the custom instruction selection problem that was discussed in Section 10.2. It can, however, be used as a subroutine by a larger procedure that solves the custom instruction selection problem by enumerating different subsets of custom instructions to synthesize. If custom instructions are added and removed from the subset one by one, then the common supergraph can be updated appropriately by applying the heuristic each time a new instruction is added. Whenever a custom instruction is removed, its mapping to the supergraph is removed as well. Any supergraph vertex that has no custom instruction DAG vertices that map to it can be removed from the supergraph, thereby reducing area costs. In this manner, the search space for the custom instruction selection problem can be explored. The best supergraph can be synthesized, and the corresponding subset of custom instructions can be included in the extensible processor design.

CHAPTER 11

INSTRUCTION MATCHING AND MODELING

Sri Parameswaran, Jörg Henkel, and Newton Cheung

Creating a custom processor that is application-specific is an onerous task upon a designer, who constantly has to ask whether the resulting design is optimal. To obtain such an optimal design is an NP-hard problem [1], made more time consuming given the numerous combinations of available parts that make up the processor. To illustrate this point, let us assume that there are 10 differing processor combinations. In addition to these combinations of processors, let us assume that 25 differing instructions can be added to the base processor, either individually or as a combination, to improve the performance of the application. Given such a scenario, the total design space would consist of more than 300 million possible designs, only a handful of which would be useful in a real system.

To overcome such a difficult design choice, differing design flows have been proposed. One such design flow is given in Fig. 11.1. In this design flow, an application program and associated data are profiled. The profiling is performed upon numerous configurations of the processor. Each configuration consists of the base processor with one or more coprocessors. For example, a configuration might be a base processor and a floating point processor, while another might be a base processor with a digital signal processing coprocessor, while yet another might have both the floating point and digital signal processing coprocessors along with the base processor. Once the profiling has been performed on all of these, a heuristic is used to select one of the configurations for further optimization.

Having selected a configuration, the profiled program is examined closely for sections of the code that are executed frequently. It is possible that one of these sections can be converted into a single instruction (replacing numerous standard instructions—for example, a cubic root instruction might well replace a set of base processor instructions that would take much longer to execute). However, it is more likely that a smaller subset of the section of code would provide the best tradeoff in terms of the necessary design objectives (such as speed, area, and power consumption).

Thus, a process is used to find the code segments within a section of code that can be converted to instructions. As will be seen

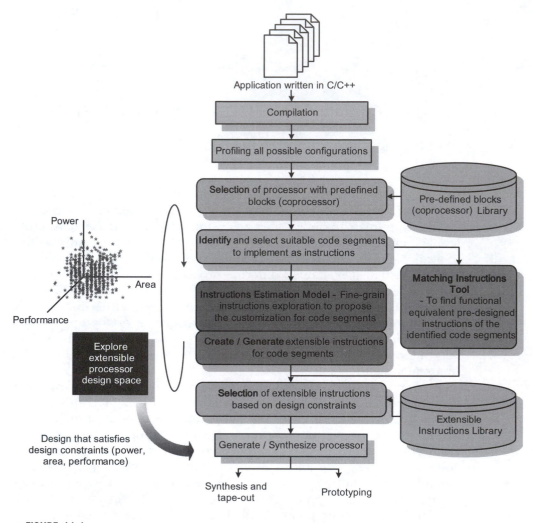

■ **FIGURE 11.1**

A design flow scenario for configurable processors.

later, many code segments can be identified as possible candidates for transformation to instructions. However, to convert these code segments to instructions, synthesize them, simulate them, and evaluate them would take an inordinate amount of time, which the modern designer can ill afford. Thus, in the design flow given in Fig. 11.1, the next step critically examines a code segment, and from that code segment, by utilizing a heuristic, extracts the code segments that would perform best for the given design criteria.

Having extracted the code segments and ranked them in descending order in which they meet the desired design criteria, the designer goes on to convert them to instructions. Since the procedure to convert a code segment in an efficient way to an instruction can be fairly time consuming, a library of instructions is kept. This library is consulted initially to see whether there exists a suitable predesigned instruction that would match the given code segment. If one exists, it can be utilized without having to create a new one; otherwise, a new one has to be synthesized. Even after the synthesis, to verify that the created instruction is actually the same as the code segment, there needs to be a matching phase.

It is likely that there will be multiple instruction manifestations for a single code segment. For example, if there are eight 16-bit numbers that needed to be added together, it is possible to either use a single instruction that adds eight registers of 16 bits in a sequential fashion or to have two 64-bit registers that are then added together in parallel. To predict which one of these combinations would be best, some form of modeling is required.

With the model in place, the identified code segments can be either be selected from a library of instructions or synthesized into instructions. Once the instructions are ready, the processor can be synthesized and evaluated. If it meets the necessary design criteria, then it can be accepted; otherwise, the parameters can be relaxed and another configuration chosen to reiterate over the design flow.

In this chapter, we take two aspects of the design flow and expand further. Section 11.1 describes the how a code segment and an instruction can be matched by utilizing BDDs. In Section 11.2, modeling of instructions is described, and finally in Section 11.3, we conclude.

11.1 MATCHING INSTRUCTIONS

11.1.1 Introduction to Binary Decision Diagrams

A reduced ordered binary decision diagram (ROBDD, but often simply referred to as "BDD") is a canonical data structure that uniquely represents a Boolean function with the maximal sharing of substructure [2]. The BDD data structure is based on the maximal sharing of substructure in a Boolean function; hence, BDD is not as prone to exponential resource blowup as other representations. To further minimize the memory requirements of BDDs, dynamic variable ordering was introduced by Rudell [3]. Dynamic variable ordering is often applied to change the order of the variable continuously (without changing the original function being represented), while the BDD application is running in order to minimize memory requirements. There are many derivatives of BDDs, such as the multivalued BDD (MDD), which has more than two branches and potentially has a better ordering, and the free BDD (FDD), which has a different variable ordering and is not canonical. Using BDDs to solve combinational equivalencechecking problems was proposed by

Madre et al. [4, 5]. They showed that BDDs can be used to verify the equivalence of combinational circuits and to determine where the two circuits are not equivalent.

The advantage of using BDDs for combinational equivalence checking is that if two functions have the same functionality (but the functions have different circuit representations), then their BDDs will still be identical. On the other hand, the disadvantage of BDDs is that during the verification of the BDDs, the memory requirement of complex modules such as multiplication is huge, causing the verification time to slow down significantly. In addition, the verification time for checking whether or not two functions are functionally equivalent consists of creation time of BDDs, dynamic variable ordering time, and checking equivalent time. Fig. 11.2 shows the verification time distribution for three extensible instructions with three code segments. The checking equivalent time often less than 50% of the total verification time.

For example, Fig. 11.3(a) shows the high-level language representation for a code segment ($S = (a + b) * 16$) and an extensible instruction ($S = (a + b) << 4$); both are functionally equivalent. Fig. 11.3(b) shows the BDD representation of the code segment and the extensible instruction. Since there are 32 bits in each variable (a, b, S), the BDDs of variable S (bit 11 to bit 4) are shown. One of these (bit 5 of variable S) is expanded out for clarity. Note that the c_i in the BDDs in Fig. 11.3(b) is the carry in of each bit. The BDD representation of extensible instruction is identical to the BDD representation of the code segment, which indicates that both the code segment and the extensible instruction are functionally equivalent.

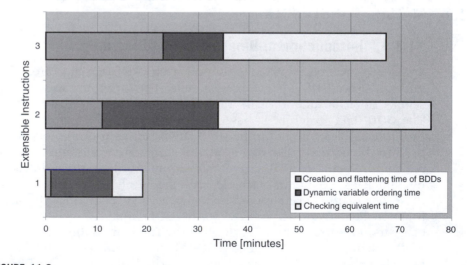

▪ **FIGURE 11.2**

Verification time distribution.

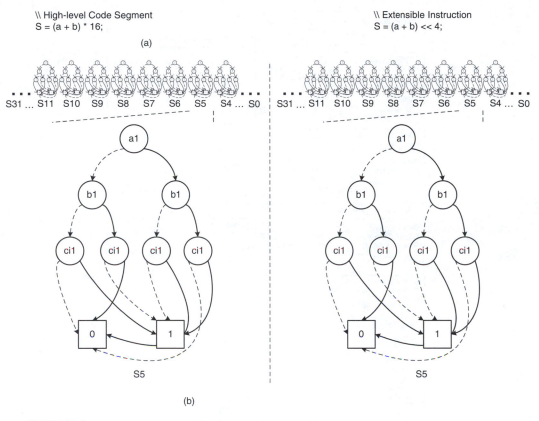

\\ High-level Code Segment
S = (a + b) * 16;

(a)

\\ Extensible Instruction
S = (a + b) << 4;

(b)

Code segment and extensible instruction and the BDD representations.

11.1.2 The Translator

Fig. 11.4 illustrates the translator flow with an example. The input of the translator is the application written in C/C++. The goal of the translator is to convert the application written in C/C++ to a set of code segments in Verilog HDL using a systematic approach. The translator consists of four steps: separate, compile, convert, and map. In addition, there is an assembler instruction hardware library written in Verilog HDL in the translator, which is a handmade library for the target processor. We refer to these instructions in hardware as "base hardware modules." These hardware modules are used for technology mapping in the translator.

The application written in C/C++ is first separated into a set of frequently used code segments written in C/C++. In other words, the complete application written in C/C++ is first profiled and then segmented, according to a ranking criteria described in [6].

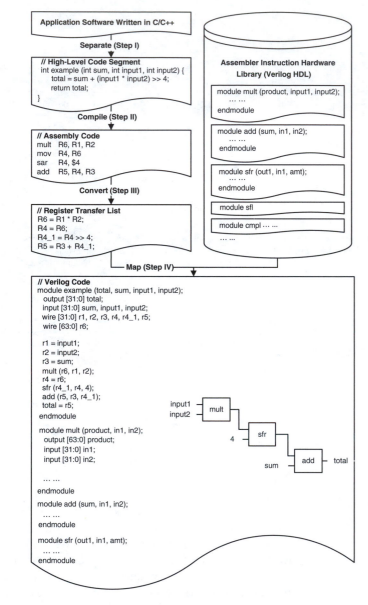

Application Software Written in C/C++

Separate (Step I)

// High-Level Code Segment
```
int example (int sum, int input1, int input2) {
    total = sum + (input1 * input2) >> 4;
    return total;
}
```

Compile (Step II)

// Assembly Code
```
mult   R6, R1, R2
mov    R4, R6
sar    R4, $4
add    R5, R4, R3
```

Convert (Step III)

// Register Transfer List
```
R6 = R1 * R2;
R4 = R6;
R4_1 = R4 >> 4;
R5 = R3 + R4_1;
```

Map (Step IV)

Assembler Instruction Hardware Library (Verilog HDL)

```
module mult (product, input1, input2);
    ... ...
endmodule
```

```
module add (sum, in1, in2);
    ... ...
endmodule
```

```
module sfr (out1, in1, amt);
    ... ...
endmodule
```

```
module sfl
```

```
module cmpl ... ...
    ... ...
```

// Verilog Code
```
module example (total, sum, input1, input2);
    output [31:0] total;
    input [31:0] sum, input1, input2;
    wire [31:0] r1, r2, r3, r4, r4_1, r5;
    wire [63:0] r6;

    r1 = input1;
    r2 = input2;
    r3 = sum;
    mult (r6, r1, r2);
    r4 = r6;
    sfr (r4_1, r4, 4);
    add (r5, r3, r4_1);
    total = r5;
endmodule

module mult (product, in1, in2);
    output [63:0] product;
    input [31:0] in1;
    input [31:0] in2;

    ... ...
endmodule

module add (sum, in1, in2);
    ... ...
endmodule

module sfr (out1, in1, amt);
    ... ...
endmodule
```

■ **FIGURE 11.4**

An example of a translation to Verilog HDL in a form that allows matching through the combinational equivalence checking model.

The C/C++ code segment is first translated into assembler code that achieves the following objectives:

- It uses all of the optimization methods available to the compiler to reduce the size of the compiled code.

- It converts the translated code into the same data types as the instructions in the library.

- It also unrolls loops with deterministic loop counts in order to convert the code segment to a combinational implementation.

An example of this step (code segment to assembler) is shown in step II of Fig. 11.4. The software code segment in the example contains addition, multiplication, and shift right operations (*mult* for multiplication, *move* for move register, *sar* for shift right, and *add* for addition). The reason the assembler code contains a *move* instruction is that the *mult* produces a 64-bit product, and hence the *move* instruction is used to reduce the size of the product to 32-bit data.

The assembler code is then transformed into a list of register transfer operations. The translator converts each assembler instruction into a series of register transfers. The main goal of this conversion step is to convert any nonregister transfer-type operations, such as *pop* and *push* instructions, into explicit register transfer operations. In this step, our tool renames the variables in order to remove duplicate name assignments automatically. Duplicate names are avoided, as Verilog HDL is a static single assignment form language [7]. In the example given in Fig. 11.4, this is shown as step III. In this example, the translator converts each assembler instruction into a single register transfer. The register transfer operations show the single assignment statement of each register, R4, R4_1, R5, and R6, where R4_1 is the variable renamed by our tool.

After the assembler code is converted to register transfer operations, the next step is the technology mapping (step IV of Fig. 11.4). In this step, the register transfer operations are mapped to the base hardware modules given in the predesigned assembler instruction hardware library. This library is manually designed to minimize the verification time of the functions. Once each register transfer has been mapped to a base hardware module, the translator creates a top-level Verilog HDL description that interconnects all the base hardware modules. The Verilog HDL, shown in Fig. 11.4, is based upon the code segment and the register transfer operations. There are three input variables (sum, input1, and input2), one output variable (total), seven temporary connection variables (r1, r2, and so forth) and three hardware modules (addition, multiplication, and shift right) in this example. The top-level Verilog HDL declares the corresponding number of variables and contains the mapped code of the register transfer operations. The technology mapping step provides a system-level approach to converting register transfer operations to a combinational hardware module. One of the drawbacks to this approach is that control flow operations such as branch and jump instructions might not directly map into a single base hardware module. Those instructions map to more complex hardware modules.

11.1.3 Filtering Algorithm

The second phase of the matching tool is the filtering algorithm. The input of the filtering algorithm is two Verilog HDL files: *extensible instruction* (written in Verilog HDL and given as the input of the matching tool) and *code segment* (written in C/C++ and translated to Verilog HDL). The goal of the filtering algorithm is to eliminate the unnecessary and complex Verilog HDL file into the combinational equivalence checking model. The idea of the filtering algorithm is to reduce the manipulation time from Verilog HDL to BDDs in the combinational equivalence checking model by inserting a low-complexity algorithm to filter unnecessary Verilog HDL files before the checking model.

Verilog HDL files can be pruned as nonmatch due to

- Differing numbers of ports (the code segment might have two inputs, while the extensible instruction only one)

- Differing port sizes

- Insufficient number of base hardware modules to represent a complex module (for example, if the code segment just contained an XOR gate and an AND gate, while the extensible instruction contained a multiplier (complex module), then a match would be impossible)

```
Algorithm Filtering (v₁, v₂) {
    if (Σ input(v₁) != Σ input(v₂)) return filtered;
    if (Σ output(v₁) != Σ output(v₂)) return filtered;
    if (Σ |input(v₁)| != Σ |input(v₂)|) return filtered;
    if (Σ |output(v₁)| != Σ |output(v₂)|) return filtered;
    for all modules v₂ do {
        if (modules(v₂) == complex_module)
            cm_list = implement(modules(v₂));
    }
    for all element i in cm_list do {
        if (cm_listᵢ ⊆ Σ modules(v₁)) return potentially_equal;
    }
    return filtered;
}
```

▪ **FIGURE 11.5**

Algorithm *filtering* for eliminating the number of extensible instructions into the combinational equivalence checking model.

The reason to check whether the code segment has an insufficient number of base hardware modules to represent the complex modules in the extensible instruction is that complex modules in the Verilog HDL file require extremely large BDDs (i.e., using 1 Gb RAM) to represent in the combinational equivalence checking model. In addition, the manipulation time (from Verilog to BDDs) for these Verilog HDL file is very large. By checking for modules, we can quickly understand that certain modules will never match.

11.1.4 Combinational Equivalence Checking Model

After filtering out unrelated instructions to the given code segment in the library, our tool checks whether the Verilog HDL converted from the software code segment is functionally equivalent to an instruction written in Verilog HDL. The checking is performed using Verification Interfacing with Synthesis (VIS) [8]. This part of the work could have been carried out with any similar verification tool.

We first convert both Verilog HDL files into an intermediate format (BLIF-MV, which VIS operates on) by a stand-alone compiler VL2MV [9]. The BLIF-MV hierarchy modules are then flattened to a gate-level description. Note that VIS uses both BDDs and their extensions, the MDDs, to represent Boolean and discrete functions. VIS is also able to apply dynamic variable ordering [3] to improve the possibility of convergence.

The two flattened combinational gate-level descriptions are declared to be combinationally equivalent if they produce the same outputs for all combinations of inputs and our tool declares the code segment and the extensible instruction to be functionally equivalent.

11.1.5 Results

The target ASIP compiler and profiler used in our experiments is the Xtensa processor's compiler and profiler from Tensilica, Inc. [10]. Our extensible instruction library is written in Verilog HDL as the assembler instruction library.

To evaluate the matching tool, we conducted two separate sets of experiments. In the first, we created arbitrary diverse instructions and matched them against artificially generated C code segments. These segments either: (1) matched exactly (i.e., they were structurally identical); (2) were only functionally equivalent; (3) the I/O ports match (i.e., code segment passes through the filter algorithm but is not functionally equivalent); or (4) did not match at all. This set of experiments was conducted to show the efficiency of functional matching as opposed to finding a match through the simulation-based approach. In the simulation-based approach, the C code segment is compiled and is simulated with input vectors to obtain output vectors. The output vectors are compared with the precomputed output result of the extensible instruction. The simulation was conducted with 100 million data sets each (approximately 5%–10% of the full data set with two 32-bit variables as inputs of the code segment). The reason for choosing 100 million as the size of the data set is the physical limitations of the hard drive. Each data set and each presimulated result of the instructions require approximately 1Gb of memory space. If more than n ($n = 1$ million, or 1% of the data set, for our experiments) differences occur in the simulation results, computation is terminated, and we state that a match is nonexistent.

The second set of experiments used real-life C/C++ applications (Mediabench) and automatically matched code segments to our predesigned library of extensible instructions. We examined the effectiveness of the filtering algorithm by comparing the complete matching time, including and excluding the filtering step. We selected the following applications, *adpcm encoder*, *g721 encoder*, *g721 decoder*, *gsm encoder*, *gsm decoder*, *mpeg2 decoder*, from Mediabench site [11] and complete *voice recognition* system [12].

Table 11.1 summarizes the results for matching instructions from the library to code segments in six different, real-life multimedia applications. We compare the number of instructions matched and time of matching extensible instructions with a reasonably experienced human ASIP designer and simulation-based approach. The ASIP designer

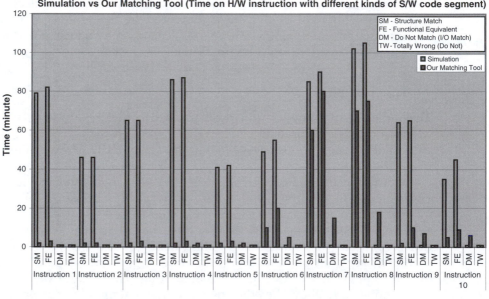

▪ **FIGURE 11.6**

Results in terms of computation time for the instruction matching step: *simulation* versus *our automated tool* (part 1).

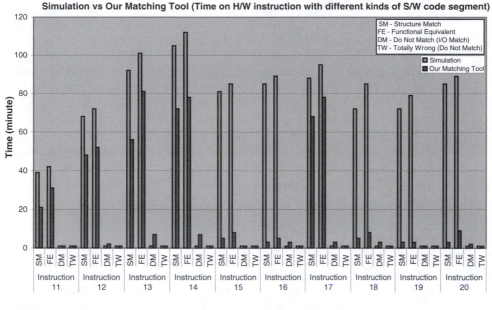

Results in terms of computation time for the instruction matching step: *simulation* versus *our automated tool* (part 2).

TABLE 11.1 ■ Number of instructions matched, matching time used, and speedup gained by different systems

Application Software	Speedup [×]	No of Instructions Matched	Simulation Time [hour]	Our Tool (no filt.) Time [hour]	Our Tool Time [hour]
Adpcm enc	2.2	3	80	25	10
g721 enc	2.5	4	75	20	8
g721 dec	2.3	4	74	20	9
gsm enc	1.1	4	105	40	25
gsm dec	1.1	4	95	35	15
mpeg2 enc	1.3	4	115	21	18
voice rec	6.8	9	205	40	25

selects the code segments manually and simulates code segments using 100 million data sets. The first column of Table 11.1 indicates the application. The second column shows the speedup achieved by the ASIP designer and our matching tool. The third and forth columns represent the number of instructions matched and the matching time used by the ASIP designer and simulation-based approach, respectively. The next two columns show the number of instructions matched and the time used by our tool (without the filtering algorithm). Finally, the last two columns display the same characteristics by the matching tool. Our automated tool is on average 7.3 times (up to 9.375 times) faster than manually matching extensible instructions. We show the effectiveness of the filtering algorithm, which reduces the equivalence checking time by more than half (compare column six and eight). In addition, we show the speedup of the embedded application that could be achieved through the automatic matching, which is 2.47 times on average (up to 6.8 times). Note also that the identical matches were made by both the human designer and our matching tool.

In summary, this section outlined a method to match instructions that is superior to existing simulation-based methods. The next section gives a method to model instructions such that their power, speed, and performance can be rapidly obtained.

11.2 MODELING

Instructions can be customized in numerous ways. One way would be to select and parameterize components like arithmetic operators. Additionally, designers can judiciously select from the available components and parameterize them for specific functionality. Parallelism techniques can be deployed to achieve a further speedup. For example, in one of the commercial vendors' tools, there are three known techniques: (1) very long instruction words (VLIW); (2) vectorization; and (3) hard-wired operations. Each of them have varying tradeoffs in performance, power and so forth [13–15]. Plus, these techniques can be used in conjunction with one another. Designers can also schedule the extensible instruction to run in multicycles. Thus the design space of extensible instructions is extremely complex.

Fig. 11.8 shows four instructions (sequences) that can be designed in order to replace a single code segment in the original software-based application. A code segment with four vectors a, b, c, and d, is summed up to generate a vector, z, in a loop with a loop iteration count of 1,000. If an extensible instruction is to replace the code segment, then a summation in series, or a summation in parallel, using 8-bit adders can be defined. These are shown in Fig. 11.8(b) and 11.8(d) respectively. Designers can also group four sets of 8-bit data together and perform a 32-bit summation in parallel, which is shown in Fig. 11.8(e). Note that this implementation loops only 250 times. In Fig. 11.8(c), an instruction using a 64-bit adder and a 32-bit adder can also be implemented, requiring just 125 loop iterations. On top of that, each of these designs, while functionally equivalent, will have differing characteristics in power, performance, and so forth. To verify the area overhead, latency, and power consumption of each instruction is a time-consuming task and is not a tractable method for exploring the instruction design space. *For a good design it is crucial to explore as many of the design points as possible. This requires fast and accurate estimation techniques.*

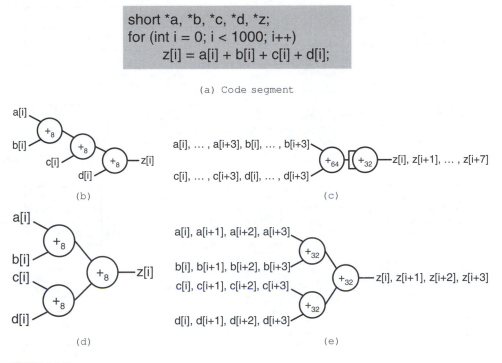

```
short *a, *b, *c, *d, *z;
for (int i = 0; i < 1000; i++)
    z[i] = a[i] + b[i] + c[i] + d[i];
```

(a) Code segment

(b)

(c)

(d)

(e)

■ **FIGURE 11.8**

Example: Four ways to design an instruction that replaces a single code segment.

In this section, we present estimation models of extensible instructions for area overhead, latency, and power consumption using *system decomposition* [16] and *regression analysis* [17]. As we will see, we achieve high accuracy, which enables us to control our previously presented techniques for semi-automatic instruction selection for extensible processors.

11.2.1 Overview

An overview of the method to derive estimation models is shown in Fig. 11.9. Extensible instructions with a set of customization parameters and synthesis and simulation results (including the results for each subsystem, such as the decoder and top logic) are inputs. The outputs are the estimation models of the extensible instructions. An extensible instruction represented by a large set of customization parameters is complex and therefore hard to analyze. Hence, we apply the system decomposition theory to decompose an instruction into its independent structural subsystems: decoder, clock gating hardware, top-logic, customized register, and combinational operations. Each such subsystem is represented by a subset of customization parameters. A customization parameter belongs to a subsystem if and only if a change in the customization parameter would affect

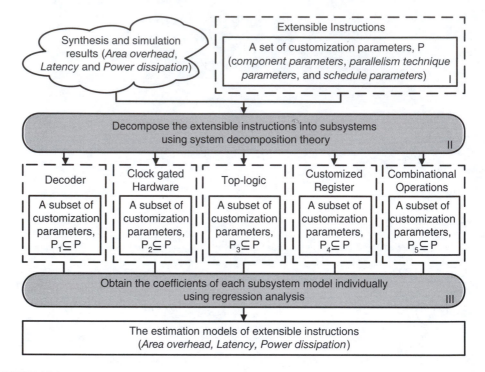

■ FIGURE 11.9

An overview for characterizing and estimating the models of the extensible instructions.

synthesis and simulation results of the subsystem. In addition, one and the same customization parameter can be contained in multiple subsystems. We then use regression analysis in each subsystem to determine: (1) the relationship between synthesis and simulation results and the subset of the customization parameters and (2) the coefficients of the customization parameters in the estimation models. In addition, the decomposition of subsystems is refined until the subsystem's estimation model is satisfactory. The estimation models for the subsystems are then combined to model extensible instructions for the purpose of estimating its characteristics. This procedure is applied separately for area overhead, latency, and power consumption.

11.2.2 Customization Parameters

Customization parameters are properties of the instruction that designers can customize when designing extensible instructions. They can be divided into three categories: (1) component parameters, (2) parallelism technique parameters, and (3) schedule parameters.

Component parameters characterize primitive operators of instructions. They can be classified based on their structural similarity, such as: (1) adder and subtractor $(+/-)$; (2) multiplier $(*)$; (3) conditional operators and multiplexers $(<, >, ? :)$; (4) bitwise and reduction logic $(\&, |, \ldots)$; (5) shifter $(<<, >>)$; (6) built-in adder and subtractor from the library (*LIB_add*) (we used these custom built components to show the versatility); (7) built-in multiplier (*LIB_mul*); (8) built-in selector (*LIB_csa*); (9) built-in *mac* (*LIB_mac*); (10) register file; and (11) instruction register. The bitwidths of all primitive operators can be altered, too.

Parallelism parameters characterize various levels of parallelism during instruction execution. As for parallelism techniques, there are three: *VLIW* allows a single instruction to execute multiple independent operators in parallel; *vectorization* increases throughput by operating multiple data elements at a time; and *hard-wired operation* takes a set of single instructions with constants and composes them into one new custom complex instruction. The parallelism technique parameters include (1) the width of the instruction using different parallelism techniques, which model the additional hardware and wider busses for paralleling the instructions; (2) the connectivity of the components (register file, instruction register, operations and so forth), which represents the components that are commonly shared; (3) the number of operations in series; (4) the number of operations in parallel; and (5) the number of operations in the instruction.

Schedule parameters represent the scheduling for instruction execution, such as multicycling. The schedule parameters are (1) the number of clock cycles it requires to execute an instruction; (2) the maximum number of instructions that may reside in the processor; and (3) the maximum number of registers that may be used by an instruction. Table 11.2 shows the notations and the descriptions for customization parameters.[1] Note that the 16-bit (or even smaller bitwidth) register file is not categorized due to the fact that the Xtensa processor is a 32-bit processor. It is often the case that a 16-bit (or smaller) register file is concatenated as 32-bit (or larger) register file. Hence, notations from Table 11.2 are used to refer to customization parameters.

11.2.3 Characterization for Various Constraints

Area Overhead Characterization

Unless the subsystems share common hardware, the area overhead of extensible instructions can be defined as the summation of the individual subsystem's area overhead.

The *decoder*, the *clock gating hardware*, and the *top-logic* are built-ins and are actually shared among extensible instructions, which has to be

---

[1] To handle scalability and to limit the design space, we only consider the following bitwidths: 8/16/32/64/128 for the operators with suffix i, and 32/64/128 for register files with suffix j in Table 11.2.

TABLE 11.2 ▪ Customization parameters of extensible instructions

	Customization parameters	Descriptions
Components	Num_{add/sub_i}	Number of i-bit addition/subtraction operators
	Num_{mul_i}	Number of i-bit multiplication operators
	Num_{cond}	Number of condition operator and multiplexors
	Num_{logic}	Number of bitwise and reduction logics
	Num_{shift_i}	Number of i-bit shifters
	$Num_{LIB\_add_i}$	Number of i-bit built-in adders
	$Num_{LIB\_csa_i}$	Number of i-bit built-in selectors
	$Num_{LIB\_mul_i}$	Number of i-bit built-in multipliers
	$Num_{LIB\_mac_i}$	Number of i-bit built-in macs
	Num_{regf_j}	Number of j-bit width register files
	Num_{ireg}	Number of instruction registers
Parallelism tech.	Wid_{vliw}	Width of the VLIW instructions
	Wid_{vector}	Width of the vectorization instructions
	Wid_{hwired}	Width of the hard-wired instructions
	Con_{regf_j}	Connectivity of j-bit register files
	Con_{ireg}	Connectivity of instruction registers
	Con_{oper_i}	Connectivity of operations
	Num_{oper_i}	Number of i-bit operations in total
	Num_{ser}	Number of operations in serial
	Num_{para}	Number of operations in parallel
Sche.	Num_{mcyc}	Number of cycles scheduled
	Num_{minst}	Number of instructions included
	Use_{reg_j}	Usage of the j-bit register files

taken in consideration. The customization parameters for these subsystems are (1) Con_{oper}; (2) Con_{regf}; (3) Con_{ireg}; (4) Num_{inst}; and (5) Num_{minst}.

Also, a customized register can be shared among the extensible instructions in the processor. The area overhead of the *customized register* is based on the size and the width of the registers. Therefore, the customization parameters are (1) Num_{regf}; and (2) Num_{ireg}.

The *combinational operations'* area overhead is not shared with other instructions and is dependent only upon the operations within the instruction. The customization parameters for combinational operations are (1) $Num_{add/sub}$; (2) Num_{mul}; (3) Num_{cond}; (4) Num_{logic}; (5) Num_{shift}; (6) Num_{LIB_add}; (7) Num_{LIB_mul}; (8) Num_{LIB_csa}; and (9) Num_{LIB_mac}.

Latency Characterization

The latency of extensible instructions can be defined as the maximum delay of each subsystem in the critical path when that specific extensible instruction is executed. A major part of the critical path is contributed to by combinational operations. Other subsystems either have very little effect on the latency or do not lie on the critical path.

The customization parameters for latency of the *decoder, clock gated, top-logic*, and *customized register* are similar to the area overhead characterization. The reason is that these subsystems mainly revolve around the connectivity between each other (i.e., fan-ins/fan-outs), while the internal latency is relatively constant within these subsystems.

In the *combinational operations*, the latency depends not only on structural components, but also on parallelism-technique parameters and schedule parameters. Component parameters represent latency of independent operators. Parallelism-technique parameters describe latency of internal connectivity, the number of stages in the instruction, and the level of parallelism; and schedule parameters represent the multicycles instructions.

Power Consumption Characterization

The characterization of power consumption is similar to the constraints described earlier.

The customization parameters of *decoder* and *top-logic* relate to the connectivity between the subsystems, and therefore are dependent upon (1) Con_{oper}; (2) Con_{regf}; (3) Con_{oper}; (4) Num_{minst}; and (5) Num_{mcyc}.

For *clock gating hardware*, the customization parameters include the connectivity and complexity of operations and the scheduling: (1) Num_{oper}; (2) Num_{regf}; (3) Num_{inst}; (4) Num_{mcyc}; (5) Num_{ser}; and (6) Num_{para}. The last two parameters specify the power consumption of the clock tree in the extensible instruction.

For *customized register*, the power consumption refers to the number of customized registers that the instruction used. The customization parameters are (1) Num_{regf}; (2) Num_{ireg}; and (3) Use_{reg}.

For *computational operations*, the power consumption characterization is further categorized into number of stages in the instruction and the level of parallelism in the stage. The reason for capturing power consumption when operations execute in parallel and when multicycle instructions are present is that stalling increases energy dissipation significantly.

11.2.4 Equations for Estimating Area, Latency, and Power Consumption

$A(opea)$ is the area overhead estimation of combinational operations and is defined as:

$$
\begin{aligned}
A(opea) = \Sigma_{i\in 8,16,32,64,128} \{ & A_{add/sub_i} Num_{add/sub_i} + \\
& A_{LIB\_mul_i} Num_{LIB\_mul_i} + A_{LIB\_mac_i} Num_{LIB\_mac_i} + \\
& A_{LIB\_add_i} Num_{LIB\_add_i} + A_{LIB\_csa_i} Num_{LIB\_csa_i} + \\
& A_{mul_i} Num_{mul_i} + A_{shift_i} Num_{shift_i} \} + \\
& A_{cond} Num_{cond} + A_{logic} Num_{logic}
\end{aligned} \tag{11.1}
$$

$T(opea)$ is the latency estimation of the combinational operations and is defined as:

$$
\begin{aligned}
T(opea) = \frac{Num_{ser}}{Num_{mcyc} \times Num_{oper}} \times \Sigma_{i \in 8,16,32,64,128} \\
\{T_{add/sub_i}Num_{add/sub_i} + T_{LIB\_add_i}Num_{LIB\_add_i} + \\
T_{LIB\_csa_i}Num_{LIB\_csa_i} + T_{LIB\_mul_i}Num_{LIB\_mul_i} + \\
T_{LIB\_mac_i}Num_{LIB\_mac_i} + T_{mul_i}Num_{mul_i} + \\
T_{shift_i}Num_{shift_i}\} + T_{cond}Num_{cond} + \\
T_{logic}Num_{logic} + T_{VLIW}Wid_{VLIW} + \\
T_{vector}Wid_{vector} + T_{hwired}Wid_{hwired} + T_{ser}Num_{ser} + \\
T_{para}Num_{para} + T_{mcyc}Num_{mcyc} + T_{minst}Num_{minst}
\end{aligned}
\tag{11.2}
$$

Finally, $P(opea)$ is the estimation of the power consumption of the instruction and is defined as:

$$
\begin{aligned}
P(opea) = \Sigma_{i \in 8,16,32,64,128}\{P_{add/sub_i}Num_{add/sub_i} + \\
P_{LIB\_mul_i}Num_{LIB\_mul_i} + P_{LIB\_mac_i}Num_{LIB\_mac_i} + \\
P_{LIB\_add_i}Num_{LIB\_add_i} + P_{LIB\_csa_i}Num_{LIB\_csa_i} + \\
P_{mul_i}Num_{mul_i} + P_{shift_i}Num_{shift_i}\} + \\
P_{cond}Num_{cond} + P_{logic}Num_{logic} + \\
P_{VLIW}Wid_{VLIW} + P_{vector}Wid_{vector} + \\
P_{hwired}Wid_{hwired} + P_{ser}Num_{ser} + \\
P_{para}Num_{para} + P_{mcyc}Num_{mcyc} + P_{minst}Num_{minst}
\end{aligned}
\tag{11.3}
$$

For detailed derivations of these equations, please see the Appendix to this Chapter.

11.2.5 Evaluation Results

In our first experiment, we examine the accuracy of the estimation models under differing customizations: VLIW, vectorization, hard-wired operation, multicycle, and sequences (multiple) of extensible instructions. Fig. 11.10 illustrates the experimental methodology used to evaluate the models. Table 11.3 shows the mean absolute error for the area overhead, latency, and power consumption of the estimation models in these categories. The mean absolute error (area overhead, latency, and power consumption) in hardwired operation is lower than those instructions using VLIW and vectorization techniques. The reason is that the estimation for VLIW and vectorization techniques instructions depends on a larger number of customization parameters, and, hence, higher error rates are observed. In terms of schedules, the mean absolute error is relatively close to the average mean error. The mean absolute error of the estimation models for all automatically generated instructions is only 2.5% for area overhead, 4.7% for latency, and 3.1% for power consumption.

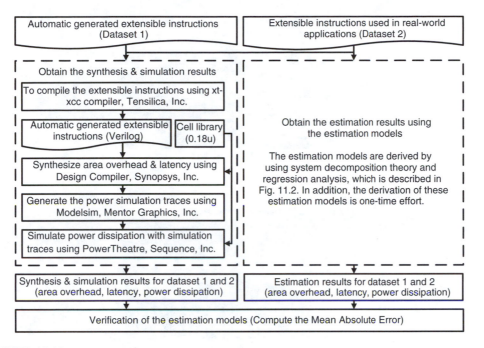

■ **FIGURE 11.10**

Experimental methodology.

TABLE 11.3 ■ The mean absolute error of the estimation models in different types of extensible instructions

Inst. Types	Area Overhead Error			Latency Error			Power Consumption Error		
	Mean Abs	Max	Min	Mean Abs	Max	Min	Mean Abs	Max	Min
VLIW	3.5	6.7	0.3	4.2	6.7	1.4	3.8	6.9	0.6
Vect.	2.8	7.5	1.4	3.5	7.0	0.3	2.5	6.8	1.7
Hard	2.6	4.4	0.2	2.5	5.4	0.3	2.0	5.1	0.8
MulCyc	4.2	7.6	0.0	4.1	8.4	0.2	4.5	7.9	0.3
MulInst	3.2	6.7	0.4	2.5	6.8	1.7	2.8	6.2	0.8
Overall	2.5	6.2	1.0	4.7	9.1	1.1	3.1	6.1	0.7

The mean absolute error for previously unseen multiple instructions ranges between 3% and 6% for the three estimation models. In addition, Fig. 11.11 summarizes the accuracy of estimation models for extensible instructions with unseen individual real-world application's extensible instructions (dataset 2). The maximum estimation error is 6.7%, 9.4%, and 7.2%, while the mean absolute error is only 3.4%, 5.9%, and 4.2% for area overhead, latency, and power consumption, respectively. The estimation errors are all far below the estimation errors of the commercial estimation tools (typically around 20% absolute error at gatelevel) we verified our models against. As such, we can conclude that our models are

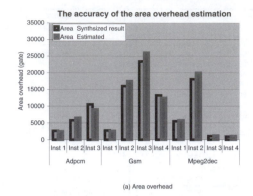

(a) Area overhead

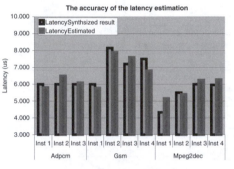

(b) Latency

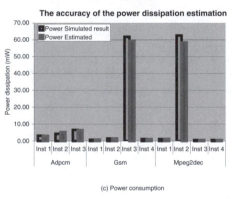

(c) Power consumption

■ **FIGURE 11.11**

The accuracy of the estimation models in real-world applications.

accurate enough for the purpose of high-level design space exploration for extensible instructions.

Our estimation models are by far faster than a complete synthesis and simulation process with commercial tools. The average time taken by *Design Compiler* and *Power Theatre* to determine the customization of an extensible instruction can be up to several hours, while our estimation models only require a few seconds in the longest case. This is another prerequisite for extensive design space exploration.

In summary, it is possible to accurately model instruction and precisely predict which of the instructions will meet the requisite criteria such as latency, power consumption, and area. Further, it is also possible to explore a very wide range of implementations very quickly to rapidly search through a large design space.

11.3 CONCLUSIONS

This chapter has shown two automatic methods to accelerate the process of designing ASIPs. The first of these shows a formal method to match instructions that is not only fast but is also accurate. The second method shows a way to model instructions so that alternate implementations of instructions can be evaluated rapidly before being synthesized. Both these methods form part of a single design flow, which is described in the introduction section of the chapter. Numerous challenges still remain to the rapid creation of ASIPs. These include taking power into consideration when selecting processor configurations and instructions, further reducing the time taken to match instructions by parallelizing matching algorithms, and modeling instructions in two separate steps, so that technology mapping is independently modeled, allowing models to be retargeted quickly as new standard cell libraries become available.

APPENDIX: ESTIMATING AREA, LATENCY, AND POWER CONSUMPTION

In this appendix, we derive the estimation equations for area, latency, and power consumption. The term used in this section are given in the text.

A.1 Area Overhead Estimation

As discussed previously, the area of extensible instructions, $A(inst)$, can be defined by using system decomposition:

$$A(inst) = \sum_{i \in \{dec,clk,top,reg,opea\}} A(i) \tag{A.1}$$

or as:

$$A(inst) = \sum_{i \in \{dec,clk,top,reg,opea\}} A(i) \tag{A.2}$$

where $A(i)$ is the area overhead estimation of all affected subsystems. Applying regression analysis on each subsystem and its customization parameter subset, the area overhead estimation of subsystems is derived and are described as follows.

The decoder has five customization parameters (according to Table 11.2). Using regression analysis, the relationship of the estimation model is seen to be linear, and the area overhead estimation, $A(dec)$, is hence defined as:

$$A(dec) = \Sigma_{i \in 32,64,128} A_{regf_i} Con_{regf_i} + A_{ireg} Con_{ireg} + \\ \Sigma_{i \in 8,16,32,64,128} A_{oper_i} Con_{oper_i} + A_{mcyc} Num_{mcyc} + \\ A_{minst} Num_{minst}$$

(A.3)

where A_{regf_i}, A_{ireg}, A_{oper_i}, A_{mcyc}, and A_{minst} are the respective coefficients.

For clock gating hardware, the area overhead estimation $A(clk)$ can be defined as:

$$A(clk) = \Sigma_{i \in 32,64,128} A_{regf_i} Con_{regf_i} + A_{ireg} Con_{ireg} + \\ A_{minst} Num_{minst}$$

(A.4)

$A(top)$ is the area overhead estimation of a top-logic and is defined as:

$$A(top) = \Sigma_{i \in 8,16,32,64,128} A_{oper_i} Con_{oper_i} + \\ \Sigma_{i \in 32,64,128} A_{regf_i} Con_{regf_i} + A_{ireg} Con_{ireg} + \\ A_{mcyc} Num_{mcyc} + A_{minst} Num_{minst}$$

(A.5)

The area overhead estimation of customized register, $A(reg)$, is defined as:

$$A(reg) = \Sigma_{i \in 32,64,128} A_{regf_i} Num_{regf_i} + A_{ireg} Num_{ireg}$$

(A.6)

$A(opea)$ is the area overhead estimation of combinational operations and is defined as:

$$A(opea) = \Sigma_{i \in 8,16,32,64,128} \{ A_{add/sub_i} Num_{add/sub_i} + \\ A_{LIB\_mul_i} Num_{LIB\_mul_i} + A_{LIB\_mac_i} Num_{LIB\_mac_i} + \\ A_{LIB\_add_i} Num_{LIB\_add_i} + A_{LIB\_csa_i} Num_{LIB\_csa_i} + \\ A_{mul_i} Num_{mul_i} + A_{shift_i} Num_{shift_i} \} + \\ A_{cond} Num_{cond} + A_{logic} Num_{logic}$$

(A.7)

A.2   Latency Estimation

As described in Section 11.2.3, the latency of extensible instructions is the maximum delay of each subsystem in the critical path of the extensible instruction. Therefore, the latency estimation, $T(inst)$, is defined as:

$$T(inst) = max_{i \in \{dec,clk,top,reg,opea\}} T(i)$$

(A.8)

where $T(dec)$ is the latency estimation of the decoder, which is defined as follows:

$$T(dec) = \Sigma_{i \in 32,64,128} T_{regf_i} Con_{regf_i} + T_{ireg} Con_{ireg} + \\ \Sigma_{i \in 8,16,32,64,128} T_{oper_i} Con_{oper_i} + T_{mcyc} Num_{mcyc} + \\ T_{minst} Num_{minst} \tag{A.9}$$

The latency estimation of the clock gated, $T(clk)$, is shown:

$$T(clk) = \Sigma_{i \in 32,64,128} T_{regf_i} Con_{regf_i} + T_{ireg} Con_{ireg} + \\ T_{mcyc} Num_{mcyc} + T_{minst} Num_{minst} \tag{A.10}$$

$T(top)$ is the latency estimation of the top logic and is shown:

$$T(top) = \Sigma_{i \in 32,64,128} T_{regf_i} Con_{regf_i} + T_{ireg} Con_{ireg} + \\ \Sigma_{i \in 8,16,32,64,128} T_{oper_i} Num_{oper_i} + T_{mcyc} Num_{mcyc} + \\ T_{minst} Num_{minst} \tag{A.11}$$

$T(reg)$ is the latency estimation of the customized register and is shown:

$$T(reg) = \Sigma_{i \in 32,64,128} \{ \frac{1}{Num_{regf} + Num_{ireg}} \times \\ T_{regf_i} Num_{regf_i} + T_{ireg} Num_{ireg} \} \tag{A.12}$$

$T(opea)$ is the latency estimation of the combinational operations and is shown:

$$T(opea) = \frac{Num_{ser}}{Num_{mcyc} \times Num_{oper}} \times \Sigma_{i \in 8,16,32,64,128} \{ \\ T_{add/sub_i} Num_{add/sub_i} + T_{LIB\_add_i} Num_{LIB\_add_i} + \\ T_{LIB\_csa_i} Num_{LIB\_csa_i} + T_{LIB\_mul_i} Num_{LIB\_mul_i} + \\ T_{LIB\_mac_i} Num_{LIB\_mac_i} + T_{mul_i} Num_{mul_i} + \\ T_{shift_i} Num_{shift_i} \} + T_{cond} Num_{cond} + \\ T_{logic} Num_{logic} + T_{VLIW} Wid_{VLIW} + \\ T_{vector} Wid_{vector} + T_{hwired} Wid_{hwired} + T_{ser} Num_{ser} + \\ T_{para} Num_{para} + T_{mcyc} Num_{mcyc} + T_{minst} Num_{minst} \tag{A.13}$$

A.3 Power Consumption Estimation

As per previous sections in this appendix, the power consumption can be modeled as:

$$P(inst) = \sum_{i \in \{dec,clk,top,reg,opea\}} P(i) \tag{A.14}$$

where

$$P(dec) = \Sigma_{i \in 32,64,128} P_{regf_i} Con_{regf_i} + P_{ireg} Con_{ireg} + \\ \Sigma_{i \in 8,16,32,64,128} P_{oper_i} Con_{oper_i} + P_{mcyc} Num_{mcyc} + \\ P_{minst} Num_{minst} \tag{A.15}$$

$$P(clk) = \Sigma_{i \in 32,64,128} P_{regf_i} Con_{regf_i} + P_{ireg} Con_{ireg} + \\ P_{mcyc} Num_{mcyc} + P_{minst} Num_{minst} \tag{A.16}$$

$$P(top) = \Sigma_{i \in 32,64,128} P_{regf_i} Con_{regf_i} + P_{ireg} Con_{ireg} +$$
$$\Sigma_{i \in 8,16,32,64,128} P_{oper_i} Con_{oper_i} + P_{mcyc} Num_{mcyc} + \tag{A.17}$$
$$P_{minst} Num_{minst}$$

$$P(reg) = \Sigma_{i \in 32,64,128} \left\{ \frac{Use_{reg_i}}{Num_{regf_i} + Num_{ireg}} \times \right.$$
$$\left. (P_{regf_i} Num_{regf_i} + P_{ireg} Num_{ireg} + P_{minst} Num_{minst}) \right\} \tag{A.18}$$

$$P(opea) = \Sigma_{i \in 8,16,32,64,128} \left\{ P_{add/sub_i} Num_{add/sub_i} + \right.$$
$$P_{LIB\_mul_i} Num_{LIB\_mul_i} + P_{LIB\_mac_i} Num_{LIB\_mac_i} +$$
$$P_{LIB\_add_i} Num_{LIB\_add_i} + P_{LIB\_csa_i} Num_{LIB\_csa_i} +$$
$$P_{mul_i} Num_{mul_i} + P_{shift_i} Num_{shift_i} \} +$$
$$P_{cond} Num_{cond} + P_{logic} Num_{logic} + \tag{A.19}$$
$$P_{VLIW} Wid_{VLIW} + P_{vector} Wid_{vector} +$$
$$P_{hwired} Wid_{hwired} + P_{ser} Num_{ser} +$$
$$P_{para} Num_{para} + P_{mcyc} Num_{mcyc} + P_{minst} Num_{minst}$$

As shown in the results section of this chapter, these equations have proved to be reasonably accurate and can be used quite readily in a rapid ASIP generation design flow.

CHAPTER 12

PROCESSOR VERIFICATION

Daniel Große, Robert Siegmund, and Rolf Drechsler

The preceding chapters focused on designing and modeling of embedded processors. As ensuring the correct functional behavior has become a major factor for successful circuit and system design, the present chapter addresses this issue for embedded processors and gives an overview on existing verification techniques. We start with the classical approaches based on simulation using directed and constrained random testing, with a particular focus on assertion-based verification. Then semiformal techniques are studied, and we end with proof techniques, such as property and assertion checking. For each method, the advantages and disadvantages are discussed. The most powerful approaches are formal proof techniques. In form of a case study, the verification of a RISC CPU is carried out considering hardware as well as software aspects. Finally, the state of the art in verification of embedded processors and peripherals in the context of multimillion gate SoC designs in an industrial environment is presented. The tradeoffs between manageable gate counts and verification effort to be spent using current simulation-based, semiformal, and formal tools are analyzed.

12.1 MOTIVATION

In the last several years, embedded processor design has become more and more important. Since embedded processors are used in safety-critical systems, e.g., in automotive components and medical devices, functional correctness is paramount. In the meantime, it has been observed that verification becomes the major bottleneck, i.e., up to 80% of the design costs are caused by verification. Due to this and the fact that pure simulation cannot guarantee sufficient coverage, formal verification methods have been proposed. These techniques ensure 100% functional correctness and, thus, they achieve the best quality. However, there are complexity problems in case of large designs. So semiformal approaches have been suggested to bridge the gap between pure simulation and formal verification.

In of this chapter Section 12.2, an overview on the major verification approaches is provided. Links to further literature are given, where the interested reader can get more information. Section 12.3 discusses formal verification of a RISC CPU. After the underlying formal technique has been described, an integrated hardware/software verification approach is presented. Based on this approach the hardware, interface, and programs of a RISC CPU are formally verified. In Section 12.4 the verification challenges for customizable and configurable embedded processors are discussed. The state of the art in verification of embedded processors and peripherals in an industrial setting at AMD is described in Section 12.5. The verification methodology based on the combination of simulation and formal techniques is introduced. Then it is discussed in the industrial context and applied to the verification of an on-chip bus bridge. Finally, the chapter is summarized.

12.2 OVERVIEW OF VERIFICATION APPROACHES

In this section we provide an overview on the three major functional verification techniques: *simulation*, *semiformal techniques*, and *proof techniques*. Since these terms are sometimes used in different contexts and with varying meanings, this chapter describes the main ideas and uses the terminology following [1]. Before the details are presented, the principles of these techniques are explained in a simplified way. Fig. 12.1 depicts the main idea of each technique. To demonstrate the different approaches, such as *device under verification* (DUV), one single AND gate is used. At the top of the figure, simulation is shown. Simulation executes the design with discrete values as inputs. Then the outputs are compared with the expected results. In case of semiformal verification, variables are applied for some inputs and diagrams are constructed based on synthesis operations—e.g., *binary decision diagrams* (BDDs) [2] can be employed. In contrast, proof techniques use variables for all inputs and thus for internal signals, the corresponding functions are computed. As proof techniques cover the complete search space, they achieve higher quality than simulation or semiformal techniques. However, resources and run time can grow quickly with increasing design sizes.

In the following, the three techniques are described in more detail. Afterward, coverage metrics are briefly discussed, which allows us to determine the verification progress. Since a complete overview cannot be given, references for further readings are provided.

12.2.1 Simulation

The most frequently used method for verification is still simulation [3]. In *directed simulation*, explicitly specified stimulus patterns are applied over a number of clock cycles to the design in order to stimulate a certain functionality, and the response is compared with the expected result. The

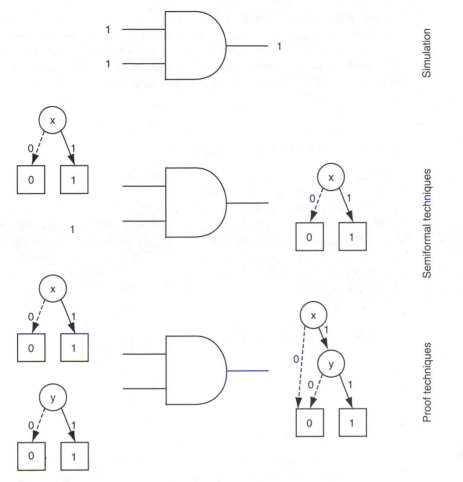

■ **FIGURE 12.1**

Illustration of different verification techniques.

main advantage of simulation is the very high execution speed. However, directed simulation only checks single scenarios. Since these scenarios have to be generated manually, this is a very time-consuming task.

To overcome this limitation, *random pattern simulation* is used to generate random stimulus pattern for the design. For example, random addresses and data are computed to verify communication over a bus.

To reduce the amount of time for the specification of simulation scenarios, *constrained-based random stimulus generation* has been introduced. For the simulation, only those stimulus patterns are generated that satisfy given constraints. Thus, the random stimulus generation process is controlled. The resulting stimuli allow us to test scenarios that are difficult to identify with directed or random pattern simulation.

Beyond the generation of stimulus pattern, a convenient way to specify expected or unexpected behavior is required. For this task, so-called *assertions* are used. In the simplest form, assertions are logical relationships between signals. This layer can be extended to allow for expressing temporal relations, too. Checking assertions during simulation is called *assertion checking* [4].

12.2.2   Semiformal Techniques

Semiformal techniques are a combination of simulation and proof techniques. As already described in Fig. 12.1, *symbolic simulation* is based on simulation with variables instead of discrete values [5]. Thus, one run of symbolic simulation covers many runs of pure simulation. As in assertion checking, this technique can be used to verify assertions. But a wider range of scenarios is tested. For example, starting from an interesting state of a simulation trace, all possible sequences of a fixed length can be checked for violation of an assertion. This corresponds to a "local proof," i.e., the assertion holds for all considered bounded sequences.

In addition, other formal methods (see Section 12.2.3) besides symbolic simulation can be combined in a hybrid approach. Following these methods, semiformal techniques bridge the gap between simulation and proof techniques.

12.2.3   Proof Techniques

Proof techniques exploit formal mathematical methods to verify the functional correctness of a design. For the proof process, different techniques have been proposed. Most of them work in the Boolean domain, like, e.g., BDDs [2] or *satisfiability* (SAT) solvers [6].

The typical verification scenarios, where formal proof techniques are applied, are

- *Equivalence checking* (EC)

- *Property checking* (PC), also called *model checking* (MC)

The goal of EC is to ensure the equivalence of two given circuit descriptions. Since EC is not in the scope of this chapter, we refer the reader for further information to [7].

In contrast to EC, where two circuits are considered, for PC a single circuit is given and properties are formulated in a dedicated "verification language." It is then formally proven whether these properties hold under all circumstances. Often, as properties, assertions specified for simulation can also be used.

While "classical" CTL-based model checking [8] can only be applied to medium-sized designs, approaches based on *bounded model checking* (BMC) as discussed in [9] give very good results when used for complete blocks with up to 100K gates.

Details on BMC and its application in a verification scenario for a RISC CPU are provided in Section 12.3.

12.2.4 Coverage

To check the completeness of the verification process, an estimation of the covered functionality is required. For simulation, an estimate of covered functionality can be given on the base of *code coverage*, which measures how well the *register transfer level* (RTL) code was exercised during simulation. For code coverage, several coverage metrics exist:

- *Line coverage*: How many lines of code have been executed?

- *State coverage*: How many states in an FSM have been reached?

- *Toggle coverage*: Which signals have been toggled during simulation?

- *Path coverage*: Which logical paths in the code were covered?

Code coverage, however, cannot measure which functionality has really been stimulated, as it has no semantic information on the RTL code. Therefore, code coverage is usually complemented by *functional coverage*, which uses coverage monitors for all functional properties to be examined and checklists driven by these monitors, which record what properties have been stimulated. Then, the result of coverage analysis for each attribute is a value for the percentage coverage. In addition, areas and functionality of the design that have not been exercised or only partially exercised are provided.

While these coverage terms are well suited for simulation and semi-formal techniques in the context of formal verification, there are no established measures yet. But this is a very active research field, and an overview on coverage metrics for formal verification can be found in [10].

12.3 FORMAL VERIFICATION OF A RISC CPU

In this section, an integrated approach for formal verification of hardware and software is described. The approach is based on BMC and demonstrated on a RISC CPU.

12.3.1   Verification Approach

To make the chapter self-contained, in the following we briefly review some of the notation and formalism. Then we describe the core of the verification approach.

We make use of BMC as described in [11], i.e., a property only argues over a finite time interval, and during the proof, there is no restriction to reachable states. The general structure of the resulting BMC instance for a property p over the finite interval $[0, c]$ is given by:

$$\bigwedge_{i=0}^{c-1} T_\delta(s_i, s_{i+1}) \ \wedge \neg\, p$$

where $T_\delta(s_i, s_{i+1})$ denotes the transition relation between cycles i and $i+1$. Thus, the initial sequential property checking problem is converted into a combinational one by unrolling the design, i.e., the current state variables are identified with the previous next state variables of the underlying *finite state machine* (FSM). The process of unrolling is shown in Fig. 12.2, where S, I, and O represent the set of states, inputs, and outputs, respectively.

Typically the property p consists of two parts: an *assume part*, which should imply the *proof part*, i.e., if all assumptions hold, all commitments in the proof part have to hold as well.

Example 1   *A simple example formulated in PSL [12] is shown in Fig. 12.3. The property* test *says that whenever signal x becomes* 1, *two clock cycles later signal y has to be* 2.

The integrated verification approach that supports hardware and software is based on the unrolling process of BMC. This also enables us to consider sequences of instructions of the RISC CPU. The steps for this verification are explained in the following.

The verification of the underlying *hardware* is done by classical application of BMC. At this stage, all hardware units are formally verified by describing their behavior with temporal properties. This guarantees the

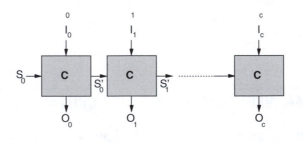

■ **FIGURE 12.2**

Unrolling.

```
1   property test
2   always
3      // assume part
4      ( x = 1 )
5      ->
6      // prove part
7      next[2] ( y = 2 );
```

■ **FIGURE 12.3**

Property test.

functional correctness of each hardware block. Basically, applying BMC for this task corresponds to block-level verification.

The *interface* is viewed as a specification that exits between hardware and software. By calling instructions of the interface, an interaction with the underlying hardware is realized. At the interface, the functionality of the hardware is available, but the concrete hardware realization is abstracted. In contrast to block-level verification, the interface verification with BMC formulates for each interface instruction the exact response of all hardware blocks involved. Besides these blocks, it is also assured that no side effects occur.

Based on instructions available through the interface, a *program* is a structural sequence of instructions. By a combination of BMC and inductive proofs [13, 14], a concrete program can be formally verified. Arguing over the behavior of a program is possible by constraining the considered sequence of instructions as assumptions in a BMC property. Thus, the property checker "executes" the program and can check the intended behavior of the proof part. Inductive reasoning is used to verify properties that describe functionality where the upper time bound varies, e.g., this can be the case if loops are used.

In the following, the basics of the RISC CPU and details on the SystemC model are given. Then the verification process for hardware and software, as outlined earlier, is described in more detail.

12.3.2 Specification

In Fig. 12.4 the main components of the RISC CPU are shown. The CPU has been designed as a Harvard architecture. The data width of the program memory and the data memory is 16 bits and the sizes are 4 KB and 128 KB, respectively. The length of an instruction is 16 bits. Due to page limitations, we only briefly describe the five different classes of instructions in the following:

- 6 load/store instructions (movement of data between register bank and data memory or I/O device, load of a constant into high and low byte of register, respectively)

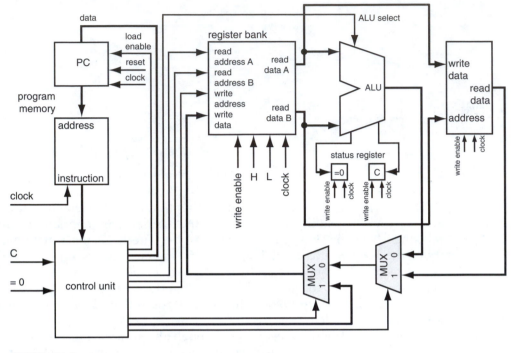

Structure of the RISC CPU, including data and instruction memory.

- 8 arithmetic instructions (addition/subtraction with and without carry, left/right rotation, and shift)

- 8 logic instructions (bit-by-bit negation, bit-by-bit exor, conjunction/disjunction of two operands, masking, inverting, clearing, and setting of single bits of an operand)

- 5 jump instructions (unconditional jump, conditional jump, jump on set, cleared carry, or zero flag)

- 5 other instructions (stack instructions push and pop, program halt, subroutine call, return from subroutine)

For more details on the CPU we refer the reader to [15].

12.3.3 SystemC Model

The RISC CPU has been modeled in the system description language SystemC [16, 17]. As a C++ class library, SystemC enables modeling of systems at different levels of abstraction, starting at the functional level and ending at a cycle-accurate model. The well known concept of hierarchical descriptions of systems is transferred to SystemC by describing a module as a C++ class. Furthermore, fast simulation is possible at an early stage of the design process, and hardware/software codesign can be carried out in the same environment. Note that a SystemC description can be compiled with a standard C++ compiler to produce an executable specification.

For details on the SystemC model of the RISC CPU, we refer the reader to [18]. To simulate RISC CPU programs, a compiler has been written that generates object code out of the assembler language of the RISC CPU. This object code runs on the SystemC model, i.e., the model of the CPU executes an assembler program. During the simulation, tracing of signals and states of the CPU are possible.

12.3.4 Formal Verification

For property checking of the SystemC model, the tool presented in [19, 20] is used. It is based on SAT solving techniques and for debugging a waveform is generated in case of a counter example. In the following, the complete verification of the hardware, interface, and programs for the RISC CPU is discussed. All experiments have been carried out on a Athlon XP 2800 with 1 GB of main memory.

Hardware

Properties for each block of the RISC CPU have been formulated. For example, for the control unit, it has been verified which control lines are set according to the opcode of the instruction input. Overall, the correctness of each block could be verified. Table 12.1 summarizes the results.[1]

The first column gives the name of the considered block. Next, the number of properties specified for a block are denoted. The last column provides the overall run time needed to prove all properties of a block. As can be seen, the functional correctness of the hardware could be formally verified very quickly with 39 properties.

Interface

Based on the hardware verification of the RISC CPU, in the next step the interface is verified. Thus, for each instruction of the RISC CPU, a property is specified, that expresses the effects on all hardware

---

[1] For the verification in the synthesized model of the RISC CPU, the sizes of the memories have been reduced.

TABLE 12.1 ■ Run time of block-level verification

Block	Number of Properties	Total Run Time in CPU Seconds
Register bank	4	1.03
Program counter	3	0.08
Control unit	11	0.23
Data memory	2	0.49
Program memory	2	0.48
ALU	17	4.41

Assembler notation: ADD R[i], R[j], R[k]

Task: addition of R[j] and R[k],
the result is stored in R[i]

Instruction format:

15	...	11	10	9	8	7	6	5	4	3	2	1	0
0	0	1	1	1	bin(i)			-	-	bin(j)		bin(k)	

■ **FIGURE 12.5**

ADD instruction.

blocks involved. As an example, we discuss the verification of the ADD instruction.

Example 2 *Fig. 12.5 gives details on the ADD instruction. Besides the assembler notation, the instruction format of the ADD instruction is also shown. The specified property for the ADD instruction is shown in Fig. 12.6. First, the opcode and the three addresses of the registers are assigned to meaningful variables (lines 1–6). The assume part of the ADD property is defined from line 11 to 12 and states that there is no reset (line 11), the current instruction is addition (line 11), and the registers R[0] and R[1] are not addressed (since these registers are special-purpose registers). Under these assumptions, we want to prove that in the next cycle the register R[i] (= reg.reg[**prev(Ri_A)**]) contains the sum of register R[j] and register R[k] (line 17), the carry (stat.C) in the status register is updated properly (line 16), and the zero bit (stat.Z) is set if the result of the sum is zero (line 18). Furthermore, we prove that the ADD instruction has no side effects, i.e., the contents of all registers that are different from R[i] are not modified.*

Analogous to the ADD instruction, the complete instruction set of the RISC CPU is verified. Table 12.2 summarizes the results. The first column gives the category of the instruction. In the second column, the number of properties for each category is provided. The last column

```
1 OPCODE := instr[15:11];
2 Ri_A := instr[10:8];
3 Rj_A := instr[5:3];
4 Rk_A := instr[2:0];
5 Rj := reg.reg[Rj_A];
6 Rk := reg.reg[Rk_A];
7
8 property ADD
9 always
10  // assume part
11  ( reset = 0 && OPCODE = "00111" &&
12    Ri_A > 1 && Rj_A > 1 && Rk_A > 1 )
13  ->
14  // prove part
15   next(
16    (reg.reg[prev(Ri_A)] + (65536 * stat.C)
17      = prev(Rj) + prev(Rk))
18    && ((reg.reg[prev(Ri_A)] = 0) <-> (stat.Z = 1))
19
20    // no side effects
21    && ( (prev(Ri_A) != 2) ->
22         reg.reg[2] = prev(reg.reg[2]))
23    && ( (prev(Ri_A) != 3) ->
24         reg.reg[3] = prev(reg.reg[3]))
25    && ( (prev(Ri_A) != 4) ->
26         reg.reg[4] = prev(reg.reg[4]))
27    && ( (prev(Ri_A) != 5) ->
28         reg.reg[5] = prev(reg.reg[5]))
29    && ( (prev(Ri_A) != 6) ->
30         reg.reg[6] = prev(reg.reg[6]))
31    && ( (prev(Ri_A) != 7) ->
32         reg.reg[7] = prev(reg.reg[7]))
33   );
```

■ **FIGURE 12.6**

Specified property for the ADD intruction of the RISC CPU.

TABLE 12.2 ■ Run time of interface verification

Instruction Category	Number of Properties	Total Run Time in CPU Seconds
Load/store instructions	6	15.16
Arithmetic instructions	8	186.30
Logic instructions	8	32.71
Jump instructions	5	6.68
Other instructions	5	7.14

```
1 /* counts from 10 downto 0 */
2        LDL R[7], 10
3        LDH R[7], 0
4 loop:
5        SUB R[7], R[7], R[1]
6        JNZ loop
```

▪ **FIGURE 12.7**

Example assembler program.

shows the total run time needed to prove all properties of a category. As can be seen, the complete instruction set of the RISC CPU can be verified in less than five CPU minutes.

Program

Finally, we describe the approach to verify assembler programs for the RISC CPU. As explained, the considered programs of the RISC CPU can be verified by constraining the instructions of the program as assumptions in the proof. These assumptions are automatically generated by the compiler of the RISC CPU. The verification of a program is illustrated in the following simple example:

Example 3 *Consider the assembler program shown in Fig. 12.7. The program loads the integer 10 into register R[7] and decrements register R[7] in a loop until it contains value 0. For this program, the property count has been formulated (see Fig. 12.8). At first, it is assumed that the CPU memory contains the instructions of the given example (lines 4–7).[2] Furthermore, the program counter points to the corresponding memory position (line 8), no memory write operation is allowed (line 9), and there is no reset for the considered 22 cycles (line 10). Under these assumptions, we prove that register R[7] is zero after 21 cycles. The time-point 21 results from the fact that the first two cycles (zero and one) are used by the load instructions, and the following 20 cycles are required to loop 10 times. The complete proof has been carried out in less than 25 CPU seconds.*

Based on proofs like those shown in the example, in a next step the verification of more complex programs can be considered. For this task, the behavior of a program can be decomposed in several parts and inductive proofs can be applied.

With the presented approach, the hardware, the interface, and simple programs of a RISC CPU have been formally verified.

In the following section, verification challenges in the context of customizable and configurable embedded processors are discussed.

---

[2] This part of the assumptions has been generated automatically by the compiler.

```
1 property count
2 always
3    // assume part
4      ( rom.mem[0] = 18186 && /* LDL R[7], 10 */
5        rom.mem[1] = 20224 && /* LDH R[7], 0 */
6        rom.mem[2] = 14137 && /* SUB R[7], R[7], R[1] */
7        rom.mem[3] = 24578 && /* JNZ 2 */
8        pc.pc = 0 &&
9        next_a[0..21]( prog_mem_we = 0 ) &&
10       next_a[0..21]( reset = 0 ))
11   ->
12   //  prove part
13   next[21] (reg.reg[7] = 0 );
```

■ **FIGURE 12.8**

Property count.

12.4 VERIFICATION CHALLENGES IN CUSTOMIZABLE AND CONFIGURABLE EMBEDDED PROCESSORS

Even though (semi)formal verification techniques are successfully applied and have become state of the art in many design flows, there are several open questions regarding their use in the context of customizable and configurable embedded processor or ASIP design. In this section, we discuss the occurring problems and point out directions for future work in this field.

Typically processors in this area have specialized functional units and instructions that speed up the considered application. The design of these processors is based on iterative architecture exploration using dedicated *architecture description languages* (ADLs; see also Chapter 4). Starting from a system description given in an ADL, a simulator, a compiler, and an HDL description are generated.

At present, customizable processors are mainly verified based on simulation, i.e., simulation test suites are generated for verification. But also in this context, the same arguments as for classical designs apply, i.e., due to increasing complexity, pure simulation is not sufficient.

As has been shown in the previous section, property checking is a well suited technique to formally verify processors. After block-level verification for each instruction of the given RISC CPU, a property has been specified and proven to hold on the design. If this procedure is transfered to the verification of customizable processors, in addition to the HDL description of the concrete processor, properties can be generated automatically. By this, e.g., the correctness of each chosen instruction can be proven. Besides such functional proofs, the correct interconnection of blocks can be shown based on formal techniques.

As a next step, the correct execution of programs can be verified following the idea presented in Section 12.3. For this task, programs in conjunction with properties can be generated based on the chosen instructions. This allows the verification of a customized processor even beyond single instructions.

Another important issue is the correctness of the generated compiler. Here two aspects should be considered. First, the compiler has to preserve the semantics of the source program, i.e., algorithms for the translation from one language into another are investigated, and algorithms for optimization have to be verified. Second, the implementation correctness of a compiler has to be considered, i.e., the question whether the translation specification has been implemented correctly. For more detailed discussion of this topic, see, e.g., [21–23].

12.5   VERIFICATION OF PROCESSOR PERIPHERALS

In the following, we discuss a verification environment used at AMD Dresden Design Center for verification of SoC designs and designs of processor peripherals.

State-of-the-art semiconductor technologies enable cost-efficient integration of CPU and peripheral system components, such as memory and display controllers as well as IO protocol controllers for various high-speed interfaces like PCIe, USB, SerialATA, and others on a single chip. The complexity of such *system-on-chip* (SoC) designs is in the multimillion logic gates range. The effort spent for functional verification of such designs accounts for up to 80% of the total SoC design effort. In particular, the verification of controller implementations for state-of-the-art high-speed protocols and the verification of correct interaction of system components in a SoC are major verification tasks.

At AMD Dresden Design Center, a verification methodology has been implemented that attempts to tackle the verification challenges imposed by complex SoC designs. In the following two sections, this methodology is discussed in detail. Afterward a case study on the verification of an on-chip bus bridge is described.

12.5.1   Coverage-Driven Verification Based on Constrained-Random Stimulation

The verification methodology is based on the following principles that are explained in more detail next:

- Coverage-driven verification: specification of functional properties to be verified as quantifiable coverage goals

- Assertion-based verification: specification of functional corner cases in the form of assertions and quantifiable assertion coverage
- Constraint-based random stimulus generation for simulation
- Verification of functional corner cases using formal techniques

Fig. 12.9 shows the components of the verification environment at AMD Dresden Design Center, which applies this verification methodology. As explained in Section 12.2, directed simulation is based on explicitly specified tests. Each test would verify a certain functional property by

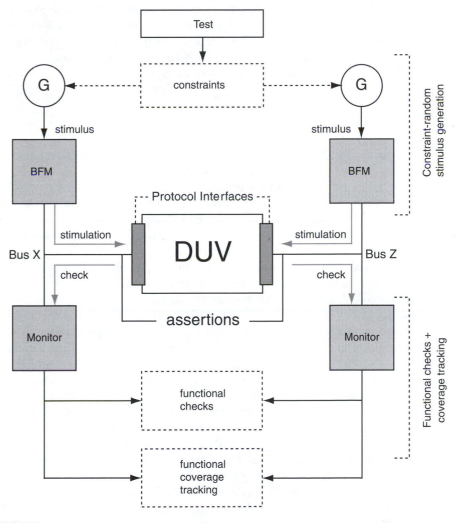

■ FIGURE 12.9

Verification methodology for protocol-oriented SoC components based on constrained-random stimulation and functional coverage monitoring.

applying defined sequences of stimulus pattern to all the DUV interfaces by means of *bus functional models* (BFMs) and observing the DUV response at each interface by means of *bus monitors* (see Fig. 12.9). The output of the monitors is compared using specific functional checkers, which, for example, check that a bus cycle on one DUV interface resulted in a bus cycle with the same address and payload data on another DUV interface. A test pass/fail criteria is then applied in order to decide whether the property holds, where a test is referred to as "passed" if none of the functional checkers reported a failure.

In contrast to directed simulation, coverage-driven verification is based on random pattern simulation. The functional properties of the system are described in the form of quantifiable coverage goals, which are measured by means of coverage monitors. Quantifiable coverage goals mean that the coverage of a certain functional property can be expressed by a rational number $C_p \in (0,1)$. Then, the total coverage for the DUV is also quantifiable as the weighted sum over the coverage for all properties:

$$C_{DUV} = \sum_p C_p w_p,$$

where $w_p \in (0,1)$ describes how the coverage of a functional property contributes to the total DUV coverage.

The set of all possible stimulus patterns to be applied to the DUV is described in the form of constraints (the solution set of the set of constraints). Stimulus patterns include both legal stimuli according to a certain interface protocol as well as illegal stimuli for which the design has to be proved robust, e.g., it must not hang even if the interface protocol is violated. For each simulation, stimulus patterns are picked randomly from the pattern set by means of stimulus generators G and are applied to the DUV interfaces by means of the BFMs (see Fig. 12.9). The functional checkers verify that the DUV responses match expected values. A set of coverage monitors track exercised scenarios and related DUV responses, if they successfully passed the functional check. As more and more scenarios are randomly chosen and exercised, the number of covered functional properties grows and the DUV coverage increases. Given a proper setup of the verification environment and a uniform distribution of the stimulus pattern obtained from the constraint-driven random generators, this process can be run almost autonomously until a sufficient functional coverage is reached, e.g., $C_{DUV} = 0.95$. Fig. 12.10 shows the typical development of functional coverage over the number of test runs. After the initial run of a complete test suite (in this case consisting of 15 tests), functional coverage is usually in the range of 35% to 45%. Over the course of random testing, it asymptotically approaches a maximum value, which is usually below the predefined coverage goal. (We will discuss strategies to close this coverage gap in the next section.) In contrast to directed testing, which usually requires the coding of

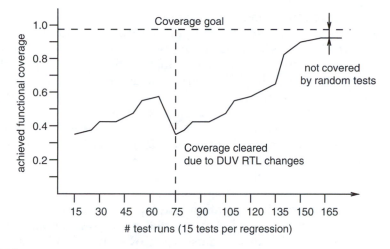

■ **FIGURE 12.10**

Development of achieved functional coverage over time.

thousands of tests in order to achieve full functional coverage of the DUV, only a very limited number of different tests (usually 15–25 per interface protocol), which set up different DUV configurations and vary the constraining of the stimulus pattern generators, are necessary.

In our verification environment, functional coverage goals are defined in terms of so-called coverage items. Each coverage item specifies a functional property to be verified. There are two kinds of coverage items:

- *Scalar items*: Scalar items cover functional properties observed by a dedicated coverage monitor. In case of a specific protocol interface, these might be the various commands defined by the protocol.

- *Cross-coverage items*: Cross-coverage items form the cross product between two or more scalar items and cover functional properties whose observation requires two or more coverage monitors running in parallel.

As part of the proposed verification methodology, in the following section assertion-based verification of corner cases is described. Especially the reasons to adopt assertion-based verification for corner cases and the resulting improvements in the verification quality are discussed.

12.5.2 Assertion-Based Verification of Corner Cases

The verification methodology described previously could potentially achieve full functional coverage by exercising all sequences of system states, given that enough simulations are run. In practice, however, simulation time is limited by computing resources and tape-out deadlines. A verification methodology based on pure simulation can only exercise a number of stimulus patterns, which in the best case grow linear with simulation time. This is in sharp contrast to the potentially exponential number of possibly exercisable, functionally interesting patterns. Therefore, verification based on pure simulation is prone to miss important but hard-to-exercise corner cases. Such corner cases include FIFO over- and under- runs, arbitration failures, deadlock scenarios, arithmetic overflows, and so forth. Fig. 12.10 illustrates the problem by means of the coverage gap: it shows that the defined coverage goal is usually not reached by constrained-random testing in a given number of simulation runs.

Assertion-based verification attempts to tackle this problem. An assertion is a functional property that must always hold. An example is that a FIFO in the design must never overrun or underrun. The task of the verification environment is to exercise scenarios to violate the property and fire the assertion, thereby revealing a design bug, or to exercise the property and cover the successful evaluation of the assertion without a firing.

To evaluate assertion checkers embedded in the DUV or placed at its interfaces, formal techniques for assertion checking have emerged over recent years. These techniques use a hybrid approach based on simulation, symbolic simulation, and model checking. To cope with the design complexity, formal tools traverse the state space only for a maximum (bounded) number of clock cycles, starting at an "interesting" initial state vector provided by the simulation engine. Current tools can handle about 1 million gates designs and are able to check about 100 clock cycles starting from the initial state vector.

12.5.3 Case Study: Verification of an On-Chip Bus Bridge

The verification methodology described in the previous sections was applied to the design of an on-chip bus bridge, which is part of an SoC recently developed at AMD. The bus bridge interfaces two on-chip buses (named X bus and Z bus) that use different protocols. The bus bridge is depicted in Fig. 12.11. The left-hand bus has separate command and data channels for target and DMA operation (where target operation refers to device register accesses and DMA operation refers to data movement from and to host memory). In one data cycle, 64 bits of data can be transferred. The right-hand bus has command and data channels for target and DMA interleaved and is capable of transferring only 32 bits in one data cycle. Both buses are capable of data bursting, e.g., a bus address cycle is followed by a number of consecutive data cycles. To improve performance for DMA read operations, the bridge furthermore implements speculative data prefetching. Both target and DMA operation can run in parallel.

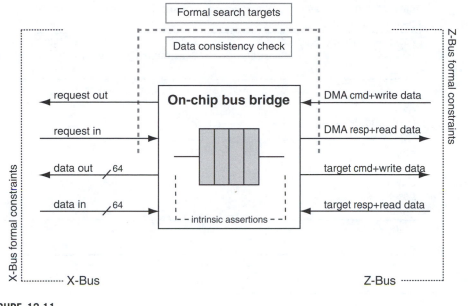

■ FIGURE 12.11

On-chip bus bridge.

The verification challenge for this bridge manifests in verifying that all DMA and target data is transferred correctly and consistently between the buses, that the bus protocol rules are obeyed by the bridge protocol controllers, and that the prefetch logic stops bursting at cache line boundaries.

The bridge design itself is dominated heavily by control logic with data paths mainly consisting of FIFOs and multiplexers. Inherent to this class of designs is a tight coupling of data path and control via the FIFO fill level signals. Therefore, discovering functional bugs is a matter of stimulating various FIFO fill levels and checks for bridge malfunctions, e.g., incorrect data is presented at one of the bus interfaces, data is lost, data is doubled. In a first step, the constrained-random simulation approach depicted in Fig. 12.9 is used to perform the verification of a set of defined functional properties, e.g., data consistency checks at both sides of the bridge. For this purpose, data transfers on the buses are stimulated by means of constrained random generators that map these data transfers to legal but randomly chosen sequences of bus cycles that adhere to the bus protocol rules. Two bus monitors observe the data traffic at either side of the bridge and check that the actual payload data contents transferred on both buses match.

To obtain a better verification of the interface data consistency property of the bridge design, which largely depends on correct handling of FIFO fill levels in the bridge, assertion-based verification is applied in

TABLE 12.3 ■ Type and quantity of assertions used for assertion-based verification of an on-chip bus bridge

Assertion Checker	X Bus Constraints	Z Bus Constraints	Formal Property (Data Consistency Check)
Maximum	2	2	-
Assert	10	0	-
Multi_enq_deq_fifo	10	20	-
Overflow	1	0	-
Fifo	4	8	-
Always	3	2	-
Never	7	7	-
Assert_timer	1	1	-
Minimum	1	1	-
Multi_enq_deq_fifo	-	-	1
Assert	-	-	2
Formal netlist complexity:		182,151 gate equivalents	

the second step. For this purpose, a formal tool targeted at assertion-based verification was used. This formal tool is driven by simulation to capture a number of system states from which a formal search for violations of assertions is spawned. To apply the tool, both X bus and Z bus protocol properties must be captured in the form of assertion checkers and passed to the tool as design constraints in order to suppress false negatives. Furthermore, the data consistency property must be formulated using another set of assertion checkers. Table 12.3 gives an overview of the assertion checkers used for constraining the protocols and for defining the interface data consistency property. The functionality of these assertion checkers corresponds to the checkers defined in the SystemVerilog Assertion library [24].

The data consistency checker itself comprises a multiple enque deque FIFO and two general-purpose assertions. Its purpose is to verify that all data presented during a DMA write on the Z bus appear correctly in the presented order and value at the X bus. No assumption is made about the latency of data transfers from the Z bus to the X bus. The two assertion blocks shown in Fig. 12.12 describe valid data cycles on either the X or the Z bus. Every time they are successfully evaluated, they are used to generate the enque/deque strobes for the multiple enque deque FIFO, which actually performs the data consistency check. With each enque strobe, 32 bits of data sampled from the Z bus are stored in the FIFO checker as one entry. Each 64-bit data transfer on the X bus triggers the two deque strobes, which cause a FIFO deque by two

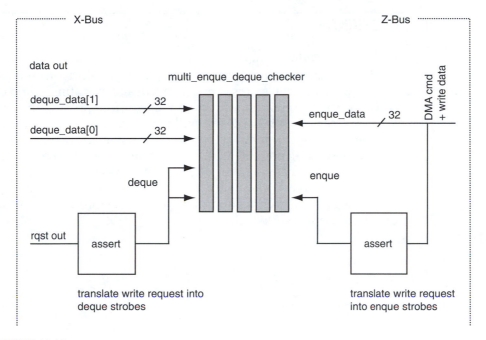

■ FIGURE 12.12

Data consistency checker for DMA write operations.

TABLE 12.4 ■ Run time for various number of proven clock cycles for formal target "data consistency checker"

Initial State Vectors Applied	Proven Clock Cycles	Assertions Fired	Run Time
90	4	-	34s
90	5	-	1min32s
90	6	-	3min10s
90	7	1	8min32s

entries. The dequeued data are checked against the data presented on the X bus data lines.

Table 12.4 shows the run time of the formal tool in order to obtain a proof that all assertions hold within a number of clock cycles starting from a given system state, or that an assertion is fired within a number of clock cycles. As can be seen, run time grows rapidly with the number of formally checked clock cycles. By using the formal tool, we were able to discover a data inconsistency bug caused by a data FIFO overrun in the bridge. This bug would probably not have been discovered using pure simulation.

12.6 CONCLUSIONS

In this chapter, three major verification techniques have been described: simulation, semiformal techniques, and proof techniques. They differ with respect to resources needed and resulting completeness of verification. Based on a formal verification approach, the functional correctness of both hardware and software for a RISC CPU has been shown. The verification of a bus bridge as part of a complex SoC developed at AMD was possible with a combination of different verification techniques. The combination of simulation-based and semiformal approaches leads to a high-quality verification result. Thus, the tight integration of different verification methods is very promising for academic and industrial work.

CHAPTER 13

SUB-RISC PROCESSORS

Andrew Mihal, Scott Weber, and Kurt Keutzer

The now-common phrase "the processor is the NAND gate of the future" begs the questions: "What kind of processor?" and "How to program them?" In the previous chapters, the focus was placed on RISC-based processors augmented with instruction extensions as the natural building block. The presumption is that programming will be done in C. This chapter challenges this viewpoint. Our opinion is than even a RISC processor is too coarse-grained for typical embedded applications, and that C is insufficient for programming a multiprocessor architecture. As an alternative, we explore the design and deployment of tiny sub-RISC processors as the "NAND gate of the future." With Tiny Instruction-set Processors and Interconnect (TIPI) processing elements, we can achieve better performance than RISC elements with fewer resource requirements. Also, we can deploy concurrent applications with a programming methodology more productive than C.

13.1 CONCURRENT ARCHITECTURES, CONCURRENT APPLICATIONS

As the previous chapters have shown, programmable processors are a compelling basic block for embedded systems. They are a coarser-grained component than RTL state machines that can be used to build large systems with higher designer productivity. Their software programmability allows them to be used across application generations, avoiding the risk of rapid obsolescence that ASICs face.

In the network processing application domain, there is already a wide variety of network processor architectures that use programmable processors as basic blocks [1]. These machines are typically a heterogeneous mix of RISC-like processing elements (PEs). This style of architecture is often called a *programmable platform*. Despite the diversity of these architectures, one thing they all have in common is that they are notoriously difficult to program.

What makes programming so hard? One proposal is that the programming difficulties are the result of poor methodologies for modeling and implementing application concurrency. Network processing applications have multiple types of concurrency on several levels of granularity.

For example, there is traditional process-level concurrency, such as the ability to process independent streams of packets in parallel. Another level of granularity is concurrency within a task, or data-level concurrency. Tasks frequently perform computations on several different fields within a single packet header, and these can often be done in parallel. At the lowest level of granularity, there is concurrency on the level of individual bits. Network processing uses custom mathematical operations (such as checksum calculations) on custom data types (such as irregular packet header fields). This is datatype-level concurrency. To meet performance goals, all of these flavors of concurrency must be considered in the implementation.

The common programming paradigm for network processors is that programmers must write code for each PE in the architecture individually. Typically this is done using assembly language, or a C language dialect with compiler intrinsics. C does not provide good abstractions for application concurrency, especially process-level concurrency. Programmers are forced to manually translate high-level application concepts into a set of interacting C programs.

In an application domain where concurrency is a primary concern, this ad hoc treatment leads to numerous troubles. Producing even one correct implementation is slow and error-prone, and effective design space exploration is out of the question due to increasing time-to-market pressure. Even minor modifications to the application require lengthy software rewrites.

The solution that has been proposed for this problem is to raise the level of the programming abstraction through the use of a domain-specific language (DSL). DSLs provide formalisms for concurrency that improve designer productivity and make it easier to build correct applications. For network processing, Click is a popular choice [2]. Projects such as SMP-Click [3], NPClick [4], CRACC [5], StepNP [6], and Soft-MP [7] start with a programming abstraction derived from Click and target implementation on platform architectures.

These techniques are successful at improving designer productivity, but they also expose a more fundamental problem. After programmers create formal models of concurrent applications, it becomes clear that there is an inherent mismatch between the concurrency requirements of the application and the capabilities of the architecture. This problem is the *concurrency implementation gap*, shown in Fig. 13.1.

Like network processing applications, network processors also exhibit multiple types of concurrency on several levels of granularity. However, these flavors of concurrency are different from those found in the applications. At the process level, multiple PEs run programs in parallel, but network-on-chip communication protocols do not match the *push* and *pull* communication semantics in Click. RISC-inspired PEs have general-purpose instruction sets and work on standard byte- and word-sized data. Unfortunately, common bit-level header field manipulations are not a part

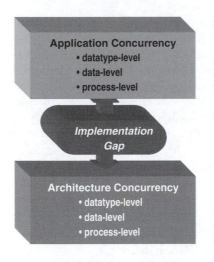

■ **FIGURE 13.1**

The concurrency implementation gap.

of standard RISC instruction sets. Irregular packet header fields do not match 32-bit-wide datapaths and ALUs. A significant part of the deployment effort is spent working around differences such as these while trying to maintain a high level of performance.

Clearly, programmable platforms face two challenges. One issue is how to model and deploy concurrent applications. The other issue is that the architectures built to date are not a good match for the applications. Although we use network processing and network processors as examples in this chapter, these issues apply to embedded applications and programmable platforms in general. If processors are to find success as the "basic blocks of the future," we must attack both of these problems.

This motivates us to rethink the way we design programmable platforms in addition to the way we program them. In this chapter, we present a novel class of processing element architectures called Tiny Instruction-set Processors and Interconnect (TIPI). TIPI PEs are a finer-grained component than RISC processors, allowing architects to build programmable platforms that provide customized support for an application's process-level, data-level, and datatype-level concurrency at low cost. This works to make the implementation gap smaller. To program these PEs, we provide a deployment methodology called Cairn. Cairn provides formal abstractions for capturing application concurrency and a disciplined mapping methodology for implementation on TIPI multiprocessor platforms. This makes the implementation gap easier to cross. Together, these two techniques make it possible to realize all of the benefits that programmable platforms are supposed to have over ASIC systems.

In the next section, we explore alternative PE architectures and motivate sub-RISC PEs as a basic block. Section 13.3 describes the design methodology and toolset for constructing TIPI PEs. In Section 13.4, the Cairn deployment methodology is introduced. Next, these techniques are used to build and program a novel architecture for packet processing applications. The ClickPE, described in Section 13.5, provides support for the process-level, data-level, and datatype-level concurrency found in a basic Click IPv4 forwarding application. This PE can be programmed using Click as a high-level input language, without requiring manual coding in C or assembler. In Section 13.6, we implement the ClickPE on a Xilinx FPGA and conclude with performance results.

13.2   MOTIVATING SUB-RISC PEs

In choosing programmable processors as a replacement basic block for state machine-based RTL design methodologies, there are several criteria we are trying to fulfill:

- *Improving designer productivity:* RTL methodologies give architects fine control over the details of a design. This is good for building high-performance systems, but it does not scale. The next architectural abstraction will accept coarser granularity in exchange for a more productive design process.

- *Customizability:* The basic block should be customizable to meet the needs of a particular application or application domain. Therefore it should not be a static entity, but rather a template that defines a family of possible blocks. We desire features that support the application's process-level, data-level, and datatype-level concurrency.

- *Degrees of reusability:* In some cases, architects are willing to make a design more application-specific in exchange for higher performance. In other cases, the ability to work across a broad application domain is more important than raw performance. The next basic block should permit architects to find the sweet spot along this spectrum by exploring tradeoffs between application-specific features and general-purpose features.

- *Architectural simplicity:* The basic block should be easy to design, verify, and compose with other basic blocks. Complexity should only be introduced (and paid for) when the application requires it. If the basic block is itself complicated, then even simple problems will lead to complex solutions. For a customizable basic block, this

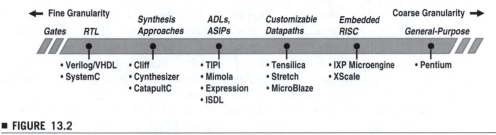

■ **FIGURE 13.2**

Spectrum of basic block candidates.

means being able to omit unnecessary features as well as to add application-specific features.

■ *Exportability:* We must be able to communicate the capabilities of the basic block to programmers so that they can deploy applications on it. Developers require a programming abstraction that finds the right balance of visibility into architectural features that are necessary for performance and opacity toward less important details. If programmers cannot figure out how to use the architecture, then the benefits of the new basic block are lost.

RISC processors are familiar to both hardware designers and software developers, so they appear to be an obvious choice. However, they are only one of many different kinds of programmable elements. Fig. 13.2 arranges the important candidates along a spectrum according to level of granularity. Existing RTL design methodologies are included as a point of reference on the left. We are interested in a basic block with a coarser granularity than state machines.

At the opposite end of the spectrum, there are large general-purpose processors such as the Pentium. A basic block such as this is very expensive in terms of area and power, and it does not include any application-specific features.

The ideal basic block will lie somewhere in between these two extremes. In this section, we will explore each candidate in detail and argue the merits and shortcomings of each relative to the goals given earlier.

13.2.1 RISC PEs

The trouble with using a RISC PE as a basic block can be summarized in one seemingly contradictory statement: they are simultaneously too simple and too complex for the applications we are interested in. The "too simple" argument follows from the customizability criterion given earlier. A traditional RISC processor provides features that work for all applications but are exceptional for no application. There are few, if any, opportunities to customize the architecture or to explore tradeoffs between application-specific features and general-purpose features. The "too complex" argument begins at the architectural simplicity criterion. In trying to be applicable to any application, RISC PEs will inevitably include features that will never be used by certain applications.

These arguments apply to each of the three levels of concurrency: process-level, data-level, and datatype-level. As an example, consider the difference between deploying a network processing application on a RISC PE versus using an approach like Cliff [8]. Cliff is a structural synthesis approach that builds a custom FPGA design starting from a Click application model. A RISC PE, on the other hand, must make do with the available instruction set. This comparison will reveal that there is a large design space gap between what is possible with an application-specific datapath versus what a RISC PE has to do.

Datatype-Level Concurrency

Datatype-level concurrency refers to the way data is represented as a bit vector and the arithmetic and logical computations that are carried out on the data. In a Click application, the fundamental data type is a packet header, and *elements* describe bit-level computations on these headers.

The DecIPTTL element is common in IP forwarding applications and will serve as an example. This element decrements the IP time-to-live header field (an 8-bit value) and incrementally adjusts the checksum to match using 1's complement arithmetic.

Cliff can produce an FPGA design that exactly matches the bit widths, and the 1's complement arithmetic expressed in the DecIPTTL element, as shown in Fig. 13.3(a). A RISC PE, on the other hand, lacks instructions for 1's complement arithmetic. This datatype-level concurrency must be emulated using logical shifts, bit masks, and 2's complement arithmetic as shown in Fig. 13.3(b). The RISC PE is sufficient for implementing DecIPTTL, but it is neither the cheapest nor the fastest way to implement it.

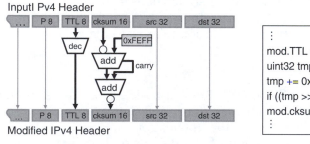

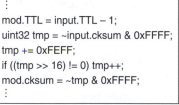

(a) Custom Logic   (b) C for RISC PE

■ **FIGURE 13.3**

Implementing datatype-level concurrency.

Data-Level Concurrency

The CheckIPHeader element demonstrates a mismatch in data-level concurrency. This element treats an IPv4 header as an array of ten 16-bit fields and performs a 16-bit 1's complement checksum operation. In the Cliff design shown in Fig. 13.4(a), the datapath is wide enough for the entire header, and all ten fields can be immediately passed into an adder tree.

A typical RISC PE will not have enough parallel adders or enough memory ports to start processing 160 bits of data right away. The C pseudocode shown in Fig. 13.4(b) specifies the checksum calculation sequentially. If the PE is a VLIW-type machine, it is up to the compiler to try to reverse-engineer the data-level concurrency. A PE with dynamic data-level concurrency (e.g., a superscalar machine) is an even more expensive solution: it has extra hardware to calculate dynamically what we know a priori.

Process-Level Concurrency

Click's *push* and *pull* connections are an abstraction for a control-follows-data style of process-level concurrency. In push communications, the upstream element performs computation on a packet header and then passes the flow of control to the downstream element. This is different from a function call because the upstream element does not wait for a return value and it does not continue. In pull communications, the downstream element initiates the transfer of control. Control is passed to the upstream element, and the downstream element blocks until the upstream element provides a packet header.

Cliff can build control structures at the multiprocessor level that mirror these semantics exactly. Each Click element in the application is implemented as a miniature datapath, as described earlier. These datapaths are controlled by finite state machines that communicate using a

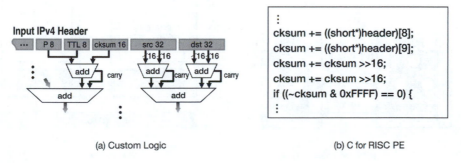

(a) Custom Logic   (b) C for RISC PE

■ FIGURE 13.4

Implementing data-level concurrency.

three-way handshake. This is shown in Fig. 13.5(a). Push and pull style communications events are implemented using these interelement control signals. This represents a tight coupling between the control logic of elements at the multiprocessor level.

A RISC PE has a more self-sufficient model of control, where the normal behavior is to forge ahead through a sequential program. Aside from interrupts, there is no mechanism for multiple PEs to directly influence each other's flow of control. A possible RISC implementation is shown in Fig. 13.5(b). Here, there are two RISC PEs that are connected to an on-chip network. The Click computations are implemented as processes, and an operating system provides access to the communication hardware resources through a message-passing interface. The processes use blocking and nonblocking send and receive functions to emulate Click's push and pull semantics in software. Since the PEs are loosely coupled in their control logic, it is more difficult and expensive to implement certain styles of process-level concurrency.

There is quite a large design gap between RISC PEs and what is possible with application-specific datapaths. In each of the three categories of concurrency, this implementation gap has a noticeable effect on performance. Tensilica reports that there is significant performance to be gained by adding support for the application's datatype-level concurrency to a RISC PE's instruction set [9]. To work around the mismatch in data-level concurrency, Sourdis et al. [10] outfit PEs with header field extraction engines to efficiently move data into register files. For process-level concurrency, SMP-Click finds that the synchronization and communication between PEs is a major performance bottleneck [3]. The mismatch between the application and the architecture is so strong in this case that the authors resort to a mapping strategy that minimizes inter-PE communication.

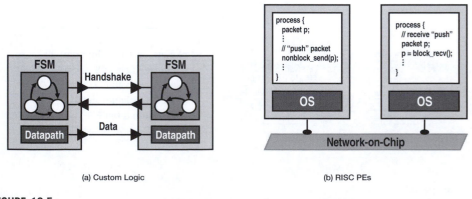

(a) Custom Logic   (b) RISC PEs

▪ FIGURE 13.5

Implementing process-level concurrency.

For these reasons, we maintain that a finer-grained programmable component is a better choice than RISC PEs. We desire more potential for application-specific concurrency, especially process-level concurrency. Whenever possible, we want to remove general-purpose features that are not used by our applications. A sub-RISC PE will be able to provide higher performance at a lower cost.

13.2.2 Customizable Datapaths

The next finer-grained PE candidate is a configurable datapath such as the Tensilica Xtensa or the Stretch architecture. They permit architects to add custom instructions to a RISC core to implement application-specific arithmetic and logic computations. This provides opportunities to match datatype-level concurrency. However, the RISC core itself is mostly fixed, and it is not possible to trim away certain unwanted general-purpose functions.

These PEs are also inadequate in terms of process-level and data-level concurrency. Traditional sequential control logic consisting of a program counter and an instruction memory is assumed. The number of register files, the number of register file read ports, and the number of memory ports are only customizable to a limited extent.

13.2.3 Synthesis Approaches

The main difficulty with synthesis approaches such as Cliff is that they do not create a programmable system. These approaches overshoot the goal in terms of customization, sacrificing reusability.

A second problem is that synthesis provides limited opportunities for design space exploration. To go from a high-level input language to an RTL implementation, the tool has to make a large number of design decisions, most of which are not visible or controllable by the architect. If the tool's design cannot meet performance goals, architects are left with undesirable choices, such as editing the generated RTL by hand.

13.2.4 Architecture Description Languages

Fortunately, there is a good basic block candidate that gives architects more programmability than synthesis approaches and provides more opportunities for customization than RISC PEs. These are the processing elements produced by architecture description language (ADL) design methodologies. In a nutshell, an ADL PE is a datapath whose capabilities can be exported as an instruction set.

Like the customizable datapaths discussed previously, ADLs permit architects to build custom instructions for application-specific arithmetic and logic operations. However, the customizability does not end at the level of making extensions to a core architecture. Architects can design the entire architecture and can therefore decide what features to leave out as well as what to include. One can design a simple architecture with a few instructions or a complex architecture with hundreds of instructions. The instructions can be tuned for a particular application, or for a domain of applications, or they can be general purpose. Traditional architectural styles, such as five-stage pipelines with a single register file and memory port, can be easily broken in favor of schemes that support application-specific data-level concurrency. This is attractive because it leads to PEs that are "simpler" and "more complex" than RISC PEs.

ADLs are also a clear winner in terms of improving designer productivity. To design a programmable element using an RTL language, architects must specify both the datapath and the control logic manually. The hardware implementation must be kept consistent with the description of the instruction set and the software development tools. This covers instruction encodings, semantics, integration with an assembler, a simulator, and many other things. Unfortunately, RTL languages do not have an abstraction for describing these details. In practice, instruction sets are usually English-language specifications. The problem of maintaining consistency is a verification nightmare. This impedes the ability to explore architectural options.

ADLs solve this problem by using a three-part approach. First, they provide formalisms for specifying instruction sets. This counters the lack of precision and analysis difficulties of English-language specifications. Second, they enforce particular architectural design patterns. Architects can no longer freely make arbitrary collections of finite state machines. Instead, design is at the level of components like register files, multiplexors, and functional units. The focus is on designing the instruction-level behavior of the PE. Third, there is a mechanism for ensuring the consistency between the hardware structure and the instruction set. ADLs can be roughly grouped into two categories based on how they attack this last problem.

Generating the Architecture from the Instruction Set

In ADLs such as ISDL [11] and nML [12], designers model an instruction set, and then a correct-by-construction synthesis tool generates a matching architecture and the appropriate software development tools. This approach has drawbacks similar to the synthesis approaches. The tool makes decisions about the structure of the architecture that are not under the control of the architect.

To remedy this, ADLs such as LISA [13] and EXPRESSION [14] were developed. Here, designers use a single unified description for the ISA, the datapath structure, and a mapping between the two. This gives architects more control over the details of the implementation. However, designers must specify an instruction set and an encoding by hand. Another downside is that the architect is expected to conform to a traditional architectural style. These ADLs are similar to the customizable datapaths in that they assume a sequential control scheme with a program counter and instruction memory. To construct application-specific process-level concurrency schemes, we want more flexibility in this area.

Extracting the Instruction Set from the Architecture

The opposite approach is to have the designer build the architecture, and then automatically extract the instruction set and the software development tools. MIMOLA is an example of an ADL in this category [15]. The theory is that for unusual architectures with complex forwarding paths, multiple memories, multiple register files, and specialized functional units, it can be more difficult to describe the instruction set than it is to just build a structural model of the architecture.

This method is clearly advantageous for the sub-RISC PEs we wish to design. However, there are still issues with application-specific process-level concurrency. MIMOLA expects a traditional control scheme in order to produce a compiler. Designers must manually identify key architectural features such as the program counter and the instruction memory. Thus the deployment methodology is only applicable to a subset of the architectures that can be designed.

TIPI Processing Elements

To build sub-RISC PEs with application-specific control logic, we must explore an area of the design space that is not adequately covered by existing ADLs. This space is shown in Fig. 13.6. Many of the basic blocks we have considered, from the hardwired datapaths created by synthesis approaches to VLIW- or RISC-like customizable datapaths, can be called ASIPs. For some of the finest grained hard-wired designs on the left in the

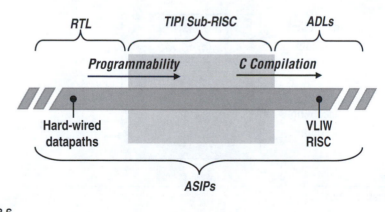

■ **FIGURE 13.6**

Spectrum of ASIP basic blocks.

figure, RTL languages offer sufficient designer productivity. However, as we start to add programmability to the architectures, designers quickly become overwhelmed in the specification of details of the control logic for the architecture. RTL languages are no longer sufficient. ADLs make it easier to describe programmable machines, but their focus is on architectures that are amenable to traditional compilation techniques (e.g., sequential C code).

When we consider application-specific process-level concurrency, and multiprocessor machines with tightly coupled control logic between PEs, it is clear that there is a gap in the design space. We have already identified that C is not a good language for modeling some of the concurrency in embedded applications. Therefore it is perfectly natural to consider architectures that are different from traditional machines that run C programs. These are architectures that are too complex to design with RTL languages in terms of productivity, but are also not targeted by existing ADLs.

This motivates the design of a new ADL. We call our approach TIPI. TIPI builds upon the ADLs that extract the instruction set from the architecture. To this we add a mechanism for expressing the programmability of architectures that lack traditional control logic.

This is accomplished by using a *horizontally microcoded, statically scheduled* machine as the core abstraction. Designers build structural models of datapaths, but leave control signals (such as register file address inputs and multiplexor select signals) disconnected. These signals are automatically driven by a horizontal microcode unit that decodes an instruction stream that is provided by a source external to the PE. This is shown in Fig. 13.7. This abstraction applies to the whole range of architectures that can be designed.

At the multiprocessor level, a TIPI PE is an element that can both receive control from and send control to other PEs. Designers build

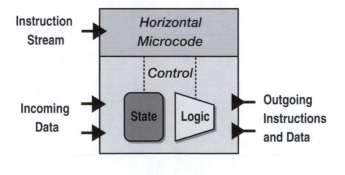

■ **FIGURE 13.7**

TIPI control abstraction.

compositions where the instruction input for each PE is connected to the outputs of other PEs. The possibility of tightly coupling control at the multiprocessor level enables the construction of hardware support for application-specific process-level concurrency.

Programming such a machine is a matter of arranging for the proper instruction streams to arrive at each PE. This is quite different from writing individual C programs for each PE. The concept of a multiprocessor machine where the PEs pass control messages among themselves forms the basis for a new application deployment methodology, which will be explored in Section 13.4.

With these pieces put together we can build programmable basic blocks that meet all of the goals given at the beginning of the section:

- *Improved productivity:* Instead of making compositions of state machines with RTL languages, architects work at the level of structural datapaths composed of register files, memories, multiplexors, and functional units. Complex control logic may be omitted. An automatic analysis process extracts the capabilities of the datapath in the form of an instruction set, preventing the need for consistency verification. A correct-by-construction code generation approach produces synthesizable RTL implementations and simulators from these high-level architectural models. These techniques are central to the MESCAL methodology [16]. This disciplined approach simplifies the design space exploration problem.

- *Customizability:* TIPI PEs can be customized to support all three levels of concurrency. Architects can add function units for application-specific arithmetic and logic computations. Beyond this, architects can create unusual pipelines with varying numbers of register files and memory ports for better data-level concurrency. Most importantly, TIPI PEs permit the construction of novel multiprocessor control schemes for process-level concurrency.

- *Varying degrees of reusability:* Architects can explore a range of architectural alternatives, from datapaths that provide a large number of general-purpose features to datapaths that focus on application-specific performance.

- *Architectural simplicity:* Deciding what to leave out of an architecture is just as important as deciding what to add. A minimal TIPI PE can have as few as two or three instructions. Architects only pay for complexity when the application demands complexity.

- *Exportability:* The automatically extracted instruction set describes the possible cycle-by-cycle behaviors of the design. This exposes the PE's capabilities for data-level and datatype-level concurrency.

For process-level concurrency, there is the abstraction that a PE can send and receive control from other PEs in a multiprocessor composition.

13.3 DESIGNING TIPI PROCESSING ELEMENTS

In this section, we cover the design of TIPI processing elements and the construction of multiprocessor architectures. This part of the system design flow is shown in Fig. 13.8.

The TIPI architecture design flow is an iterative methodology. Designers begin on the left side of the diagram by building structural models of PE datapaths. The capabilities of each datapath are characterized by running an automatic operation extraction algorithm. At this point, architects can iterate over the design and improve the architecture until it implements the desired functions, as shown by the dotted line feedback path. TIPI also produces a cycle-accurate, bit-true simulator for a PE, so timing simulation can also be included in the design loop.

Once one or more PE architectures have been designed, architects can build programmable platforms by making multiprocessor compositions of PEs. A transaction-level simulator can be automatically generated from a multiprocessor model. TIPI can also produce a synthesizable Verilog model that can be passed to third-party FPGA or ASIC design tools, or simulated using a third-party Verilog simulator.

In Section 13.4, we will extend this design flow to include application deployment. The following sections describe each of the architectural design steps in detail.

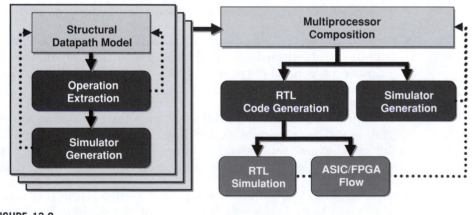

■ FIGURE 13.8

TIPI design flow.

13.3.1 Building Datapath Models

TIPI PEs are constructed by assembling *atoms* from a library. This library contains common things such as register files, multiplexors, memories, pipeline registers, and arithmetic and logical functional units. Most TIPI atoms are type polymorphic in terms of bit width, which simplifies the design process. Atoms are written in a feed-forward dataflow language based on Verilog. This allows architects to specify arbitrary datatype-level arithmetic and logic computations.

Fig. 13.9 is an example of a simple architecture. This datapath has two register files. One is connected to a decrement functional unit, and the other is connected to an equality comparison unit and a checksum updating unit.

A TIPI datapath is a nondeterministic description of computations that are atomic with respect to architectural state and outputs. One example of this nondeterminism is that many of the ports in the design are left unconnected. The read and write address ports on both register files, one input to the comparison functional unit, and the select signal to the multiplexor are all unspecified in this design.

The TIPI abstraction states that these signals are implicitly connected to a horizontal microcode control unit. Their values will be controlled by statically scheduled software on a cycle-by-cycle basis. The register file read and write addresses and multiplexor select signals become part of the instruction word. Unspecified data inputs, such as the input to the comparison functional unit, become immediate fields in the instruction word. Also missing are traditional control logic elements such as a program counter and instruction memory. Instead, this design receives instruction words from an external source as in Fig. 13.7.

Obviously there are restrictions on what combinations of control bits represent valid and useful settings for the datapath. The next step

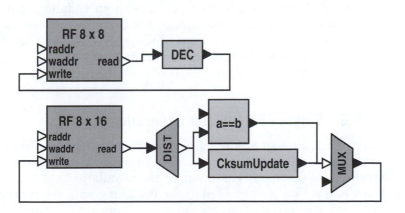

■ **FIGURE 13.9**

Example TIPI architecture model.

in the design process is to analyze the datapath and determine what these are.

13.3.2   Operation Extraction

TIPI provides an operation extraction algorithm that analyzes a datapath design and finds the set of valid execution paths that can be configured statically by software. The resulting *operations* are the state-to-state behaviors of the datapath.

A complete description of the operation extraction process is given in [17]. Fundamentally, operation extraction works by converting a structural datapath model into a system of constraints, and then finding satisfying solutions to the constraint problem. Each atom has firing rule constraints that specify the valid ways to use the component in a cycle. An algorithm based on iterative Boolean satisfiability is used to find configurations where all of the atoms' rules are satisfied.

To represent the programmability of the PE compactly, TIPI defines an operation to be a *minimal* solution to this constraint system. A minimal solution is a datapath configuration that is not simply a combination of smaller solutions. Continuing with the example of Fig. 13.9, Fig. 13.10 shows the set of extracted operations. There are three operations for this datapath: decrement, compare immediate, and checksum update. A *nop* operation also exists in which the datapath remains idle for a cycle, but this is not shown in the Figure. The valid datapath configuration shown in Fig. 13.10(d) is a solution, but it is not an operation because it is simply the union of the decrement and checksum update operations.

TIPI captures these nonminimal solutions by creating a *conflict table* for the PE. This is given in Fig. 13.11. The conflict table shows us that it is possible for software to issue the decrement and checksum update operations simultaneously in one cycle without violating any of the PE's constraints. On the other hand, it is not possible to issue the compare and checksum update operations simultaneously. Together, the operation set and the conflict table are a compact way of describing the valid ways in which software can use the datapath.

13.3.3   Single PE Simulator Generation

After operation extraction, TIPI can produce a cycle-accurate bit-true simulator for the PE. This is a C++ program that compiles and runs on the host machine. The input to the simulator is a list of data and instructions that are to be fed to the PE, along with the timing information that says on what cycles the inputs appear. The simulator models the cycle-by-cycle functional behavior of the datapath. The GNU GMP library is used to model the arbitrary bit widths in the datapath [18].

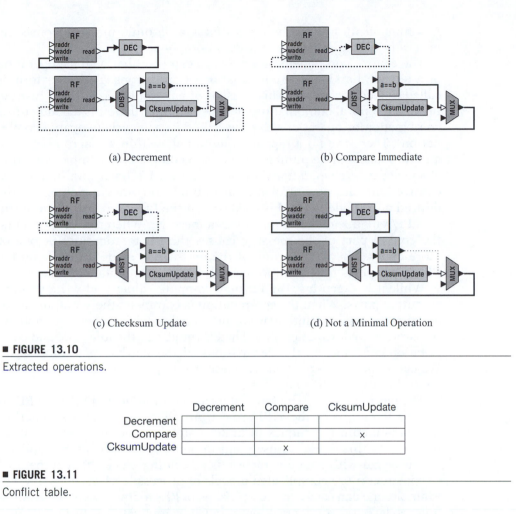

(a) Decrement

(b) Compare Immediate

(c) Checksum Update

(d) Not a Minimal Operation

■ **FIGURE** 13.10

Extracted operations.

	Decrement	Compare	CksumUpdate
Decrement			
Compare			x
CksumUpdate		x	

■ **FIGURE** 13.11

Conflict table.

Performance results for our C++ simulator are given in [19]. On average, the simulator is 10 times faster than an equivalent Verilog simulation using a commercial simulator. If the program trace to be simulated is known a priori, TIPI also provides the option to create a statically scheduled simulator. With this technique we can obtain a speedup of 100 times more than the Verilog simulation. Thus, simulation is not a bottleneck in the iterative design approach even when large programs are used.

13.3.4   TIPI Multiprocessors

At the multiprocessor level, a TIPI PE is modeled as an opaque block as shown in Fig. 13.7. It is an element that can send and receive data and control to and from other PEs. Communication between PEs in a multiprocessor is governed by a *signal-with-data* abstraction.

A communication event between PEs is an atomic message consisting of a control component and a data component.

The control component of a message can be either a single operation or a temporal and spatial combination of operations. The latter form is called a *macro operation*. This consists of an atomic sequence of instructions (a temporal combination), where each instruction is a conflict-free combination of operations to be executed in one cycle (a spatial combination). There are no jumps or conditionals within a macro operation, and they take a finite number of cycles to complete. For implementation efficiency, macro operations are stored in a PE's horizontal microcode control unit and a control message simply provides a pointer to the desired microcode. A RAM is used so that the PE can be reprogrammed.

A PE remains idle until it receives a message. When this event occurs, the control part of the message causes the PE to compute for one or more cycles. The data portion of the message is made available on the PE's input ports for the entire period that the computation occurs.

While it is executing, the PE may compute values and write them to its output ports. When the computation is complete, these output values are bundled together into new atomic signal-with-data messages and are transmitted to destination PEs. The PE returns to the idle state and computation continues at the destinations of the newly created messages. We call these messages *transfers* because they represent the transfer of control from one PE to another.

The mechanism by which transfers are communicated between PEs is outside the scope of the TIPI project. Architects can use best-practices network-on-chip techniques to build a suitable communication infrastructure. In this work, we use point-to-point connections for simplicity.

The signal-with-data abstraction is a building block that can be used to construct more complicated models of process-level concurrency. For example, one can easily create a pipeline of PEs to implement the control-follows-data style of control found in Click applications. In Fig. 13.5(a), Cliff uses handshaking signals to pass control between basic blocks. In a TIPI multiprocessor, we can achieve the same effect by sending push and pull transfers between PEs.

It is also possible to duplicate the behavior of a sequential von Neumann machine using transfers. At the multiprocessor level, one makes a feedback loop between a PE's outputs and inputs so that the PE may send transfers to itself. The PE computes its own next control word every cycle. This is exactly how a RISC processor behaves. In every cycle, in addition to doing an ALU or memory access operation, the processor also updates its program counter. This value is passed through an instruction fetch and decode unit to obtain the control bits that are used in the next cycle.

Fig. 13.12 shows a typical configuration of a TIPI PE in a multiprocessor context. This PE can send transfers to itself as well as to other PEs. An arbitrator is included to merge incoming transfers from multiple

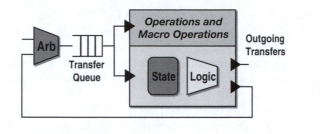

■ **FIGURE 13.12**

Example of a TIPI PE in a multiprocessor context.

sources. The signal-with-data abstraction does not specify priorities or any specific arbitration scheme. The only requirement is that arbitration be fair and that transfers are never dropped. Deadlock is possible in multiprocessor compositions with feedback loops. A queue may be provided to buffer transfers. This is consistent with a globally asynchronous, locally synchronous (GALS) multiprocessor abstraction.

13.3.5 Multiprocessor Simulation and RTL Code Generation

TIPI produces transaction-level simulators for multiprocessor models. This type of simulator captures the bit-true functionality of PEs and the creation and consumption of transfers. The timing details of sending transfers between PEs is not modeled. Like the single-PE case, these simulators are C++ programs that compile and run on the host workstation.

TIPI can also produce a synthesizable Verilog implementation of the multiprocessor system. If a bit-true, cycle-accurate simulation of the entire system is desired, the Verilog can be simulated using third-party tools. Otherwise, the RTL code can be sent directly to a standard ASIC or FPGA flow.

13.4 DEPLOYING APPLICATIONS WITH CAIRN

With TIPI PEs, we can create architectures that are good matches for the application's concurrency. Process-level, data-level, and datatype-level concurrency are all considered. However, the fact that we are building unusual architectures means that we cannot use the usual compilation techniques to deploy application software. TIPI PEs are not standalone elements that execute sequential programs from an instruction memory; they are peers that pass the flow of control among themselves using a signal-with-data abstraction. A C-based programming abstraction will not work for this style of architecture.

Simply programming at the level of TIPI operations is not sufficient. This is micromanaging the datapaths at the pipeline level. Like writing assembly language, this is error prone, difficult to debug, and has poor productivity overall. Designers would much rather work at a higher level, where application concepts like Click push and pull connections are the primary abstractions.

Approaches like NPClick provide a "programming model" that combines architectural concepts with DSL-like features that are intuitive for the application. This is a delicate balance between opacity of unimportant architectural details, visibility of features that are crucial to performance, and understandability to domain experts. For one application and one architecture, this is an acceptable compromise. However, it is not flexible to architectural design space exploration or evolving architectural families (e.g., IXP1200 to IXP2800). Also, it is difficult to add additional abstractions for multifaceted applications.

These shortcomings motivate a more general approach. A fundamental tenet of Cairn is that there is no single abstraction or language that covers all aspects of system design. Instead, designers should have access to different abstractions for different facets of the system.

Cairn is built around not one but three major abstractions, as shown in Fig. 13.13. These abstractions are designed to support a Y-chart-based design methodology. This is shown in Fig. 13.14. Application developers and hardware architects can work in parallel, using separate abstractions for their tasks. The application and the architecture are brought

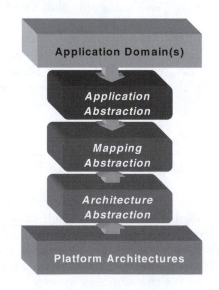

▪ **FIGURE 13.13**

Three abstractions for system design.

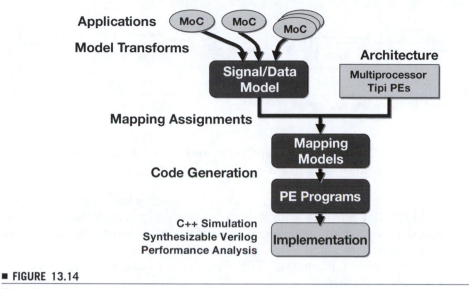

Applications

Model Transforms

Architecture

Mapping Assignments

Code Generation

C++ Simulation
Synthesizable Verilog
Performance Analysis

■ **FIGURE 13.14**

Cairn Y-chart design flow.

together in a mapping step to create an implementation. Here, another abstraction is used to manage the concurrency implementation gap.

On the architecture side, designers use TIPI to construct PEs. TIPI operations capture the capabilities of individual PEs, and the signal-with-data abstraction captures the process-level concurrency of the platform as a whole. In this section, we describe the Cairn application and mapping abstractions.

13.4.1 The Cairn Application Abstraction

Inspired by systems like Ptolemy II, Cairn uses *models of computation* and the concept of *actor-oriented design* as the basis for application abstractions [20]. An actor is a modular component that describes a particular computation, such as a checksum calculation. A model of computation is a set of rules that governs the control and communication within a composition of actors. Entire applications are described by building compositions of actors with different models of computation governing different subgroups of actors. It is important to be able to use different models of computation to model different facets of the application. No single abstraction is likely to be a good choice for the entire application.

To use high-level application models for implementation, and not just early exploration, the models must make a precise statement about the requirements of the application. One problem with the Ptolemy approach is that the actors are written in Java. Actors frequently make use of JVM features that are undesirable to implement on embedded

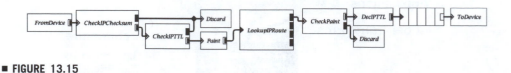

■ FIGURE 13.15

Simplified Click IP forwarding application.

platforms for reasons of efficiency and performance. Also, the Java language cannot strictly enforce the rules imposed by a model of computation.

A solution to this is to use an actor description language such as Cal [21]. Cal provides a formalism for designing actors based on firing rules. However, Cal actors can still use control structures like conditional statements and loops. As described in Section 13.3, "Designing TIPI Processing Elements," TIPI PEs do not necessarily support such control mechanisms.

Therefore, Cairn uses an actor language where the firing rules are restricted to be primitive mathematical functions. More complex computational models, such as general recursion, are built using compositions of actors. The Cairn actor language also supports arbitrary bit types for describing datatype-level concurrency within an actor.

For network processing applications, Cairn implements a model of computation based on Click. In this model, actors are equivalent to Click elements. Interaction between actors uses Click's push and pull style communication semantics. This abstraction is good for modeling the data plane of a networking application.

Cairn provides a graphical editor for assembling actor-oriented models. Just as in the C++ implementation of Click, designers assemble actors from a preconstructed library. To date, Cairn has actors for only a small portion of the Click library. Converting Click elements from C++ to Cairn is straightforward, but an important concern is to correctly model the data-level and datatype-level concurrency of the element. These characteristics are difficult to see in the C++ implementation. However, it is not expected that most users will need to extend the actor library.

A simplified IPv4 routing application model using the Click model of computation is shown in Fig. 13.15. This figure shows one input and one output chain where packets enter and leave a router. Normally, several such chains would be connected to the central route lookup actor. This actor classifies packets according to a routing table and sends them along the proper output chains. The graphical syntax in this model matches that given in the Click documentation.

Designers can perform functional simulation and debugging of the application model as it is being built. Cairn provides an automatic code generation tool that converts the application model into a bit-true C++ simulator that runs on the development workstation. Design space exploration can be done at the application level to make sure that functional requirements are met.

13.4.2 Model Transforms

As in Cal, Cairn uses model transforms to prepare high-level application models for implementation. Model transforms perform syntax checking and type resolution, and check that a composition of actors is valid under the rules of a particular model of computation. The transforms then replace the abstract communication and control semantics with concrete implementations based on transfers.

Push and pull connections are transformed into transfer-passing connections. Each firing rule of each actor is inspected to determine where the flow of control can come from and where the flow of control goes next. This information can be determined statically from the structure of the Click diagram. The model transform makes extensions to the firing rules so that they compute outgoing transfers in addition to performing packet header computations.

For example, in Fig. 13.15, the FromDevice and CheckIPChecksum actor are connected by a push link. The FromDevice actor has one firing rule that receives a packet from off-chip and writes the header onto its output port. After this, computation is supposed to continue by executing the CheckIPChecksum actor's firing rule. FromDevice's firing rule is extended to compute the corresponding control word for CheckIPChecksum. This control word and the header data make up a transfer that will be sent to the PE that implements CheckIPChecksum.

The reappearance of the signal-with-data abstraction at the application level is not a coincidence. The concept of blocks that communicate by passing transfers describes a fundamental capability of the architecture. By describing the application in the same way, we can make a formal comparison between the capabilities of the architecture and the requirements of the application.

13.4.3 Mapping Models

Designers use a mapping abstraction to assign application components onto PEs in the architecture. In Cairn, this is done by making mapping models. Every PE in the architecture has one mapping model. After the appropriate model transform is applied to the application model, the resulting transformed actors are simply assigned to various PEs using a drag-and-drop interface. One-to-one and many-to-one mappings are supported. Mappings are made manually, but in the future an automated optimization process may be used to help find a good assignment [22].

A mapping model is a high-level description of the software that is to run on each PE. The goal is to make the PE behave like the union of the actors that are mapped to it. When a PE receives a transfer, it will begin executing a program that corresponds to one of the firing rules for one of the mapped actors. Since a firing rule is a primitive mathematical function, these programs will be finite-length schedules of TIPI operations without loops or jumps.

Thus, each program can be described as a TIPI macro operation. Programming a TIPI PE is the process of converting actor firing rules into macro operations and placing the resulting machine code into the PE's microcode memory.

13.4.4   Code Generation

Designers do not convert actors into machine code by hand. This would be slow and error prone, making it impossible to perform effective design space exploration.

Instead, an automatic code generation tool is provided. Actor firing rules are processed one at a time. The tool takes as input a netlist of the PE architecture, a netlist representing the firing rule computations, and a total cycle count constraint that bounds the length of the program schedule. A combination of symbolic simulation and Boolean satisfiability is used to find a solution.

A solution is not guaranteed to exist, since TIPI PEs can be highly irregular and simply not capable of doing the specified computations. If this occurs, there is a serious mismatch between the application and the architecture. Designers must explore alternative mappings or try making modifications to the PE architecture.

Once the mapping models are made and code generation is performed, designers have a complete model of the hardware and software of the system. The implementation process now picks up at the "Multiprocessor Composition" block of Fig. 13.8.

The concept of having independent models of the application, the architecture, and the mapping is the core of the Cairn approach. A clear model of the application's requirements and the architecture's capabilities helps designers understand what needs to be done to cross the implementation gap. This comparison also provides insight into how the application or the architecture can be changed to make the implementation gap smaller. Potential improvements can be explored using the feedback paths in the Y-chart. The application and the architecture can be modified separately, and different mapping strategies can be tried and evaluated quickly. Performing this design space exploration is critical to finding an implementation that meets performance goals.

13.5 IPv4 FORWARDING DESIGN EXAMPLE

To demonstrate the capabilities of TIPI architectures, we construct a novel processing element for network processing. The ClickPE architecture takes into account the process-level, data-level, and datatype-level concurrency expressed in Click applications. We will show that hardware support for the application's concurrency can be implemented at low cost. By shrinking the concurrency implementation gap, we can also achieve high performance.

13.5.1 Designing a PE for Click

How does an architect approach the problem of designing a new PE? There are three categories of knowledge that guide the design process.

Application Knowledge

First, we look at Click models like the router in Fig. 13.15 and analyze the computations they perform. The Click model divides packets into headers and bodies in the ingress side. Headers are passed through elements like Paint, CheckPaint, CheckIPChecksum, DecIPTTL, CheckIPTTL, and LookupIPRoute. Headers and bodies are reassembled on the egress side.

The ClickPE should be able to support all of these operations with good performance. In this chapter, we focus on the elements that perform computations on packet headers. We do not consider the problem of queuing packets or packet header/body separation.

Architectural Knowledge

Second, we look at the target implementation platform and see what features are available to construct the ClickPE. We wish to implement our PE on a Xilinx Virtex 2 Pro FPGA. Since this is a configurable platform, the ClickPE should be a customizable element.

One important customization is extensibility. It should be possible to extend the ClickPE to add support for additional Click elements. As a complement to this, the ClickPE should also be reducible. It should be possible to trim it down to support just one or two Click elements with greater performance and smaller area requirements.

Prior Experience

Third, we look at other custom PEs for network processing to see what is possible and what the potential pitfalls are:

- *Longest prefix match computation:* A common observation is that the LookupIPRoute element is a bottleneck. Several architectures have been proposed to speed up this operation. Henriksson and Verbauwhede propose a pipelined architecture based on a 2-bit trie [23]. There, the key to performance is to saturate accesses to the routing table stored in external SRAM. This has the potential to do one route lookup every clock.

Taylor et al. [24] use eight PEs to saturate access to an external routing table. The PEs perform a more complicated routing algorithm that saves significant memory over the trie approach.

The Lulea algorithm [25] is compelling for the ClickPE because it offers several ways to exploit datatype-level concurrency. One operation in this algorithm is a reduction add, which counts the number of set bits in a 16-bit word. Also, the Lulea algorithm can be modified to take advantage of the 18-bit wide BRAM blocks on the Xilinx FPGA, thus permitting a compressed routing table to be stored on-chip.

▪ *Manipulating header fields:* A second observation is that it is good to be able to quickly modify irregularly sized bit fields in packet headers. Sourdis et al. [10] build PEs with header field extraction and modification engines to do this efficiently.

▪ *Custom control flow:* The Linkoping architecture [26] recognizes that standard sequential control flow is not the best match for packet processing. This PE offers an application-specific data-driven control scheme. In one cycle, the PE can perform a field matching operation and compute the next control word for itself.

To guide the integration of all of these ideas, we rely on architectural design patterns [27] and try to follow a minimalist approach.

13.5.2 ClickPE Architecture

The resulting ClickPE architecture is shown in Fig. 13.16. This schematic is the top level of a hierarchical design that matches the hierarchical arrangement of fields in an Ethernet packet header.

At the top level, the datapath is 344 bits wide. This allows the PE to process a complete packet header in parallel. The 344 bit number is the sum of a 20-byte IPv4 header, a 14-byte Ethernet header, and a 9-byte Click header for annotations such as Paint values and timestamps. There is a 344-bit register file at this level, and network-on-chip I/O ports for sending and receiving headers as the data portion of signal-with-data transfers.

The operations at this level of the ClickPE are to break up the header into IPv4, Ethernet, and Click sub-headers and pass these components to SubPEs for further processing. Additional operations provide for

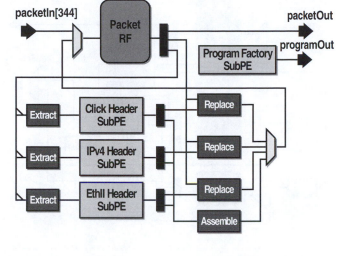

■ **FIGURE 13.16**

ClickPE architecture schematic.

reassembling modified packet fields and storing the result in the packet register file.

The program factory SubPE is used to calculate control words for transmission to other PEs as a transfer. Control can be generated from an immediate value in the instruction word, by a lookup within a small memory, or by a hardware if-then-else operation. Each of the other SubPEs can perform comparison operations to drive the if-then-else operation. These connections are not shown in the diagram for clarity.

The SubPEs do the bulk of the header processing. The IPv4 SubPE is diagrammed in Fig. 13.17; the remaining SubPEs are similar in structure. There is a primary register file for storing entire IPv4 packet headers. The SubPE has operations to break out the fields into smaller register files and reassemble them. This hierarchy of register files allows computation to occur on different parts of the header in parallel.

The IPv4 SubPE also has functional units for doing checksum calculations, testing the values of fields, decrementing fields, and incrementally adjusting the checksum. All of these operations are inspired by the elements in the Click router model. This combination of functional units and register files of various widths supports the datatype-level and data-level concurrency in the application.

13.5.3 ClickPE Control Logic

The ClickPE is also interesting for what it omits. There is no program counter or instruction fetch and decode logic. Instead, the control template shown in Fig. 13.12 is used. The ClickPE can calculate transfers

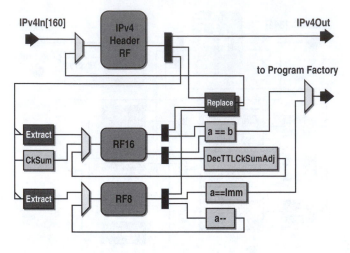

■ **FIGURE 13.17**

IPv4 SubPE architecture schematic.

for itself or it can receive transfers from another PE in the platform architecture.

By omitting the standard RISC control logic, we are simultaneously simplifying the architecture and realizing a model of process-level concurrency that is a better match for the requirements of the application. And still, the ClickPE is a programmable architecture. The programmability lies in how one defines the macro operations that are performed when transfers are received.

13.5.4 LuleaPE Architecture

The LuleaPE is a trimmed-down version of the ClickPE designed to perform the Lulea longest prefix match algorithm. It is shown in Fig. 13.18. This architecture trades off reusability (since it supports fewer Click elements) for higher performance and lesser resource utilization. This PE is meant to perform one stage of the three-stage Lulea algorithm. In Section 13.6 we will implement a pipeline of three LuleaPEs to do the entire algorithm.

The Lulea approach involves performing several table lookups using parts of the IPv4 destination address as indices. In [25], the sizes of these tables are constrained by the byte boundaries of a traditional processor's memory space. The FPGA offers embedded memories that are 18 bits wide; therefore we modify the Lulea algorithm to take advantage of this. Several of the lookup tables can be made smaller as a result, allowing us to store a larger routing table with fewer BRAM blocks. These lookup tables are implemented as functional units directly in

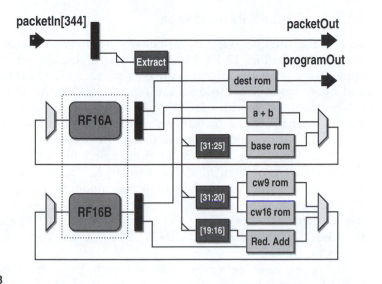

■ **FIGURE 13.18**

LuleaPE architecture schematic.

the PE datapath. They can be accessed concurrently, thereby improving data-level concurrency.

Another modification from the original algorithm is in the reduction add operation. This was originally done using a lookup table. The LuleaPE provides a hardware operation for this, eliminating extra memory references and better matching datatype-level concurrency.

Several of the ClickPE's wide register files are absent in the LuleaPE. This is because the LuleaPE only requires read access to the packet header. No temporary storage for modified headers is required. The LuleaPE reads the header data directly from the packetIn port, where it arrives as the data component of an incoming transfer. This saves extra cycles that the ClickPE needs to move data to SubPEs.

13.6 PERFORMANCE RESULTS

To measure the performance and implementation costs of the ClickPE and LuleaPE architectures, we use TIPI's simulator generation and RTL generation tools. A cycle-accurate simulator is used to measure how fast the PEs can run Click application tasks. The generated Verilog code is synthesized and implemented on a Xilinx Virtex 2 Pro FPGA to determine a feasible clock rate and hardware resource requirements.

13.6.1 ClickPE Performance

To test the ClickPE, we implement an architecture based on the template shown in Fig. 13.12. This represents a single ClickPE as it would appear in a larger multiprocessor configuration. The external incoming and outgoing transfer sources are mapped to FPGA I/O pins. A round-robin arbitrator provides fair access to a 16-deep transfer queue that is shared by the external transfer source and the internal feedback path.

The ClickPE is programmed using the example Click application shown in Fig. 13.15. This application is first modeled using the Cairn design tools. The Click model transform is applied, and the resulting actors are assigned to the PE's mapping model. A straightforward deployment strategy for a multiprocessor router is to use one ClickPE for packet ingress and egress on a single network port. Therefore we assign the CheckIPChecksum, CheckIPTTL, Paint, CheckPaint, and DecIPTTL actors to the PE.

The Cairn code generation tool converts these mapped actors into individual programs that the PE can run. Table 13.1 shows the results of the code generation process. These values are the optimal cycle counts for executing each mapped actor's functionality.

For this experiment, we included additional hardware in the IPv4 SubPE for performing the L1 stage of the Lulea lookup algorithm. This is not shown in Fig. 13.17 for clarity. The lookup tables were sized for a 10,000-entry routing table.

FPGA resource usage is given in Table 13.2. For comparison, a medium-sized FPGA in the Virtex 2 Pro family (2VP50) has 23,616 slices and 232 block RAMs.

TABLE 13.1 ■ Cycle counts for mapped Click elements

Click Element	ClickPE Cycle Count
Paint	5
CheckPaint	4
CheckIPChecksum	12
DecIPTTL	10
CheckIPTTL	5
Lulea L1 Lookup	10

TABLE 13.2 ■ FPGA implementation results

Architecture	Slices	BRAMs	Clock Rate
ClickPE	2841	12	110 MHz
Lulea L1, L2, L3 Pipeline	2487	219	125 MHz

An incoming packet is processed by the Paint, CheckIPChecksum, and CheckIPTTL actors for a total of 22 cycles of computation. An outgoing packet requires 14 cycles for the CheckPaint and DecIPTTL operations. At 110 MHz, the ClickPE can therefore process more than 3 million packets/sec. Assuming minimum-sized 64-byte Ethernet packets with a 64-bit preamble and a 96-bit interframe gap, this corresponds to a line rate of 2 Gbps. This is ideal for a full-duplex Gigabit Ethernet port.

13.6.2 LuleaPE Performance

To test the performance of the LuleaPE, we constructed a pipeline of three PEs to match the three stages of the Lulea lookup algorithm. This is shown in Fig. 13.19. The L1 stage matches the first 16 bits of the IPv4 destination address. The L2 stage matches the next 8 bits, and the L3 stage the last 8 bits. After the Click model transform is applied, each stage is mapped to the corresponding LuleaPE. If a longest prefix match is found in the L1 or L2 PEs, the remaining PEs simply push this result through to the end of the pipeline instead of doing further computation.

The routing tables across the three LuleaPEs together contain 100,000 entries, compared to the 10,000 entries for the ClickPE by itself. Of these 100,000 entries, 10,000 are 16 bits or less, 89,000 are distributed between 17 and 24 bits, and the remaining 10,000 are longer than 24 bits. This distribution mimics that found in a typical core network router, and is much larger than the tables used in NPClick (10,000 entries) and Cliff (a small 12-entry table). The routing tables use 435 KB of on-chip BRAM memory.

Results of the Cairn code generation process are shown in Table 13.3. The L2 and L3 PEs require larger lookup tables than the L1 PE. The extra two cycles these PEs use to process a packet are due to extra pipelining in these large RAMs.

For packets that match prefixes of 16 bits or less, the three-PE pipeline can process up to 10.5 Gbps of traffic. A random mix of incoming packets will result in a processing rate of between 8.4 and 10.5 Gpbs of traffic.

FPGA implementation results for the LuleaPE pipeline are included in Table 13.2. As expected, the LuleaPE is both smaller and faster than the ClickPE. Three LuleaPEs together occupy about the same number

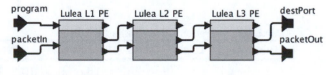

■ **FIGURE 13.19**

Three LuleaPE pipeline for route lookup.

of slices as a single ClickPE. The BRAMs are used exclusively for storing the routing tables. Although the FPGA offers dual-ported BRAMs, we are only using one read port. The second port remains available for making updates to the routing tables.

13.6.3 Performance Comparison

Comparison of these performance numbers to other methodologies is complicated by the differences in the target architectures. Furthermore, we have not constructed a complete stand-alone router with Ethernet interfaces and proper handling of packet bodies. This work focuses on the header processing path. We believe that the same techniques demonstrated in this chapter can be applied to the remaining facets of the IPv4 forwarding application to create a full implementation with good performance results. Supporting additional Click elements on the ClickPE will require making extensions to the architecture.

SMP-Click achieves 493K packets/sec on a dual-processor Xeon Linux workstation. Synchronization and cache coherency between PEs are major bottlenecks in this implementation. With the LuleaPE pipeline multiprocessor architecture, we are able to better match process-level concurrency and greatly exceed this packet rate.

NPClick implements a full 16-port Fast Ethernet router at rates between 880–1360 Mbps on the IXP1200. The ClickPE performs better than the IXP microengines, but the IXP architecture also interfaces with actual Ethernet MACs. The NPClick code spends about 50% of its time moving packet data on and off the IXP. A full Cairn design should surpass this limitation by using more PEs for packet I/O. The raw lookup performance of our LuleaPE pipeline currently matches the lookup rate of a hand-coded design for the IXP2800 [28], so we expect to be competitive with this architecture but with higher designer productivity.

CRACC's DSLAM application requires a much lower packet rate than a core router. It achieves 3 Mbps downstream and 512 Kbps upstream with a single embedded microprocessor.

StepNP can achieve a 10-Gpbs throughput, but this requires 48 ARM processors, each with 16 hardware threads and a 500-MHz clock rate. These hardware requirements are significant, and performance is dependent on network-on-chip latency.

TABLE 13.3 ▪ Performance results for LuleaPE

Lulea Stage	LuleaPE Cycles	Packets/sec	Rate
L1	8	15.6 M	10.5 Gbps
L2	10	12.5 M	8.4 Gbps
L3	10	12.5 M	8.4 Gbps

The SoftMP approach achieves 2.4 Gbps on an FPGA with 14 Micro-Blaze cores at 100 MHz. This design requires 11,000 slices.

Finally, the Cliff router is able to handle 2 Gbps of traffic in about 4,000 slices. This includes the logic to interface to two Gigabit Ethernet MACs, but not the MACs themselves. The comparable Cairn design would have two ClickPEs and the pipeline of three LuleaPEs, requiring 8,200 slices plus additional logic for MAC interfaces. For the extra area, we gain programmability and lookup performance headroom to scale to several more Ethernet ports.

13.6.4 Potentials for Improvement

In both the ClickPE and the LuleaPE, transfers are processed one at a time. This leads to relatively poor utilization of the datapath resources since the lookup tables are only accessed once in each 8- or 10-cycle program schedule. Better performance can be achieved if the PE processes transfers in a pipelined fashion.

The Cairn code generation tool can accept an *initiation interval* constraint to produce schedules that can be pipelined. If we apply the datapath duplication design pattern and include four dual-port register files, four adders, and four reduction adders, we can find a schedule with an initiation interval of only one cycle. The lookup tables are accessed every cycle, as recommended in [23]. This provides 8 times the improvement in throughput (to 125M packets/sec or 84 Gbps) at a small extra cost in hardware and latency.

If the LuleaPE is implemented in an ASIC process instead of on an FPGA, we can expect a further improvement from an increased clock rate. A 1.3-Mb SRAM (the largest lookup table among the three Lulea PEs) should run at 500 MHz in a 90-nm technology. By saturating accesses to this memory, throughput would improve by another 4 times (to 500M packets/sec). Packet data I/O will be the bottleneck at this point. The extra processing capability can be applied to more sophisticated packet-matching techniques.

13.7 CONCLUSION

The concurrency implementation gap is a major impediment to deploying programmable platforms. Architectures that provide poor support for application concurrency requirements make this gap wide; ad hoc methodologies for programming concurrent applications make the gap difficult to cross.

Designers must address both of these issues to achieve good results. First, it is important to choose a basic block processing element that can support application-specific process-level, data-level, and datatype-level concurrency. TIPI sub-RISC PEs are a compelling choice because they provide the right balance between programmability and application specificity. This comes at a lower cost than typical processor architectures in terms of hardware resources and designer hours.

Second, designers must take a disciplined approach to deploying concurrent applications. Cairn provides multiple abstractions for the different facets of the design problem. This makes it easy to cross the implementation gap. Designers can experiment with changes to the application, the architecture, and the mapping individually. Effective design space exploration will lead to high performance.

The performance numbers for our FPGA test case implementation show the advantages of sub-RISC PEs. Implemented in an ASIC, a multiprocessor of ClickPE and LuleaPE elements can easily surpass the IXP2800 in raw packet destination lookup performance. Clearly, these PEs can make excellent building blocks for future programmable platforms.

ACKNOWLEDGMENTS

The TIPI and Cairn frameworks are funded, in part, by the Microelectronics Advanced Research Consortium (MARCO) and Infineon Technologies, and are part of the efforts of the Gigascale Systems Research Center (GSRC).

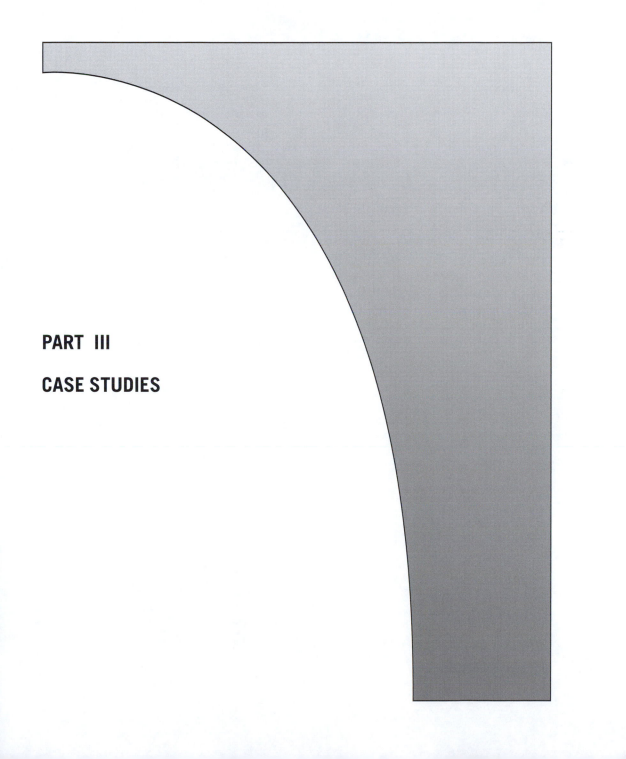

PART III

CASE STUDIES

CHAPTER 14

APPLICATION SPECIFIC INSTRUCTION SET PROCESSOR FOR UMTS-FDD CELL SEARCH

Kimmo Puusaari, Timo Yli-Pietilä, and Kim Rounioja

The wireless terminals of today are capable of rather spectacular tasks: small handheld devices provide users with communication services that seemed unimaginable just 10 years ago. The connectivity with the seemingly omnipresent network is provided by one or more wireless modem engines embedded in the device. These engines are designed to conform to requirements set by the appropriate standard in terms of, for example, performance and processing latencies. For a standard to be competitive against others, the requirements tend to be set according to the upper limit of the capabilities of the contemporary implementation technology. On the other hand, the engine designs have tight constraints in terms of both cost (owing to multimillion production volumes) and power dissipation (owing to the device being operated by a small battery). Consequently, aside from antennas and basic radio frequency processing functions, a conventional wireless modem engine is implemented using digital circuit technology as a system-on-chip consisting of a set of standard-specific, fully optimized, fixed-function accelerators controlled by one or more processors.

Universal mobile telecommunications system (UMTS) is a third-generation cellular system providing terminal users seamless mobile data and voice services. The frequency division duplex (FDD) version of UTMS is widely adopted around the world, with 82 commercial networks in service at the time of this writing [1]. UMTS is one example of the complex communication systems of today: the technical specifications related to radio aspects alone consist of some 164 documents. What is more, the standard keeps evolving and incorporating new functionality: release 7 is currently under work and also the work for the so called long-term evolution or 3.9G has started [2]. In particular, 3.9G is expected to introduce significant new functionality that further pushes the limits of technology capabilities. It is clear that developing and maintaining fixed-function accelerators and control software for such a system requires a significant work effort. Furthermore, UMTS is only one system that a practical terminal of today is required to support: the future terminals will be required to support an increasing number of complex systems.

The capability to develop efficient customized processors (ASIPs) with reasonable effort is one means to enhance the efficiency of the design work. Flexible and power-efficient ASIP modules can incorporate both control and data processing related to a specific function of the wireless system. This chapter describes development of an ASIP for a subfunctionality of a UMTS terminal. The ASIP implements the physical layer functionality necessary for synchronizing the terminal with the cellular network and measuring the basic properties of the radio link. This chapter begins by illustrating some issues related to utilization of ASIPs in wireless modem design. In the subsequent section, the application functionality is described. This is followed by describing the design and verification of the ASIP. The chapter is concluded with a description of the results and a discussion about their significance.

14.1   ASIP ON WIRELESS MODEM DESIGN

More flexibility for modems is required due to the accelerating pace of evolution of the radio standards. Modern mobile terminals incorporate wireless modems for multiple cellular and broadband radio standards. Modem complexity has increased by orders of magnitude since early 2G mobile phones. This is due to several factors: higher order of modulation, need for chip-rate processing, packet radio protocol, support for multiple radios, more complex channel coding, and antenna diversity. The user data rate has increased roughly 10,000 times, from 2G phones to future 3.9G phones [2], while the number of options and modes that each modem supports has increased. Flexibility in modem processing helps to manage this increased complexity and allows late changes to the implemented algorithms. This lowers the risk when critical new algorithms are introduced and allows more flexibility for differentiation. A similar tendency can be seen from the results shown in Fig. 2.2 in Chapter 2. In wireless modem applications, ASIPs have the potential to meet energy and area efficiency requirements while maintaining the flexibility of a general-purpose processor.

14.1.1   The Role of ASIP

Currently, mainstream wireless modem architectures combine standard processors with hardware accelerators. The main driver has been area and energy efficiency. The drawbacks are poor flexibility of the accelerator modules and high processor load related to hardware control. An ASIP solution can get closer to a mobile processor with HW accelerators in computing efficiency than the more general-purpose DSP, SIMD, and VLIW architectures. In Fig. 14.1, the normalized area and energy

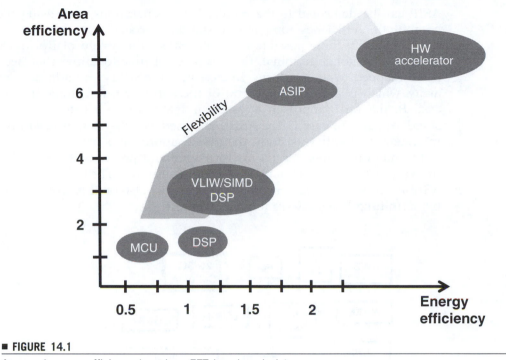

■ FIGURE 14.1
Area and energy efficiency based on FFT benchmark data.

efficiency values are shown for the FFT benchmark. FFT is utilized in modem platforms supporting broadband data rates.

Computation within wireless modems can be divided into control and data functions. Control functions maintain real-time timing of receiver and transmitter chains and interface to higher layers in the communication protocol stack. Typically, the control functions are mapped to a standard processor, which is suitable for control operations such as branching, multitasking, interrupt handling, and external interfacing. Conventionally, control functions are accelerated by selecting a high-performance standard processor. Another solution is distributing the control tasks close to where the actual data processing is carried out. This is an opportunity for an ASIP-like approach. However, distributed control also introduces new design and verification challenges and might compromise system performance.

The performance requirements for data functions in multiradio devices are approaching tera operations per second (TOPS) due to increased user data rates. Data operations are diverse, including, for example, bit manipulation, comparison, real and complex arithmetic, multiplication, division, scaling, and data reordering. Many of the operations are memory bandwidth limited, especially in low-power architectures. An

ASIP can be dedicated to the required data operations, increasing the computation efficiency compared to standard processors.

In Fig. 14.2, three possible roles of ASIPs in a system platform are shown. In a mobile terminal, there is a set of functionalities that need to be mapped to the platform. In case A, the ASIP is a multiprocessor array, which is able to run several of the subsystem functionalities. In case B, the ASIP approach is used for instruction set extension of a standard core. In case C, the ASIPs are used as subsystem accelerators in system ASICs. In this study, the case C approach is used.

In addition to meeting the performance and power budgets, design time is a critical success factor for a mobile terminal. The flexibility of an ASIP-based solution can have a positive impact also on this requirement by permitting late software fixes and reuse.

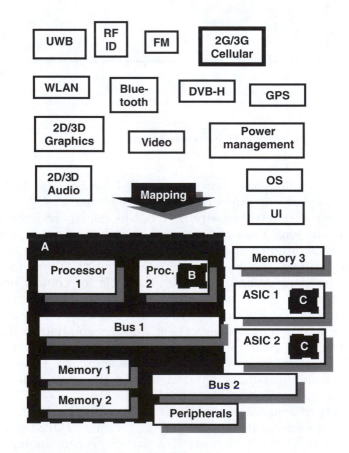

 FIGURE 14.2

Different ways of using ASIPs in a processing platform.

14.1.2 ASIP Challenges for a System House

One challenge is the practical utilization of ASIP in the hardware-software partitioning stage. As illustrated in Fig. 14.2, ASIP is one of the possible technologies that may be utilized in implementing the complete system. Benchmarking is a typical method used in partitioning. For an ASIP, benchmarking requires developing a rather detailed model of the ASIP using a process called design space exploration, which is illustrated in Section 14.4. Design space exploration is an iterative process that requires constant designer interaction. Depending on how much the ASIP toolset can automate the exploration, benchmarking an ASIP typically takes a significant amount of time.

A more advanced form of benchmarking would consider, instead of a single ASIP core, the whole wireless modem system architecture with ASIP models in place. There is a need for system-level architecture exploration to decide the number and type of ASIPs that will provide the best computing efficiency. Single kernel function acceleration can be analyzed without system-level architecture exploration, but it is important to understand the full system performance and the impact of ASIP on that. Architecture exploration with system-level simulation can give insight to the expected system-level performance and how well the timing and power budgets can be met. So the question is not only to accelerate or explore implementation alternatives for single function only, but also to ensure that the entire system operates efficiently, including a memory system, interconnects, a main controller, an interrupt controller, external interfaces, and a DMA.

In wireless modem development, time to market is very tight, as the design work is performed in parallel to standardization. This is a concern, since designing an ASIP takes additional time compared to a conventional solution. In the conventional ASIP design flow, the ASIP is designed to fit the requirements given by the software that it is intended to execute. However, in practice, time-to-market constraints require that the ASIP design begins before the software is complete. When the completed software is available, it is almost certain that the hardware components, including the ASIP, have been frozen. In practice, then, the ASIP architecture design is done based on "prototype" software and estimates. As a result of this uncertainty, the ASIP design carries a significant risk that must be managed. For example, the ASIP might be designed for a significantly greater performance target than required, as increasing the processing power later can be difficult.

Consider a system containing a pool of ASIPs. Each ASIP design is optimized for the tasks it carries out. There may be multiple generations of the system, and each may have different versions of ASIPs (either because of new optimizations or new system features). Consequently, there is a significant volume of different versions of design tools, hardware and software designs, as well as design teams to be maintained.

The overall work effort can become unreasonably high if the situation is not managed carefully.

The software development effort is a significant part of the overall ASIP cost. In practice, it is too expensive to maintain dedicated SW teams with specialists for each ASIP. Instead, the programming interface of the various ASIPs in the system must be made consistent. It is recognized that the standard ISO-C, even with the recently adopted embedded processor extensions, is not as such sufficiently expressive to serve as the programming interface for ASIPs with application-dependent operations and data types.

A typical design team in a system house consists of system, hardware, and software designers. ASIP technology requires new competences from all of them. Applying application-specific processor technology requires rather deep understanding of procedures related to instruction set processor development, which, generally speaking, does not exist in a system house. Specification, design, and verification tasks for an instruction set processor are rather different compared to those for a hardware accelerator, owing to significantly higher complexity of the module. In addition, for the ASIP to be usable in practice, a correctly functioning and efficient compiler is needed. The contemporary ASIP design tools require designer interaction for building a decent compiler; this is another type of competence that generally does not exist in the system house. In general, external expertise may be utilized to circumvent these issues.

Some of the listed challenges are related to immaturity of the ASIP as implementation technology. In any case it is clear that controlling the amount of work effort is critical for the success of ASIP.

14.1.3 Potential ASIP Use Cases in Wireless Receivers

Fig. 14.3 shows the typical wireless receiver functionality with filtering, demodulation, and channel decoding stages. Demodulation and decoding stages are sometimes also referred to as inner and outer receivers, respectively. Each stage has unique processing characteristics and requirements, as already discussed in the UMTS receiver example in Chapter 2. The filtering stage is simple and regular, the demodulation stage includes a wide variety of data and control processing, and the decoding stage includes irregular bit and word processing. The filter stage is practical to implement as a parametrizable HW module, since its flexibility requirements are limited to modifying the filter length and coefficients value. The channel decoding stage requires many read and write operations for bit and word rearrangements—this would suggest software implementation. However, the throughput requirement may be very high and render the software solution too inefficient. In the following, we postulate that the characteristics of the demodulation stage provide good opportunities for ASIP implementation.

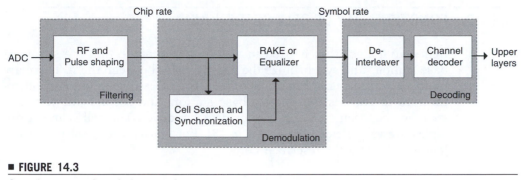

Generic stages of a wireless receiver.

Demodulation consists of various synchronization, channel equalization, and detection functions. A wide variety of data types and word widths is characteristic of the demodulation stage. The basic arithmetic operations for data path are rather simple signal processing functions, such as complex arithmetic, correlation, and integration. Input data rate, or chip rate, is typically in the order of millions of samples per second, while symbol rate is at the range of tens of kilo samples per second.

In contemporary wireless systems, the demodulation stage may incorporate rather diverse functionality. For example, simultaneous reception of multiple channels with different transmission schemes and antenna configurations results in many possible algorithm combinations that need to be supported. Therefore, demodulation is also characterized by control processing. Both data and control processing functions require high computational performance and have tight real-time constraints. For this reason, it may make sense to use modules that incorporate control processing capabilities in addition to data flow processing capabilities suitable for hardware accelerators.

As an example, a wireless demodulation stage includes a detector, such as RAKE or Equalizer, and a cell searcher (Fig. 14.3). An ASIP-based demodulator could consist of a dedicated ASIP for each such function, leading to an ASIP-pool type of system architecture. These ASIPs would be connected to each other and to system control via SoC interconnect. Synchronization and cell search functionalities have very high computation requirements in wireless modems: in the UMTS case, they require several giga operations per second (GOPS). Flexible implementation of synchronization and cell search would enable utilization of the same hardware module for other purposes as well. In this case, embedding all the low-level measurement functionality in the cell search ASIP minus 1pt allows the rest of the system to be in power-save mode, while continuous cell searching and channel measurements are active. In conclusion, investigating ASIP implementation for the cell search functionalities appears justified.

14.2   FUNCTIONALITY OF CELL SEARCH ASIP

Cell search algorithm is used in mobile terminals to find the strongest base stations and synchronize the data transfer between them. There are mainly two use cases. First, when the terminal is turned on, it synchronizes to the nearest (strongest) base station, and second, the terminal switches between cells, i.e., performs a handover. The cell search principles are very clearly presented in [3]. For clarity, the most important features are shortly described here.

14.2.1   Cell Search–Related Channels and Codes

Cell search makes use of three physical channels [3]. Primary and secondary synchronization channels (P-SCH and S-SCH) and the common pilot channel (CPICH) are presented in Fig. 14.4. Also their frame structure and timing is visible in Fig. 14.4. One frame lasts for 10 ms and is divided into 15 slots. Each slot is in turn divided into 10 symbols of 256 chips in length. P-SCH and S-SCH are only transmitted during the first symbol of each slot, whereas the CPICH is transmitted continuously. The cell-specific scrambling code is not used for synchronization channels, because the timing synchronization must be acquired before the actual code can be found. The primary synchronization code (PSC, C_p) transmitted on P-SCH is the same for each cell and slot. The secondary synchronization code (SSC) on S-SCH varies from cell to another ($C_s(i, slot\#)$). Each slot consists of one of the $i = 0\ldots15$ codes. They can form the frame-long sequence in 64 legal ways, as defined in the SSC

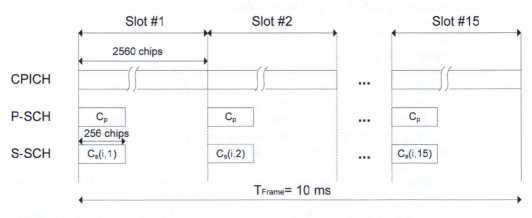

■ **FIGURE 14.4**

The channels related to cell search.

allocation table presented in the 3GPP standard [4]. The 64 sequences identify also the scrambling code (SCRC) groups. Each group includes eight scrambling codes. The codes are used to separate different base stations and different mobile terminals from each other.

14.2.2 Cell Search Functions

The purpose of the PSC search is to find the 256-bit-long PSC defined in [4] from the data stream. The code is the same for both real and imaginary components. By finding the PSC from the input data stream, the slot timing is found (Fig. 14.5). Also, the 256-bit SSCs have identical real and imaginary components. By finding the SSC sequence and the number of slots it has shifted, the frame timing is found (Fig. 14.5). Finally, the SCRC search starts from the acquired frame boundary. The search detects the "identity" of the cell by finding out which one of the eight codes of that group is used in that cell. The primary scrambling code is unique for that cell. The impulse response measurement (IRM) for P-CPICH is done to obtain information about the multipath properties of the channel. The signal can take the direct line or be reflected, for example, from buildings or hills. This is called multipath propagation [3]. The primary CPICH's scrambling code is found as a result of the initial synchronization. This code is then generated locally and correlated against the P-CPICH input data. The correlation should reveal peaks, which correspond to the signals arriving on different time instants. This is the equivalent of the delay profile.

14.2.3 Requirements for the ASIP

The ASIP for cell search (hereafter the Searcher) is required to be able to conduct the PSC, SSC, and SCRC searches and form the channel delay

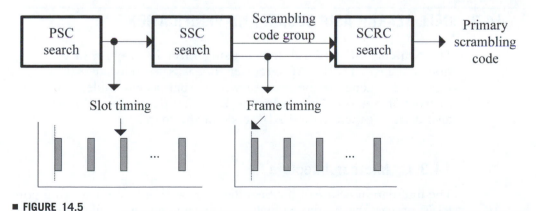

■ **FIGURE 14.5**

Proceeding of the initial synchronization.

profile based on IRM. The standards do not set any specific constraints, in which time, for example, the initial synchronization must complete. However, the input chip rate of 3.84 Mcps must be respected. We examine the case where the 4-bit input samples are additionally oversampled, setting even tighter constraints. Oversampling allows, e.g., more accurate detection of the multipaths during IRM. Of course, the searches (PSC, SSC, SCRC, and IRM) must be configurable, so that it is possible to determine over how many slots PSC search is executed or how many cell candidates are stored to memory, for example.

The samples are read from the memory in 32-bit words containing eight 4-bit words. There are four real and four imaginary samples in that bundle. This allows more efficient memory bandwidth use than separately reading the samples. Some idea of the computation requirements, e.g., in the case of the PSC search, can be obtained with the following reasoning. Within one slot, 2,560 searches must be done, where the 256-bit long PSC code (the taps) are correlated against 256 four-bit samples. Correlation consists of summing the actual or negated value of the 256 samples, depending on each tap. This is done for both real and imaginary samples. Assuming that one sign selection can be done per operation, and the adder tree takes 255 operations, the total computation requirement is in order of 3,900 MOPS. The oversampling doubles this number, which means that the requirement is about 7.8 GOPS of performance just to calculate the correlations.

The Searcher needs to be programmable in C, so a compiler must be configured for the ASIP's instruction set. In an IP block survey [5], it was concluded that the probability of finding an efficient compiler is highest for a RISC processor. Such a processor, combined with execution-accelerating HW that is integrated to the instruction set architecture, sets a solid ground for the searcher's design.

14.3   CELL SEARCH ASIP DESIGN AND VERIFICATION

The Searcher ASIP was designed using LISA 2.0 architecture description language. The LISATek toolset that was introduced in Chapter 2 was used to generate the necessary tools such as assembler, linker, and instruction set simulator from the LISA 2.0 description. LISATek was also used to generate the VHDL code of the searcher.

14.3.1   Microarchitecture

The microarchitecture of the Searcher is based on a 32-bit starting-point RISC processor with three pipeline stages and a register file with 32 registers. Other starting point architectures, such as a VLIW template, were considered, but the C compiler, which was one of the requirements, is likely to be much more efficient for a single-issue processor. The Searcher has a Harvard memory architecture with separate program and data memory. The memory accesses are 32 bits wide, and the data memory has separate read and write ports. The program memory has only a read port. Both memories are word addressable.

The design process was started by identifying the most time-consuming parts of the application. A general cell search C code was used in the application analysis. As expected, performing the matched filtering/correlation operations was the most expensive procedure within cell search. The short reasoning in the previous section about the GOPS figure supports this. Another evident bottleneck was the scrambling code generation that is required for every symbol and code in SCRC search and for every symbol in IRM. Cycling the code generator once would require about 50 ALU operations if executed sequentially, so a huge benefit can be achieved if that is supported on hardware. These were the two main bottlenecks, but other critical loop properties, such as the amplitude calculation, also affected the microarchitecture and instruction set.

Based on these analyses, the RISC processor was modified by adding a special-purpose register (SPR) file for matched filter data lines and tap lines (DataLine file). Also, special function units for matched filtering (MF_corr) and updating and modifying the DataLine file (MF_*) were added. Different amounts of registers in one DataLine were evaluated during architecture exploration (from 16 to 256), but the full 256-long register was the only one giving adequate performance. The scrambling code generator (CG) unit that has two shift registers and fixed XOR-logic as shown in [4] was also added. The ALU unit was enhanced with four instructions for split register instructions (SRI), which meant that the general-purpose register file needed an extra input port to support the two-output SRIs. The properties of the ASIP were fine-tuned during the design space exploration process depicted in Fig. 14.6. The evaluation was mainly done based on the application code cycle counts in the LISATek-generated instruction set simulator (ISS). In this case, the application code was also edited during exploration to make it more suitable for the updated architecture. The final microarchitecture of the Searcher is depicted in Fig. 14.7. The architecture also contains bypassing logic to prevent data hazards and zero overhead hardware loops to enable efficient critical loop calculations (not visible in Fig. 14.7). The introduced instruction set enhancements do not affect the instruction level parallelism. The processor only fetches one instruction per cycle. The extensions are highly application-specific units that are used via single instructions. Therefore, the only (yet extremely significant) parallelism used by the Searcher comes from the bit-level parallelism inside the execution units.

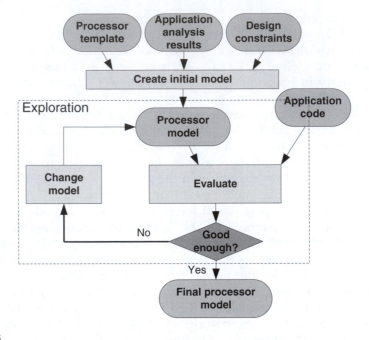

The principle of design space exploration.

14.3.2   Special Function Units

There are three types of application-specific instructions in the Searcher. The matched filter, the scrambling code generator, and the split-register-related instructions and hardware are elaborated in this section.

The functionality of the most important enhancements, the DataLine File and MF_corr decode (DC) units, is depicted in Fig. 14.8. As seen from the Fig. 14.8, the MF calculation is started already in the DC stage of the pipeline. This is possible because MF_corr implicitly reads all the 4-bit registers of the DataLine file, and no delay from addressing is caused, which would be the case in a conventional SIMD or VLIW architecture. The execution is launched with the MF_corr r<destreg#>, <mode>, <parity> instruction. The instruction defines the destination register in GPR file, the *mode* (should I samples be correlated against TapLineI or TapLineQ?) and *parity* (whether the even or odd DataLine registers are used). There are two correlators in parallel in MF_corr (DC) unit, one for real (I) and one for imaginary (Q) samples. During the DC stage, the two correlators read all the DataLine registers and calculate the results that MF_corr (EX) then packs from pipeline registers sregI and sregQ to a single GPR. Split-register instructions are then used to further process the results. Updating the DataLines' registers is a circular buffer process. The oldest value in all DataLines can be replaced

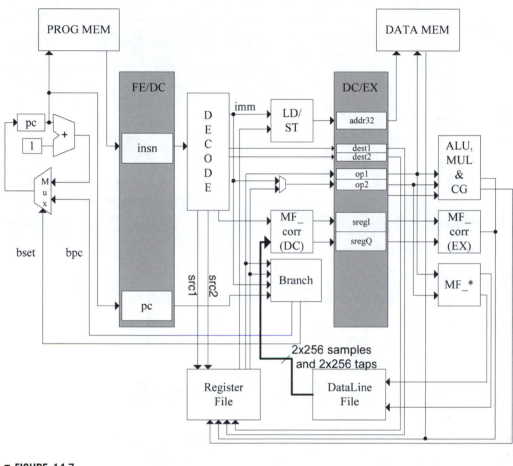

■ FIGURE 14.7

Microarchitecture of the searcher.

with a single instruction. This also means that DataLines are not shifted in filtering, but instead the TapLines are rotated, thus consuming less power.

The scrambling code generator flow diagram is depicted in Fig. 14.9. It more clearly shows the amount of XOR operations required to get one pair of bits. The searcher was extended with a CG unit that can perform the XOR operations for the implicit registers during one cycle. The area of such a unit is minimal compared to the size of the complete Searcher, so the cost-benefit ratio of the unit is favorable. Both I and Q registers also have shadow registers that can store the state of the generator, so it is not always necessary to cycle the registers from the initial values to the desired state. The CG unit has two outputs, one for the I bit and one for the Q bit. They are stored in separate registers, where they can be used to form the 256-bit scrambling code tap sequence.

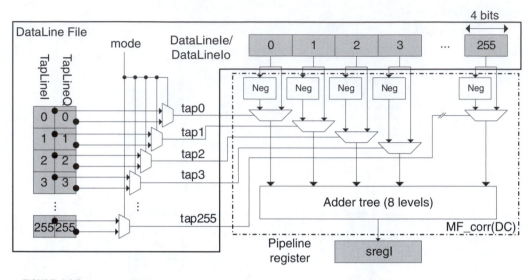

▪ FIGURE 14.8

The functionality of I-half of the MF_corr (DC) unit with DataLine file.

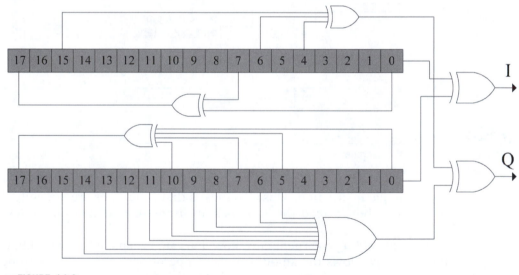

▪ FIGURE 14.9

Flow graph of scrambling code generator. There are two shift registers and several (multi-input) exclusive-or ports. The shift direction is from left to right.

The last group of application-specific instructions is the ALU (SRI), which is used to process both even and odd results from I and Q correlators in parallel. The data flows of the instructions are depicted in Fig. 14.10. The instructions efficiently implement the functionality of

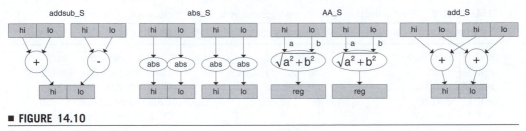

Split-register instructions.

amplitude approximation for the correlation results. Note that the *abs_S* and *AA_S* instructions have two outputs. For example, in the case of *AA_S*, the first output register holds the value of amplitude calculated based on the correlation results from even DataLineIe and DataLineQe. The second register has the amplitude based on DataLineIo and DataLineQo. *Abs_S* finds the absolute values of all input half words and must be used before *AA_S*. *Add_S* is useful, for example, in adding up the I,Q values in coherent memory. *Addsub_S* is used to add and subtract the correlator results to get the real and imaginary values for the amplitude approximation.

14.3.3 Instruction Set

The instruction set of the 32-bit RISC template had regular load/store instructions with base register plus offset indirect addressing. Control instructions make altering the program flow possible by conditional and unconditional branching and using subroutine calls. The template also had a set of common arithmetic-logic and ALU-immediate instructions. The instructions related to the application-specific enhancements were added on top of those. The instruction set was also streamlined with more general additions. The post-increment load and store operations are essential in efficiently loading the samples from memory to DataLines and in storing the amplitude results to memory. *Do* instruction is used for initializing the hardware loops. In total, there are 55 instructions in the Searcher's instruction set. The application-specific instructions are listed in Table 14.1.

It was very important to make the ASIP programmable in C. The template with which the design was started had a preliminary compiler ready that required significant enhancement. Most of the work was done in GUI, but intrinsic and inline assembly mappings were described textually. These mappings were used to directly utilize the application-specific instructions from C. To maximize the ASIP efficiency, the critical loops of the application code were hand-optimized at assembly level. The difference between writing the intrinsics code in C and assembly in this case is small. The biggest issue is that in C, the sources and the destinations are denoted by variables, whereas in assembly, registers are used.

TABLE 14.1 ■ The application-specific instructions and the corresponding C intrinsic

C intrinsic (uint = unsigned int)	Assembly instruction	Used to...
void MF_upd(short)	MF_upd r1	Update datalines
uint MF_corr0()	MF_corr r1, 0, 0	Even, normal correlation
uint MF_corr1()	MF_corr r1, 0, 1	Odd, normal correlation
uint MF_corr2()	MF_corr r1, 1, 0	Even, crossed correlation
uint MF_corr3()	MF_corr r1, 1, 1	Odd, crossed correlation
void MF_rtaps()	MF_rtaps	Rotate TapLines
void MF_utaps(uint, uint)	MF_utaps r1, r2	Update taps
void MF_zeroDI(uint)	MF_zeroDI r1	(Re)set datalines' write pointer
void ld_cg(uint, uint)	ld_cg r1, r2	Init codegen regs
dint_t cycl_cg()	cycl_cg r1, r2	Cycle codegen
void ld_cg_shad()	ld_cg_shad	Load from shadows
void st_cg_shad()	st_cg_shad	Store to shadows
dint_t addsub_S(uint,uint)	addsub_S r1, r0 = r1, r2	SRI add and substract
dint_t add_S(uint,uint)	add_S r1, r0 = r1, r2	SRI add
dint_t abs_S(uint,uint)	abs_S r1, r2 = r1, r2	SRI absolute value
dint_t AA_S(uint,uint)	AA_S r1, r2 = r1, r2	SRI amplitude approximation
int read_pin()	read_pin r1	Read input pin value
void set_ready_pin()	set_ready_pin	Set output pin to '1'
void clear_ready_pin()	clear_ready_pin	Set output pin to '0'

14.3.4   HDL Generation

The synthesizable VHDL code of the Searcher was automatically generated by the LISATek toolset. In LISA language, the operations are grouped to execution units using the "UNIT" keyword. The principle in the grouping is that all the operations within a group are executed in the same stage of the pipeline. It is also advantageous if the inputs and outputs are similar for all operations of the group. Additionally the naming of the generated VHDL files is based on the UNIT labels.

The VHDL generated from the Searcher LISA model contained 29 files, with one file for each UNIT. A file was also generated for the general-purpose register file, which also includes all the special-purpose registers, such as the DataLineIe. This is not desired behavior, because it makes the architecture of the register file unnecessarily complex, due to too-complex addressing logic. It would be better to have DataLines, TapLines, the code generator's Ireg and Qreg, and other SPRs separated from the GPR file and make them locally accessible by their respective units. This will be possible in the updated version of the tool in the future. Other generated files comprise files that combine the VHDL blocks into a complete model. Depending on the code generation configuration, the simulation models of the memories were also generated. On top of everything, a test bench was generated to wrap the model. The 29 files contained about 80,000 lines of VHDL code in them. The

Verilog code was generated for reference, and it had about 1,000 more lines.

The inability of the HDL generator tool to separate SPR registers from the general-purpose register (GPR) file led to two fairly complicated problems. First, it meant that the DataLines were treated as regular registers, and thus the same addressing logic for read and write accesses was created for each of them. For write accesses this was acceptable, because only one DataLine address (pointed to by DataIndex) is written to at a time. However, during the execution of MF_corr instruction, all registers from two of the DataLines are read. The separate addressing logic for many such registers is tremendously complex. This basically explained the tens of thousands of code lines. Naturally, the effect is also negative for the operations using the general registers. The second problem is tool related. Such a large GPR file jams the synthesis program. Thus, all the references to DataLines and TapLines were removed from the GPR file block, and a new block that includes them, DataLineFile, was manually coded. DataLineFile has more optimized enable signals and addressing that selects all registers from either the even or odd DataLines, based on the *parity* bit (see Section 14.3.2) input only. These changes also required quite extensive changes to the signal names in other matched filter related blocks, i.e., UNITs *MF_corr (DC)*, *MF_corr (EX)*, and *MF_**. In addition, the input and output signals and port maps were changed accordingly in the higher hierarchical blocks *DC* (encapsulates the decode stage of the pipeline), *Pipe* (encapsulates the whole pipeline), and *SEA* (encapsulates the whole Searcher). After these changes, the amount of code lines in VHDL files was down to about 38,000.

14.3.5 Verification

The verification of the Searcher had essentially three main steps. First, the correct functionality of the LISA model and application SW was verified using the LISATek-generated instruction set simulator. The Searcher was fed with input data generated with the software developed in [6]. It was shown that Searcher can find the initial synchronization and form the delay profile. The pair of graphs in Fig. 14.11 shows how the PSC search and IRM functions were able to locate the peaks of the two possible base station cells and the delays of the three multipaths, respectively. The multipaths are actually faintly visible already in the PSC graph. The results of all the cell search functions match with the output of the reference simulator also implemented as a part of [6].

Second, after the VHDL code was generated and manually modified, it was verified in a standard simulator, using the same application code as in ISS verification. The results and memory images matched the ISS results, and the cycle counts were also the same. The register states during code execution were also matched. The fact that the VHDL model is

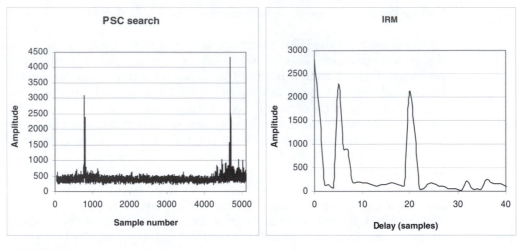

■ **FIGURE 14.11**

Results for PSC search and IRM.

an exact match with the LISA model is very important, because it makes it possible to develop software in ISS, where the simulation speeds are much higher than in a VHDL simulator. This is also helpful for a system house, in case the processor IPs are not made by us, because fewer licenses are needed.

Third, to further ensure the correct functionality of the hand-modified parts of the VHDL code, some dynamic assertion-based verification (ABV) methods were used. Assertions include the expected behavior and compare the actual results to it. If they don't match, a notification is given during VHDL simulation. Assertions were active during the application tests, and some special test code was also written to get better coverage.

14.4   RESULTS

This section represents the results for the searcher ASIP. The performance is given in terms of cycle counts from the ISS, from which the execution time can easily be derived if the clock frequency is known. This information is provided by the delay calculations of the synthesis tool, which also gives the area in logical gates and the power dissipation figures.

14.4.1   Performance

The cycle counts for the cell search functions are listed in Table 14.2. The middle column shows for how long each function was run. The values selected for the table are common for the functions. In the PSC search case, the cycle count consists mostly of the correlation loop. The maximum value search and memory arranging take about 12% of the time. In the SSC search, the correlations take only 15% of the time, and the most time-consuming phase is the finding of the right group when comparing the sequences of Appendix 1. In the SCRC search, the cycle count depends on the scrambling code number. In this case, the code number was 1184 (minimum 0, maximum 8171). The same applies for the IRM.

14.4.2 Synthesis Results

The VHDL code of the Searcher was synthesized with 90-nm technology libraries using a standard synthesis tool. The tool-generated scripts were used, with some exceptions. For example, the library references were modified to point to correct libraries and some constraints, such as the modified system clock. The synthesis results for the VHDL code are introduced in Table 14.3. A system clock of 10 ns was used.

The largest block in the Searcher is the DataLineFile, mostly because of its registers. The MF_corr(DC) is the next biggest block with about 30 kgates. The size of the Searcher could be reduced by removing half of the GPRs (reduced by about 13 kgates) and removing the multiplier (3 kgates) without much affecting the performance. If the size

TABLE 14.2 ■ Cycle counts of the cell search functions

Function	Executed over...	Cycle Count
PSC search	1 frame (15 slots)	583,041
SSC search	1 frame	270,765
SCRC search	8 symbols	123,440
IRM	8 symbols	107,884

TABLE 14.3 ■ Synthesis results

Property	Value
Gate Count	165,500 gates
DataLineFile	90,100 gates
MF_corr(DC)	29,900 gates
GPR File	26,400 gates
Other	19,100 gates
System Clock	100 MHz (10 ns)
Power Consumption	70 mW

of the program memory (5 KB of read-only memory) is added to the core gate count, the total gate count nears 200,000. The power estimate is calculated using the execution profiles. The figure was calculated assuming that 10% of the executions are MF_corr instructions (the most power-consuming) and that fetch, decode, and execute stages are always active at the same time. The register file was assumed active in 90% of the cycles.

Using the 10-ns system clock, the cycle count available for processing one frame (10 ms) is 1,000,000. When comparing this to the values in Table 14.3, it can be seen that the PSC search can clearly process the samples in real time. Three SSC searches could be run within one frame's time. SCRC and IRM don't quite meet the requirement. However, that is not critical, because SCRC can utilize the time saved by PSC and SSC (the search is done via memory, not for streaming samples), and it is unlikely that the Searcher will be required to be active all the time. If the first three cycle counts from Table 14.3 are added up, it can be seen that the whole initial synchronization can be executed in 10 ms.

The cycle counts and synthesis results can also be compared to other existing implementations. In general, the fixed ASICs have their performance mapped to the input sample rate, which means that they can process the sample stream in real time. The same real-time performance was achieved clearly with Searcher ASIP, as shown by Table 14.3. When the gate counts are compared, the ASIC implementations are usually smaller. In [7] some digital matched filter ASICs are listed with gate counts ranging from 54 to 79 kilogates. The matched filter-related structures in the Searcher consist of DataLineFile and MF_corr, which have a total of 120 kilogates. The hybrid Golay-MF in [8] has only 42.5 kilogates, and it performs the PSC, SSC, and SCRC searches. It is only a quarter of the Searcher's gate count of 165,500. For flexibility and hardware reuse reasons, Golay implementations were not possible for the ASIP. It must also be noted that in addition to PSC, SSC, and SCRC searches, the Searcher can also perform the IRM, which was not considered in the previously mentioned ASICs.

For another reference, an out-of-the-box C code (written using only ANSI-C) for PSC search was compiled and run using the ISS of Texas Instruments' TMS320C64xx eight-issue DSP. The cycle count for the PSC search over two slots was about 6,660,000. If multiplied with 7.5 (one frame time), the count is about 50 million cycles. If TMS320C64xx is run at 600 MHz (compared to 100 MHz of the searcher ASIP), the searcher still consumes only 7% of the TMS320C64xx's elapsed time. The C code was not specifically optimized for TMS320C64xx. Therefore, it is expected that the difference could be smaller if more effort was put into coding. However, this comparison and the ASIC references show that the results obtained in this work are aligned with Fig. 14.1.

14.5 SUMMARY AND CONCLUSIONS

Traditionally the functionality of wireless modems has been implemented using a control processor and a set of HW accelerators. As evolving radio standards set ever higher requirements to the modems, flexibility and programmability become essential. The area/power efficiency of the contemporary DSP processors is not at a desired level, so the gap between them and HW accelerators must be filled. A new player, an application specific instruction set processor (ASIP), is capable of this task. It can combine programmability and high performance. ASIPs also bring new challenges to a system house. For example, more SW effort is needed, and a new competence, compiler expertise, must be gained. It was shown that the use cases for ASIP can be found in the demodulator of a wireless receiver.

Therefore, an application-specific instruction set processor for UMTS-FDD Cell Search and impulse response measurements was implemented. The implemented Searcher ASIP is capable of processing the samples in real time and can be used in mobile terminals. The ASIP model was designed using LISATek tools. The model is based on a 32-bit RISC starting-point architecture and instruction set that was extended and enhanced with application-specific functional units and instructions. The most important addition included the matched filter data line registers and logic that were necessary to boost the Searcher's processing capabilities to the real-time level. Software development tools, i.e., assembler, linker, and instruction set simulator, were generated from the LISA model. VHDL code was also generated from the model. Some manual input was needed during C compiler configuration and VHDL code optimization.

The Cell Search application software for Searcher ASIP was written in C language, and the critical loops were then hand optimized in assembly. The quality of the compiled code was very good for general C code, but intrinsics caused redundant nops in the critical loops, which justified hand-optimizing the assembly. The functionality of the application code and the ASIP was verified using the instruction set simulator and VHDL simulations. The size of the application code in the program memory is about 5 kilobytes. A synthesis was performed for the VHDL model, giving the gate count of 165,500 at 100 MHz without memories. Most of the gate count is due to the matched filter registers. The comparison of the results showed that performance is equal to that of contemporary ASICs, but the gate count of the ASIP is clearly larger. DSP's performance was clearly topped by the Searcher.

Generally speaking, area and energy efficiency of ASIP implementations seem good compared to most processor architectures with various levels of parallelism. One challenge is the nonstandard programming interface of the ASIP, which can lead to difficulties while porting the code for the ASIP from an existing system. The quality of the C-compiler is vital. However, manual code optimization is usually required to achieve high computation efficiency. This is not a big problem for some applications, but code portability is an issue in most cases. New design rounds lead to scalability and code reuse issues when, for example, higher data rates or added functionality are required. It is important to support reliable version handling of previous versions of the tools. The development tools and the programming model will play imperative roles when the way is paved for the wider usage of ASIPs.

CHAPTER 15

HARDWARE/SOFTWARE TRADEOFFS FOR ADVANCED 3G CHANNEL DECODING

Daniel Schmidt and Norbert Wehn

In parts I and II of this book the main goals, problems, and motivational issues of using customizable embedded processors have been discussed thoroughly. Key phrases like Moore's Law, the "energy/flexibility gap," the demand of the market for short development times and ever-growing design productivity, and the like are known very well. But while many methodologies, like the automatic instruction-set extensions presented in Chapter 7, are still in research status, how can customizable processors be used in today's demanding wireless communications applications?

In this chapter we will describe in more detail the implementation of an important building block in the baseband processing of wireless communication systems, the channel decoder. We will show how to efficiently map the algorithms onto an application-specific multiprocessor platform, composed of low-complexity application-specific processor cores connected by a network on chip (NoC). We extend a Tensilica Xtensa base core with custom application-specific instructions and show that it clearly outperforms DSPs or general-purpose computing engines while still offering the flexibility of a programmable solution. We also show how these processing nodes can be connected by a custom tailored communication network to form a *highly parallel and scalable application-specific multiprocessor platform*. This platform allows us to run the processors at relatively low clock frequencies and, furthermore, to improve energy efficiency by using voltage scaling. This case study shows that ASIPs are clearly able to address a lot of today's design issues but also identifies problems and limitations of the available methodologies.

15.1 CHANNEL DECODING FOR 3G SYSTEMS AND BEYOND

In digital communications bandwidth and transmission power are critical resources. Thus, advanced wireless communications systems have to rely on sophisticated *forward error correction* (FEC) schemes. FEC allows for a reduction of transmission power while maintaining the quality of service (QoS) or, vice versa, to improve the QoS for a given transmission power. Shannon proved in his pioneering work [1] the noisy channel coding theorem, which gives a lower bound on the minimum bit energy to noise spectral density necessary to achieve reliable communication. However, he gave no hint on the structure of encoding and decoding schemes approaching this so-called Shannon limit. During the following more than four decades, researchers have been trying to find such codes. In 1993, a giant leap was made toward reaching this goal, when Berrou published a paper entitled *Near Shannon Limit Error Correcting Coding and Decoding: Turbo-Codes* [2], which could be considered a revolution in this field. Turbo-Codes approach Shannon's limit by less than 1 dB. The outstanding forward error correction provided by these codes made them part of 3G wireless standards like UMTS and CDMA2000, the WLAN standard, and satellite communications standards [3], to name a few. However, this improvement in communications performance comes at the expense of an increase in complexity. For example, the computational complexity of this new type of channel decoding in the 3G standard increased by an order of magnitude compared to the channel coding scheme used in the 2G standard.

Berrou's important innovation was the introduction of iterative decoding schemes of convolutional codes by means of soft information. This soft information represents the confidence of the decoder in its decision. The design of Turbo-Codes is highly demanding from a communication theory point of view. But turbo encoding is, from an implementation perspective, a low-complexity task, with the encoder mainly comprising some shift registers and an address generator unit for interleaving. However, the iterative nature of turbo decoding implies big implementation challenges. Hence, efficient implementations of turbo decoders in hard- and software have become a very active research area. The system and implementation design space is huge and has been efficiently explored on multiple abstraction levels, to yield an optimum decoder architecture [4]. On the system and algorithmic level, the designer is faced with the relations between communications performance—a typical metric in this application is the bit error rate (BER) versus signal to noise ratio (SNR)—and implementation performance (throughput, energy dissipation, flexibility, area, and code size in the case of programmable solutions). Hence, prior to any detailed implementation steps, major steps on the system and algorithmic level have to be carried out. These are detailed algorithm analysis with respect to computational complexity, data transfer characteristics, storage demands, algorithmic transformations, evaluation of the impact of quantization, iteration control, and so forth. On the architectural level, key to an optimized decoder are the selection of the appropriate level of parallelism and optimal memory/computation tradeoffs.

15.1.1 Turbo-Codes

This chapter briefly introduces Turbo-Codes. FEC is typically done by introducing parity bits. In Turbo-Codes, the original information ($\vec{x}^s$), denoted as *systematic information*, is transmitted together with the parity information ($\vec{x}^{1p}$, $\vec{x}^{2p}$). To increase the code rate, i.e., the ratio of sent bits to systematic information bits, a puncturer unit can be employed to remove (puncture out) a number of the parity bits before combining the systematic and parity information to the final encoded stream—see Fig. 15.1(a). The choice of which of the parity bits to transmit and which ones to omit is very critical and has a strong influence on the communications performance of the code. For the 3GPP standard [5], the encoder consists of two recursive systematic convolutional (RSC) encoders with constraint length $K = 4$, which can also be interpreted as

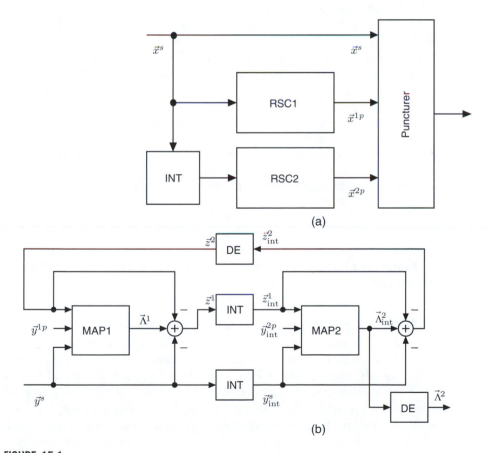

(a)

(b)

■ **FIGURE 15.1**

Turbo encoder and turbo decoder.

8-state finite state machines. The code rate is 1/2. One RSC encoder works on the block of information in its original, the other one in an interleaved order. On the receiver side, one corresponding component decoder for each of them exists. Interleaving is scrambling the processing order of the bits inside a block to break up neighborhood relations. The quality of the "randomness" of the interleaving strongly influences the communications performance. Due to this interleaving, the two generated parity sequences can be assumed to be uncorrelated.

The *maximum a posteriori* (MAP) decoder has been recognized as the component decoder of choice, as it is superior to the *soft output Viterbi algorithm* (SOVA) in terms of communications performance and implementation scalability [6]. The soft output of each component decoder ($\vec{\Lambda}$) is modified (subtracting the soft input and systematic information) to reflect only its own confidence ($\vec{z}$) in the received information bit being sent as either "0" or "1." These confidence values are exchanged between the decoders to bias the estimation in the next iteration—see Fig. 15.1(b) and Algorithm 15.1. During this exchange, the produced information is interleaved and deinterleaved respectively (INT, DE), following the same scheme as in the encoder. The iteration continues until a stop criterion is fulfilled, see [7]. The last soft output is not modified and becomes the soft output of the turbo decoder ($\vec{\Lambda}^2$). Its sign represents the 0/1 decision, and its magnitude represents the confidence of the turbo decoder in its decision.

Given the received samples of systematic and parity bits (*channel values*) for the whole block (Y_1^N, where N is the block length), the MAP algorithm computes the probability for each bit to have been sent as $d_k = 0$ or $d_k = 1$, with $k \in \{1 \ldots N\}$. The *logarithmic likelihood ratio* (LLR) of these probabilities is the soft output, denoted as:

$$\Lambda_k = \log \frac{\Pr\{d_k = 1 | Y_1^N\}}{\Pr\{d_k = 0 | Y_1^N\}}. \tag{15.1}$$

Equation 15.1 can be expressed using three probabilities, which refer to the encoder states S_m^k with $m, m' \in \{1 \ldots 8\}$:

▪ The *branch metrics* $\gamma_{m,\,m'}^{k-1,\,k}(d_k)$ calculate the probability that a transition between states S_m^{k-1} and $S_{m'}^k$ in two successive time steps $k-1$ and k took place. These metrics are derived from the received signals, the a priori information given by the previous decoder, the code structure, and the assumption of $d_k = 0$ or $d_k = 1$. Note that for a given m, $\gamma_{m,\,m}^{k-1,\,k}(d_k) = 0$ holds for all but two m', because to each state there exist exactly two possible successor states and two possible predecessor states. Consequently, there exist exactly two transitions from and to each state, one for $d_k = 0$ and one for $d_k = 1$ respectively. For a detailed description see [8].

- From these branch metrics, the probabilities α_m^k that the encoder reached state S_m^k given the initial state and the received sequence Y_1^k are computed through a forward recursion:

$$\alpha_m^k = \sum_{m'} \alpha_{m'}^{k-1} \cdot \gamma_{m', m}^{k-1, k} \qquad (15.2)$$

- Performing a backward recursion yields the probability β_m^{k-1} that the encoder has reached the (known) final state, given the state S_m^{k-1} and the remainder of the received sequence Y_k^N:

$$\beta_m^{k-1} = \sum_{m'} \gamma_{m, m'}^{k-1, k} \cdot \beta_{m'}^k \qquad (15.3)$$

Both the αs and βs are called *state metrics*.

Using these, Equation 15.1 can be rewritten as:

$$\Lambda_k = \log \frac{\sum_m \sum_{m'} \alpha_m^{k-1} \cdot \beta_{m'}^k \cdot \gamma_{m, m'}^{k-1, k}(d_k = 1)}{\sum_m \sum_{m'} \alpha_m^{k-1} \cdot \beta_{m'}^k \cdot \gamma_{m, m'}^{k-1, k}(d_k = 0)}. \qquad (15.4)$$

Algorithm 15.1 and Algorithm 15.2 summarize the turbo decoding and the MAP algorithm. The original probability-based formulation implies many multiplications and has thus been ported to the logarithmic domain resulting in the *Log-MAP algorithm* [8]. In the logarithmic domain, multiplications turn into additions and additions into maximum selections with additional correction terms denoted as max* operation:

$$\max{}^*(\delta_1, \delta_2) = \max(\delta_1, \delta_2) + ln(1 + e^{-|\delta_2 - \delta_1|}). \qquad (15.5)$$

This logarithmic transformation does not affect the communications performance. As all but two of the $\gamma_{m', m}^{k-1, k}$ in Equation 15.2 equal 0, the sum reduces to a single addition. Hence, one max* operation suffices to calculate a forward recursion. The same holds for Equation 15.3, accordingly. Arithmetic complexity can further be reduced by omitting the correction term (*Max-Log-MAP algorithm*), which leads to a slight loss in communications performance (about 0.1–0.2dB). The Log-MAP and Max-Log-MAP algorithms are common practice in state-of-the-art implementations.

Algorithm 15.1 Turbo Decoder

{Constants:}
{N: block length}
{Input:}
{$y^s(k); k \in \{1 \ldots N\}$: systematic channel values}
{$y^{1p}(k)$, $y_{int}^{2p}(k); k \in \{1 \ldots N\}$: parity channel values}
{Output:}
{$\Lambda^2(k); k \in \{1 \ldots N\}$: soft outputs of turbo decoder}
{Variables:}
{y_{int}^s: interleaved systematic channel values}
{$z^1(k)$, $z^2(k)$, $z_{int}^1(k)$, $z_{int}^2(k); k \in \{1 \ldots N\}$: soft outputs of MAP decoder}

for $k = 1$ to N **do**
$\quad z^2(k) = 0$
end for

repeat
$\quad \vec{z}^1 = \text{MAP\_decoder}(\vec{y}^s, \vec{y}^{1p}, \vec{z}^2)$
$\quad \vec{z}_{int}^1 = \text{Interleave}(\vec{z}^1)$
$\quad \vec{y}_{int}^s = \text{Interleave}(\vec{y}^s)$
$\quad \vec{z}_{int}^2 = \text{MAP\_decoder}(\vec{y}_{int}^s, \vec{y}_{int}^{2p}, \vec{z}_{int}^1)$
$\quad \vec{z}^2 = \text{Deinterleave}(\vec{z}_{int}^2)$
until stopping criterion fulfilled
$\vec{\Lambda}^2 = \vec{y}^s + \vec{z}_{int}^1 + \vec{z}_{int}^2$

15.2 DESIGN SPACE

It is well known that optimizations on system and algorithmic levels have the largest effect on implementation performance, but they also affect communications performance. Therefore, an exploration of the interrelation of different system/algorithmic transformations and their tradeoffs between communications and implementation performance is necessary (e.g., see the Log-MAP simplification in the previous section). However, due to limited space, we omit a detailed discussion of this issue and emphasize the implementation aspect. As already mentioned, key to an efficient implementation is an in-depth exploration of the parallelism inherent to the algorithm.

As can be seen from the algorithms, a turbo decoder offers potential for parallelization on various architectural levels [9]. On the *system level*, multiple decoder units may be used in parallel. However, this approach implies a large latency and low architectural efficiency. On the level below, the *turbo decoder level*, three distinct architectural layers

Algorithm 15.2 MAP Decoder

{Constants:}
{M: number of states}
{N: block length}
{Input:}
{$\vec{y}^s$, $\vec{y}^p$, $\vec{z}$: systematic and parity channel values, extrinsic information}
{Output:}
{$\Lambda_{\text{out}}(k); k \in \{1\ldots N\}$: produced likelihood values}
{Variables:}
{$\alpha_m(k); m \in \{1\ldots M\}$, $k \in \{0\ldots N\}$): forward state metrics}
{$\beta_m^{\text{predecessor}}$, $\beta_m^{\text{current}}; m \in \{1\ldots M\}$: backward state metrics}
{$\Lambda_{\text{in}}(k) = (\Lambda(y^s(k)), \Lambda(y^p(k)), \Lambda(z(k)))$: consumed likelihood values, where $\Lambda()$ is the transformation from probabilities to metrics}

$\alpha_1(0) = 0$, $\beta_1^{\text{current}} = 0$
for $m = 2$ to M **do**
    $\alpha_m(0) = -\infty$, $\beta_m^{\text{current}} = -\infty$
end for

{forward recursion}
for $k = 1$ to $N - 1$ **do**
    for $m = 1$ to M **do**
        {Let states i and j be the predecessors of state m}
        $\alpha_m(k) = recursion\_step\left(\alpha_i(k-1),\ \alpha_j(k-1),\ \Lambda_{\text{in}}(k)\right)$
    end for
end for
for $k = N$ to 1 **do**
    {backward recursion}
    for $m = 1$ to M **do**
        {Let states i and j be the successors of state m}
        $\beta_m^{\text{predecessor}} = recursion\_step\left(\beta_i^{\text{current}},\ \beta_j^{\text{current}},\ \Lambda_{\text{in}}(k)\right)$
    end for

    $\Lambda_{\text{out}}(k) = softout\_calculation(\alpha_1(k-1), \ldots, \alpha_M(k-1),$
    $\beta_1^{\text{current}}, \ldots, \beta_M^{\text{current}}, \Lambda_{\text{in}}(k))$
    for $m = 1$ to M **do**
        $\beta_m^{\text{current}} = \beta_m^{\text{predecessor}}$
    end for
end for

exist: the *I/O interface layer*, the *MAP decoder layer*, and the *interleaver network layer*. The MAP decoder layer is subdivided into different processing levels. On the *window level*, the data dependencies in the state metrics can be broken. This allows us to subdivide the data block into subblocks—called windows—that can be processed independently [10]. However, some acquisition steps are necessary at the borders of each window, which slightly increases the computational complexity. Windowing can be exploited in two ways. First, by sequentially processing different windows on the same MAP unit, the memory for the state metric values can be reused, thus reducing memory needs from block size to window size. Second, windowing also allows for a partially or fully overlapped processing of different windows using several MAP units in parallel, thus increasing throughput. Below the window level is the *recursion level*. State metrics for forward and backward recursion can be calculated in parallel or sequentially. Further parallelization is possible by unrolling and pipelining the recursion loop. Also, the max* operation can be done in parallel for several encoder states, yielding the so-called butterfly operation. To do this, the state metrics updates have to be reordered or scheduled in a way that the update parameters can be reused. This is possible because the encoder states can be grouped to pairs with a common pair of predecessor states. Through this reuse of parameters, the butterfly operation is able to calculate the max* operation for two states in parallel while cutting down memory accesses by 50% at the same time. On the *operator level* the max* calculation is composed of basic operations like additions and comparisons. The challenge is to trade off the parallelism on the various architectural levels with respect to implementation style and throughput requirements.

15.3 PROGRAMMABLE SOLUTIONS

The importance of programmable solutions for the previously mentioned algorithms was already stated in [11]: "It is critical for next generation of programmable DSPs to address the requirements of algorithms such as MAP or SOVA, since these algorithms are essential for improved 2G and 3G wireless communications." State-of-the-art DSPs offer increased instruction level parallelism by grouping several instructions to so-called very long instruction words (VLIW). Furthermore, the processing units usually support the single instruction/multiple data approach (SIMD). Hence, we are restricted to this type of parallelism and have to use the lower architectural turbo decoder levels to speed up processing.

The performance-critical parts of Algorithm 15.2 are the `recursion_step` and the `softout_calculation` functions. The `recursion_step` function is called $2 \cdot M$ times for each bit and comprises one max* operation per call (see Equation 15.2, Equation 15.3, and Equation 15.5). The `softout_calculation` function is only called once for each decoded bit; however, it comprises $2 \cdot (M - 1)$ max* operations per call (see Equation 15.4 and Equation 15.5). This results in a total of $(4 \cdot M) - 2$ max* operations per bit for each pass of the Log-MAP algorithm. Besides its computational complexity, the MAP algorithm contains a huge amount of data transfers. Although pairs of subsequent `recursion_step` function calls use a common set of parameters and can be grouped together, the needed memory bandwidth is still huge.

Most DSPs provide special instructions to efficiently implement a Viterbi decoder. However, in case of the Log-MAP the butterfly update requires branch metrics values instead of branch metrics bits, and the use of the max* operation—see Equation 15.5.

15.3.1 VLIW Architectures

We made several implementations on state-of-the-art DSPs. The used DSPs correspond to a 0.18-μm technology node. Although compiler technologies for DSP technologies have made big progress, they still produce significant overhead. Consequently, we implemented the algorithms on assembler level to fully exploit the DSP capabilities. Special emphasis was put on:

- An efficient butterfly scheduling to fully exploit the VLIW parallelism

- Loop merging of backward recursion and soft output calculation

- Efficient register allocation to enable subword parallelism and avoid unnecessary register-register transfers

- Efficient memory bank partitioning and memory layout optimization to minimize memory transfer bottlenecks

Table 15.1 shows the results for implementations on different processors using the Turbo-Code parameters of the 3GPP standard for the Log-MAP and Max-Log-MAP algorithms, respectively. While both the ST120 from ST Microelectronics [12] and the StarCore SC140 from Motorola [13] are general-purpose VLIW DSPs, the ADI TigerSharc from Analog Devices [14] is a DSP targeted on wireless infrastructure applications. Of these, it is the only processor providing dedicated instruction support for the Log-Map algorithm: A set of enhanced communication instructions enables, for example, efficient max* selections in a variety of configurations. The fourth column in this table shows the number

TABLE 15.1 ■ Turbo decoder throughput (Kbps) on various DSPs

Processor	Architecture	Clock freq. (MHz)	Cycles/bit	Throughput @ 5 Iter.
		Max-Log-Map		
STM ST120	2 ALU	200	37	540 kbit/s
SC140	4 ALU	300	16	1875 kbit/s
		Log-Map		
STM ST 120	2 ALU	200	~100	~200 kbit/s
SC140	4 ALU	300	50	600 kbit/s
ADI TS	2 ALU	180	27	666 kbit/s

of clock cycles necessary to calculate 1 bit in the MAP algorithm; the last column shows the achievable throughput, assuming five decoding iterations.

We see that a throughput of 2 Mbps, as required for a single UMTS data channel, cannot be achieved by any of the tested configurations. Only the SC140 using Max-Log-MAP even comes close. We can also see that for processors without a dedicated instruction support for the max* operation, the Log-MAP algorithm requires three times as many clock cycles per bit as the Max-Log-MAP algorithm. This is due to the calculation of the correction term of the max* operation, which requires a subtraction, the calculation of an absolute value, access to a look-up table (implementing the exp function), and an addition. This sequence requires nine additional clock cycles to the plain maximum calculation of the Max-Log-MAP algorithm, which can be done in just one clock cycle.

15.3.2   Customizable Processors

Significant attention has recently been drawn to configurable processor cores, such as those offered by Tensilica [15], ARC [16], Co-Ware [17], and others. These are based on classical RISC architectures and can be configured in two dimensions: First, with respect to the architectural features of the core, e.g., inclusion of a fast MAC unit, cache size and policy, memory size, and bus width. Second, the instruction set can be extended by user-defined instructions. This approach offers the benefit of removing application-specific performance bottlenecks while still maintaining the flexibility of a software implementation, thus blurring the borders between hardware and software. For instance, performance-critical code portions that require multiple instructions on a generic RISC architecture can be compressed into a single, user-defined instruction to obtain a significant speedup. More important, on a system level this can eliminate the need for a heterogeneous RISC/DSP or RISC/IP-block architecture, thereby simplifying the architecture, reducing the cost of the system, and shortening validation time of the total system. Key to efficiently using a configurable processor core is the methodology for defining and implementing the custom instructions as well as toolset support. For a detailed discussion of research in this area, we refer to Chapters 5, 6, 7, and 9 of this book. In our case study, we applied the Tensilica approach and implemented dedicated instructions to increase the instruction level parallelism of the Tensilica Xtensa base core.

To extend the instruction set of the Tensilica Xtensa base core, each additional instruction has to be implemented using a special extension language, called Tensilica instruction extension (TIE) language. After the designer has specified all details (including mnemonic, encoding, and semantics) of the new instruction, the TIE compiler automatically adds new instructions to the processor's RT-level description and generates a complete toolset for the extended processor. The generated hardware is integrated into the *EXE* stage of the processor pipeline, as can be seen in Fig. 15.2. Here the two stages of a classical RISC five-stage pipeline, the instruction decode and the execution stage, are shown. Boxes with solid lines represent combinatorial logic in the processor, boxes with dashed lines represent flip-flops, and bold boxes represent combinatorial logic added by the TIE compiler. As the TIE extensions to the processor pipeline appear simply as another functional unit in the *EXE* stage, the new instructions are indistinguishable from the native instructions. All new instructions are executed in a single cycle. That means that the maximum frequency of the extended processor may depend on the TIE compiler–generated hardware extensions.

As mentioned earlier, the kernel operation is the butterfly operation, which is used for forward and backward recursion. This operation processes two states concurrently. Consequently, to efficiently process a Log-MAP butterfly, we introduced a new kernel operation, called LM_ACS (*Log MAP Add Compare Select*), in which we seamlessly integrated the max* operation. A sample architecture for the LM_ACS operation is given in Fig. 15.3. The lookup table is used for the calculation of the correction

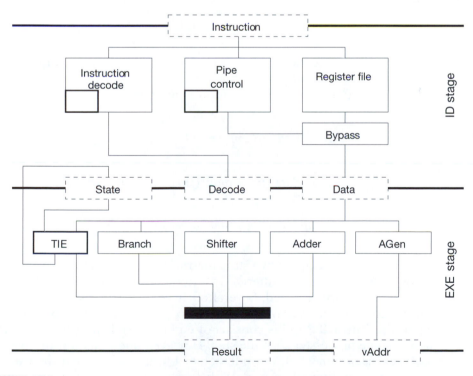

■ **FIGURE 15.2**

TIE hardware integration into Xtensa pipeline [18].

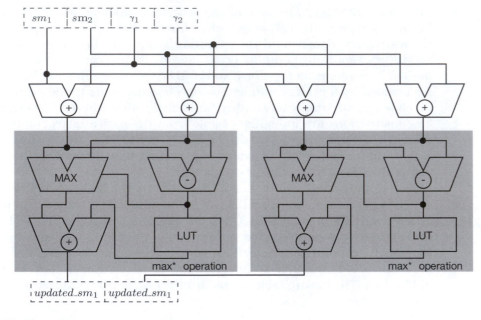

■ **FIGURE 15.3**

Sample architecture for LM_ACS operation.

term (exp) in Equation 15.5 and can be implemented as combinatorial logic with a complexity of less than a dozen gates. The format and arithmetic functionality of this operation is as follows:

$$(updated\_sm_1, updated\_sm_2) = LM\_ACS(sm_1, sm_2, \gamma_1, \gamma_2) \qquad (15.6)$$

$$update\_sm_1 = max^*(sm_1 + \gamma_1, sm_2 + \gamma_2) \qquad (15.7)$$

$$update\_sm_2 = max^*(sm_1 + \gamma_2, sm_2 + \gamma_1) \qquad (15.8)$$

The LM_ACS requires two state metrics sm_1, sm_2 and two branch metrics γ_1, γ_2 to process a butterfly. Two updated state metrics $updated\_sm_1$, $updated\_sm_2$ are returned. Several of these butterfly instructions can be grouped into one VLIW instruction to process multiple butterflies in parallel. The 3GPP trellis consists of eight states, hence, four butterflies have to be processed per time step. However, due to data transfer issues of the branch metrics, the number of butterflies that can be executed in parallel is limited to two. With two parallel butterflies, four recursion steps (either forward or backward recursions) can be processed in 1 cycle. Hence, the forward and backward recursion for the state metrics updates for an eight-state trellis can be done in 4 clock cycles. In a similar way, we implemented instructions to accelerate the soft-out calculation (5 clock cycles) and the calculation of the branch metrics (2 clock cycles). With these new instructions, 1 bit in the Log-MAP can be processed in 9 clock cycles.

An appropriate addressing scheme must ensure the correct mapping of source operands and results. The processing of forward and backward recursions is executed in a loop, which should take only a minimum number of clock cycles. As pointed out earlier, branch and state metrics have to be read from and written to the memory. Usually, a memory access takes more than 1 clock cycle to complete. Using means such as prefetching and implicit address generation, even multicycle memory accesses can be parallelized with the processing of the data. In case of writing, the latency is hidden in the processor pipeline. In addition to the previously mentioned instructions, we implemented different memory operations, which permit us to totally hide the data transfer latencies, as they can be executed in parallel to the recursion computations.

Due to some limitations in the TIE language to the functional interface of extension instructions, some of them had to be implemented several times with slightly different data transfers [19]. To pass more values to an instruction than the functional interface allows, dedicated custom registers had to be added to the processor pipeline. These can only be used by the new instructions. To access them properly, the instructions had to be implemented once for each possible input source, as the parameters could not all be passed through the functional interface. Without this workaround, some of the implemented extensions would have had to be implemented as two or more instructions, thus reducing throughput of the architecture. As a drawback of this approach, the application-specific instructions cannot be used independently, but have to be executed exactly in the way intended at design time. This limitation, of course, reduces the flexibility of the extended processor. It was also necessary to implement additional instructions to initialize the custom registers.

In principle, it is possible to further increase the instruction parallelism, and thus the performance, at the cost of decreased flexibility, i.e., tuning the instructions completely to the 3GPP constraints. However, whenever there is a future change in the parameters of the standard (e.g., the number of encoder states) that is not known at the current design time, the instructions have to be modified. In our case, we traded off performance for flexibility, providing a certain degree of flexibility for changing parameters without the need to modify instructions, but at the cost of a certain performance loss.

We synthesized the extended core using a 0.18-μm technology under worst case conditions (1.55 V, 125°C). The Tensilica RISC core without instruction set extensions runs at a frequency of 180 MHz and requires about 50 Kgates, whereas the extended core runs at a frequency of

133 MHz, requiring 104 Kgates. The resulting throughput for the Log-MAP turbo decoder algorithm under 3GPP conditions yields 1.4 Mbps. Comparing this throughput with, e.g., the STM ST120 (its gate complexity corresponds to about 200 Kgates), we get a performance gain of seven for the Log-MAP and about three for the Max-Log-MAP algorithms with half of the gate count complexity. Extrapolating the results to a 0.13-μm technology, the 3GPP throughput conditions could easily be fulfilled with this extended RISC core. This result impressively shows the advantage of the application-specific core approach.

15.4 MULTIPROCESSOR ARCHITECTURES

While the maximum throughput of 3GPP is limited to 2 Mbps, WLAN and future communication standards will demand much higher data rates. Hence, only increasing the instruction level parallelism of a single processor core is not sufficient, and we have to move to massively parallel multiprocessor architectures, called application-specific multiprocessor system on chip (MPSoC).

A simple solution is to put several processors in parallel, each of them decoding a different block. However, such a solution has a low architectural efficiency and a large latency for each individual block. Since latency is a very critical issue, such a solution is in many cases infeasible. Parallelizing the turbo decoding algorithm itself is a much better approach. For this, we have to decompose the algorithm into a set of parallel tasks running on a set of communicating processors [20]. As explained in the previous section, the MAP algorithm has a large inherent parallelism. We exploit the window level by breaking up a complete block into smaller subblocks, which can be decoded independently from each other at the cost of some acquisition steps at the border of the subblocks. Each of these subblocks is mapped onto one processing node. In this way all nodes can run independently from each other. The real bottleneck is the interleaving of the soft information, which has to be performed in the iterative loop between the individual component decoders—see Algorithm 15.1. Interleaver and deinterleaver tables contain one-to-one mappings of source addresses to destination addresses (e.g., a 3GPP-compliant table contains up to 5,114 entries). One address translation has to be read from these tables for every calculated MAP value. As long as no more than one MAP value is generated per clock cycle, interleaving can be performed on the fly through indirect addressing. However, in highly parallel architectures, more than one value is produced per clock cycle. This means several values have to be fetched from and stored to memories in the same clock cycle, resulting in memory conflicts. Note that good interleavers, in terms of communications performance, map neighboring addresses evenly to spread-out target addresses. Thus we have no locality in the interleaving process, and an efficient distribution of the interleaved values is key to high throughput turbo decoding on any parallel architecture [21].

In our multiprocessor architecture we use the extended Tensilica core as the base processing node. Each node decodes a windowed subblock. As described in the previous section, every 5 cycles a new value is produced during the combined backward recursion (2 cycles) and soft-out (3 cycles) calculation loop. In the following, this data production rate is denoted as R, i.e., $1/5$ in our case. For efficiency reasons all input data of an iteration has to be stored in a fast single cycle access memory. The Xtensa architecture provides an interface to an optional local data RAM, where a fast SRAM can be mapped into the processor address space. The input data are stored in this fast memory M_C instead of the main memory M_P, because of the required high memory bandwidth. The generated soft-out values also have to be delivered in a single cycle to the processing node to calculate branch metrics. We use the *Xtensa local memory interface* (XLMI) to link the memory to a communication device (see Fig. 15.4). XLMI is a general-purpose, single-cycle data interface designed for interprocessor communication. The communication device implements a message passing model. It is composed of status registers, a FIFO buffer, and two dual-ported memories, which are mapped into the XLMI address space. During branch metrics calculation, one of the two buffers provides the values of the last MAP calculation to the processor pipeline. During soft-out calculation, the processor stores all calculated values with the appropriate address

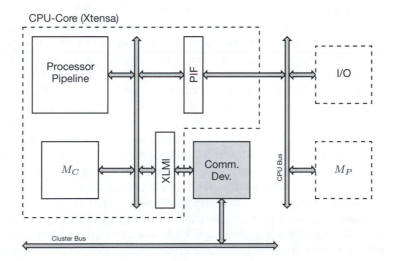

■ **FIGURE 15.4**

Architecture of a processing node.

information as multiprocessor messages in the FIFO buffer. A bus interface controller reads the messages from the FIFO and transmits them to the targeted processor via the cluster bus (see Fig. 15.4). The bus interface controller reads the message on the cluster bus and stores the corresponding value in the other of the two buffers. When a processor finishes its soft-out calculation, it actively polls the status register until a ready-bit is set. When all calculated values of all processor nodes have been delivered to the targeted processors, the MAP algorithm is completed and the ready-bit of all status registers is set. The processor may then continue with the next MAP calculation, reading the distributed values from one of the two buffers while the other one is being filled with the new calculated values. The Xtensa PIF interface is used to link the processor to the CPU bus. Fig. 15.4 shows the overall architecture of such a processing node.

Assuming a block of length N and P processing elements, each node has to process a subblock of N/P bits. The throughput requirement on the communication network is $P \cdot R$ values/cycle to completely hide the communication latency and avoid stalling of the processors due to missing data. If $P \cdot R \leq 1$, at most 1 value is generated per clock cycle and no interleaving conflicts exist. In this case a simple bus architecture is sufficient. For a 0.18-μm technology, about five processors can be connected to a single bus, not exceeding its capacity. This yields a maximum throughput of about 7 Mbps. If $P \cdot R > 1$, write conflicts can occur and a new communication strategy becomes mandatory. We make use of a packet-switched network on chip approach. Packets are composed of the ID of the target processor, the local address in the corresponding target buffer, the target buffer ID, and the soft-out value. The overall communication architecture is shown in Fig. 15.5.

P_B processors are connected to a single bus, forming a *cluster*. Let C be the number of clusters. P_B is chosen to fully exploit the bus capacity. Each cluster is connected over a bus switch to a so-called ring interleaving network, which is composed of output-queued routing nodes, or so-called RIBB cells. Each of these cells is locally connected to two neighbors, forming a ring. A RIBB cell decides whether the incoming data has to be sent to the local cluster or has to be forwarded to either the left or the right RIBB cell, respectively, by using a shortest path routing algorithm. Thus, for each input of the RIBB cell, a routing decision unit that decides in which direction the message has to be forwarded exists. Each output of the RIBB cell is linked to a buffer, the output queue. These buffers dominate the size of a RIBB cell.

Assuming an equal distribution of traffic, which is a feasible assumption for good interleavers, we can statically analyze the network and derive necessary and sufficient conditions, such that the throughput of the communication network does not degrade the data throughput [22].

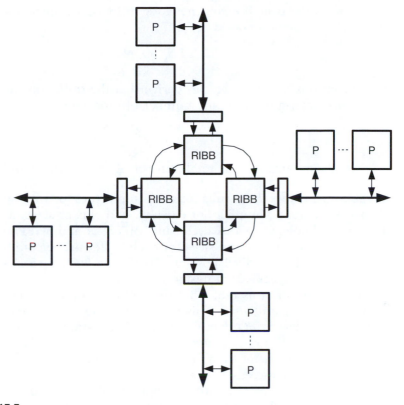

Communication structure.

- The traffic inside a cluster and to/from the outside must be completed within the data calculation period. This yields

$$\frac{1}{C^2} \cdot N + 2 \cdot \frac{C-1}{C} \cdot N \le \frac{1}{R} \cdot \frac{N}{P} \qquad (15.9)$$

From this equation we can derive that the cluster bus determines the number of processors per cluster:

$$P_B \le \frac{1}{R} \cdot \frac{C}{2C-1} \Rightarrow P_B \approx \frac{1}{2R} \qquad (15.10)$$

This equation gives the bus arbitration scheme. For larger values of C, a round robin arbitration, which grants the bus to one of its local nodes and the bus-switch alternately, is sufficient.

■ The traffic over the ring network must be completed within the data calculation period:

$$\frac{1}{8} \cdot N \le \frac{N}{P \cdot R} \qquad (15.11)$$

From this equation we can derive that the traffic on the ring network determines the total number of processors:

$$P \le \frac{8}{R} \qquad (15.12)$$

Hence in our case, the total number of processors is limited to 40.

Table 15.2 shows the results for our architecture for varying numbers of processor nodes, numbers of clusters, and cluster sizes, assuming a 0.18-μm technology, a clock cycle time of 7.5 *ns* and 3GPP conditions. To determine the minimum sizes for each buffer in the communication network, we simulated all possible 3GPP interleavers (5,114 different interleavers are specified) at design time.

We can see from this table that the architectural efficiency increases with increasing parallelism. This is due to the fact that the application memory (interleaver, I/O data memories) is constant, and the overhead of the communication network is for most cases less than 10%. The throughput grows faster than the area and, as a consequence, the architectural efficiency increases.

Fig. 15.6 compares area and throughput of architectures exploiting block level parallelism by using several decoders in parallel to those exploiting subblock level parallelism. We see that the parallelization on subblock level is much more efficient. Even more important is that the latency—not shown in this graph—is much lower, since the latency decreases proportionally with the parallelism degree.

Several possibilities exist to further increase the throughput of the communication network. Substituting the ring topology by a chordal ring or by random topologies allows for a direct connection of more clusters.

TABLE 15.2 ■ MPSoC results

Total Nodes (N)	# of Clusters (C)	Cluster Nodes (N_C)	Throughput [*Mbps*]	Area Comm. (mm^2)	Area Total (mm^2)	Efficiency ($Mbps \cdot mm^2$)
1	1	1	1.48	NA	6.42	1
5	1	5	7.28	0.21	14.45	2.19
6	2	3	8.72	0.66	16.73	2.26
8	4	2	11.58	1.25	20.91	2.40
12	6	2	17.18	2.02	28.92	2.58
16	8	2	22.64	2.88	36.98	2.66
32	16	2	43.25	7.29	70.26	2.67
40	20	2	52.83	10.05	87.47	2.62

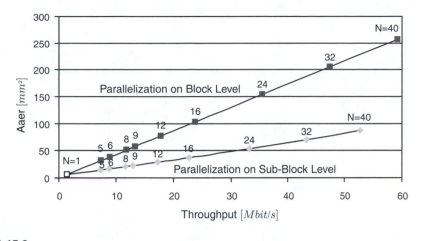

■ **FIGURE 15.6**

Block level versus subblock level parallelism.

However, all these approaches do not feature any kind of flow control to prevent buffer overflow. To avoid buffer overflow, all interleaver patterns need to be known at design time so that the buffer sizes can be adjusted to hide all access conflicts. Due to the limited space of this paper, we omit a more detailed discussion on this issue and refer the interested reader to [21] for communication strategies that exploit flow control.

15.5 CONCLUSION

In this chapter we described how to efficiently map the turbo-decoding algorithm onto an application-specific multiprocessor platform by tailoring the processing nodes and communication network to the application. We exploited the concurrency inherent in the algorithm to parallelize the implementation on various levels. The methodology we presented is not limited to turbo decoding. It can also be used to efficiently implement other iterative decoding techniques, especially LDPC codes [23]. These codes were already invented in 1963 and rediscovered in 1996 [24]. Since then, LDPC codes have experienced a renaissance, and, from a communications performance point of view, they are among the best known codes, especially for very large block lengths. The decoding process of LDPC codes has many similarities to Turbo-Codes: information is "randomly" and iteratively exchanged between component decoders. Hence, we can apply the same architectural approach to the implementation of LDPC code decoders. But this is beyond the scope of this chapter and is an active research area of its own.

CHAPTER 16

APPLICATION CODE PROFILING AND ISA SYNTHESIS ON MIPS32

Rainer Leupers

Based on an extensive analysis of the wireless communication application domain, in Chapter 2, Ascheid and Meyr propose an architecture description language (ADL) based design flow for ASIPs. While Chapter 4 has elaborated on ADL design issues in detail, in this chapter we focus on advanced software tools required for ADL-based ASIP architecture exploration and design.

The concept of ADL-based ASIP design is sketched in Fig. 16.1. The key is a unique "golden" processor model, described in an ADL, which drives the generation of software (SW) tools as well as hardware (HW) synthesis models. The retargetable SW tools allow us to map the

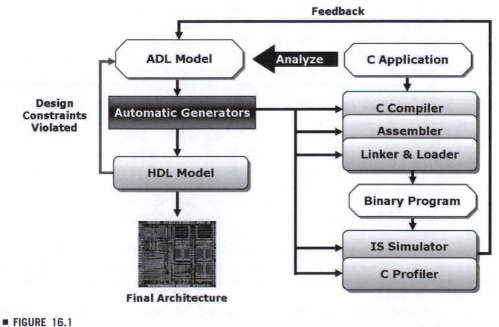

■ **FIGURE 16.1**

ADL-based ASIP design flow.

application to the ASIP architecture and to simulate its execution on an architecture virtual prototype. A feedback loop exists to adapt the ADL model (e.g., by adding custom machine instructions to the ASIP) to the demands of the application, based on simulation and profiling results. Another feedback loop back-annotates results from HW synthesis, which allows the incorporation of precise estimations of cycle time, area, and power consumption. In this way, ASIP design can be seen as a stepwise refinement procedure: Starting from a coarse-grain (typically instruction-accurate) model, further processor details are added until we obtain a cycle-accurate model and finally a synthesizable HDL model with the desired quality of results.

This iterative methodology is a widely used ASIP design flow today. One example of a commercial EDA tool suite for ASIP design is CoWare's LISATek [1], which comprises efficient techniques for C compiler retargeting, instruction set simulator (ISS) generation, and HDL model generation (see, e.g., [2–4] for detailed descriptions). Another state-of-the-art example is Tensilica's tool suite, described in Chapter 6 of this book.

While available EDA tools automate many subtasks of ASIP design, it is still largely an interactive flow that allows us to incorporate the designer's expert knowledge in many stages. The automation of tedious tasks, such as C compiler and simulator retargeting, allows for short design cycles and thereby leaves enough room for human creativity. This is in contrast to a possible pure "ASIP synthesis" flow, which would automatically map the C application to an optimized processor architecture. However, as already emphasized in Chapter 2, such an approach would most likely overshoot the mark, similar to experiences made with behavioral synthesis in the domain of ASIC design.

Nevertheless, further automation in ASIP design is desirable. If the ASIP design starts from scratch, then the only specification initially available to the designer may be the application C source code, and it is far from obvious how a suitable ADL model can be derived from this. Only with an ADL model at hand, the earlier iterative methodology can be executed, and suboptimal decisions in early stages may imply costly architecture exploration cycles in the wrong design subspace. Otherwise, even if ASIP design starts from a partially predefined processor template, the search space for custom (application-specific) instructions may be huge. Partially automating the selection of such instructions, based on early performance and area estimations, would be of great help and would seamlessly fit the above "workbench" approach to ASIP design.

In this chapter we therefore focus on new classes of ASIP design tools, which seamlessly fit the existing well-tried ADL approach and at the same time raise the abstraction level in order to permit higher design productivity.

On one hand, this concerns tools for application source code profiling. A full understanding of the dynamic application code characteristics is a prerequisite for a successful ASIP optimization. Advanced profiling tools can help to obtain this understanding. Profilers bridge the gap from a SW specification to an "educated guess" of an initial HW architecture that can later be refined at the microarchitecture level via the earlier ADL-based design methodology.

Second, we describe a methodology for semiautomatic instruction set customization that utilizes the profiling results. This methodology assumes that there is an extensible instruction set architecture (ISA) and that a speedup of the application execution can be achieved by adding complex, application-specific instruction patterns to the base ISA. The number of potential ISA extensions may be huge, and the proposed methodology helps to reduce the search space by providing a coarse ranking of candidate extensions that deserve further detailed exploration.

The envisioned extended design flow is coarsely sketched in Fig. 16.2. The two major new components are responsible for fine-grained application code profiling and identification of custom instruction candidates. In

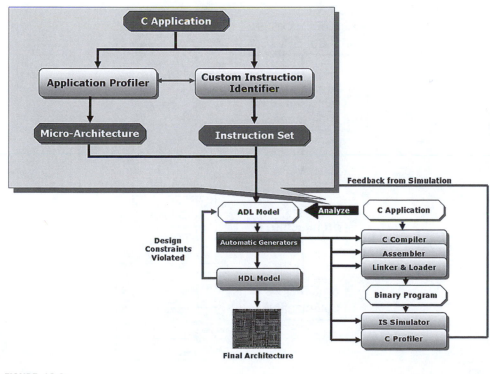

■ FIGURE 16.2

Extended ASIP design flow.

this way, they assist in the analysis phase of the ASIP design flow, where an initial ADL model needs to be developed for the given application. Selected custom instructions can be synthesized based on a given extensible ADL template model, e.g., a RISC architecture. The initial model generated this way is subsequently further refined within the "traditional" ADL-based flow. Details on the proposed profiling and custom instruction synthesis techniques are given in the following sections.

16.1   PROFILING OF APPLICATION SOURCE CODE

Designing an ASIP architecture for a given application usually follows the coarse steps shown in Fig. 16.3. First, the algorithm is designed and optimized at a high level. For instance, in the DSP domain, this is frequently performed with block diagram–based tools such as Matlab or SPW. Next, C code is generated from the block diagram or is written manually. This executable specification allows for fast simulation but is usually not optimized for HW or SW design. In fact, this points out a

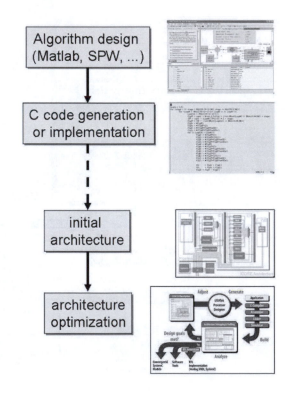

▪ **FIGURE 16.3**

Coarse steps in algorithm-to-architecture mapping.

large gap in today's ASIP design procedures: taking the step from the C specification to a suitable initial processor architecture that can then be refined at the ISA or microarchitecture level. For this fine-grained architecture exploration phase, useful ADL-based tools already exist, but developing the initial ADL model is still a tedious manual process.

16.1.1 Assembly and Source Level Profiling

Profiling is a key technology to ensure an optimal match between a SW application and the target HW. Naturally, many profiling tools are already available. For instance, Fig. 16.4 shows a snapshot of a typical assembly debugger graphical user interface (GUI). Besides simulation and debug functionality, such tools have embedded profiling capabilities, e.g., for highlighting critical program loops and monitoring instruction execution count. Such assembly-level profilers work at a high accuracy level, and they can even be automatically retargeted from ADL processor models. However, there are two major limitations. First, in spite of advanced ISS techniques, simulation is still much slower than native compilation and execution of C code. Second, and even more important, a detailed ADL processor model needs to be at hand in order to generate an assembly-level profiler. On the other hand, only profiling results can tell what a suitable architecture (and its ADL model) need to look like.

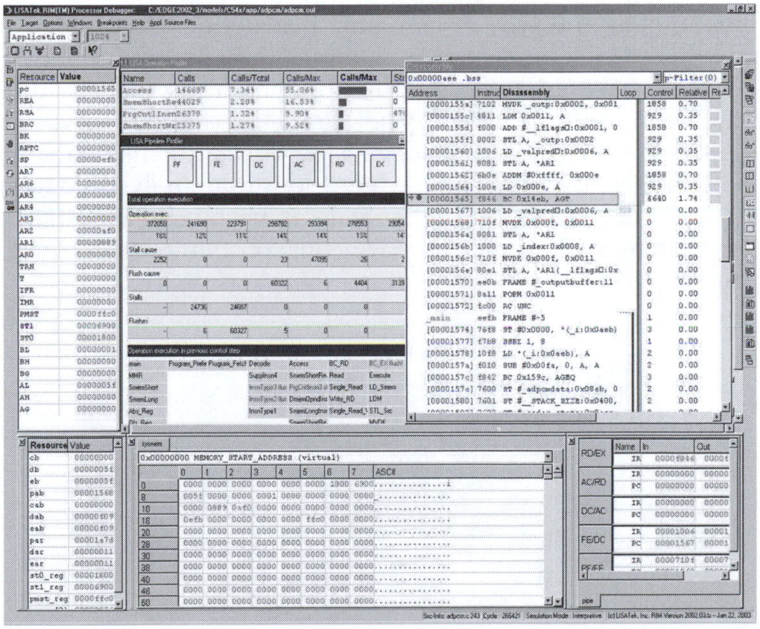

■ **FIGURE 16.4**

Typical debugger GUI with profiling capabilities.

Due to this "chicken-and-egg" problem, assembly-level profiling only is not very helpful in early ASIP design phases.

Profiling can also be performed at the C source code level with tools like GNU gprof. Such tools are extremely helpful for optimizing the application source code itself. However, they show limitations when applied to ASIP design. For example, the gprof profiler generates a table as an output that displays the time spent in the different C functions of the application program and therefore allows identification of the application hot spots. However, the granularity is pretty coarse: a hot-spot C function may comprise hundreds of source code lines, and it is far from obvious what the detailed hot spots are and how their execution could be supported with dedicated machine instructions.

A more fine-grained analysis is possible with code coverage analysis tools like GNU gcov. The output shown in Fig. 16.5 shows pairs of "execution count:line number" for a fragment of an image processing application (corner detection). Now, the execution count per source code line is known. However, the optimal mapping to the ISA level is still not trivial. For instance, the C statement in line 1451 is rather complex and will be translated into a sequence of assembly instructions. In many cases (e.g., implicit type casts and address scaling), the full set of operations is not even visible at the C source code level. Thus, depending on the C programming style, even per–source line profiling may be too coarsely grained for ASIP design purposes. One could perform an analysis of single source lines to determine the amount and type of assembly instructions required. For the previous line 1451, for instance, 5 ADDs, 2 SUBs, and 3 MULs would be required, taking into account the implicit array index scaling. However, in case an optimizing compiler, capable of induction variable elimination, will later be used for code generation, it is likely that 2 MULs will be eliminated again. Thus, while the initial analysis suggested MUL is a key operation in that statement, it becomes clear that pure source level analysis can lead to erroneous profiling results and misleading ISA design decisions.

```
372: 1447:  while (corner_list[n].info !=7)
  -: 1448:  {
371: 1449:    if (drawing_mode==0)
  -: 1450:      {
371: 1451:        p = in + (corner_list[n].y-1)*x_size + corner_list[n].x - 1;
371: 1452:        *p++=255; *p++=255; *p=255; p+=x_size-2;
371: 1453:        *p++=255; *p++=0;    *p=255; p+=x_size-2;
371: 1454:        *p++=255; *p++=255; *p=255;
371: 1455:        n++;
  -: 1456:      }
```

▪ **FIGURE 16.5**

GNU gcov output for a corner detection application.

16.1.2 Microprofiling Approach

To overcome the limitations of assembly level and source level profilers, we propose a novel approach called microprofiling (μP). Fig. 16.6 highlights the features of μP versus traditional profiling tools. In fact, μP aims at providing the right profiling technology for ASIP design by combining the best of the two worlds: source-level profiling (machine independence and high speed) and assembly-level profiling (high accuracy).

Fig. 16.7 shows the major components. The C source code is translated into a three address code intermediate representation (IR) as is done in many C compilers. We use the LANCE C frontend [5] for this purpose. A special feature of LANCE, as opposed to other compiler platforms, is that the generated IR is still executable, since it is a subset of ANSI C. On the IR, standard "Dragon Book"–like code optimizations are performed in order to emulate, and predict effects of, optimizations likely to be performed anyway later by the ASIP-specific C compiler. The optimized IR code then is instrumented by inserting probing function calls between the IR statements without altering the program semantics. These functions perform monitoring of different dynamic program characteristics, e.g., operator frequencies. The implementation of the probing functions is statically stored in a profiler library. The IR code is compiled with a host compiler (e.g., gcc) and is linked together with the profiler library to form an executable for the host machine. Execution of the instrumented application program yields a large set of profiling data, stored in XML format, which is finally read via a GUI and is displayed to the user in various chart and table types (see Fig. 16.8).

Simultaneously, the μP GUI serves for project management. The μP accepts any C89 application program together with the corresponding

	C source level (e.g., gprof)	assembly level (e.g., LISATek)	micro-profiler
primary application	source code optimization	ISA and architecture optimization	ISA and architecture optimization
needs architectural details	no	yes	no
speed	high	low	medium
profiling granularity	coarse	fine	fine

■ **FIGURE 16.6**

Summary of profiler features.

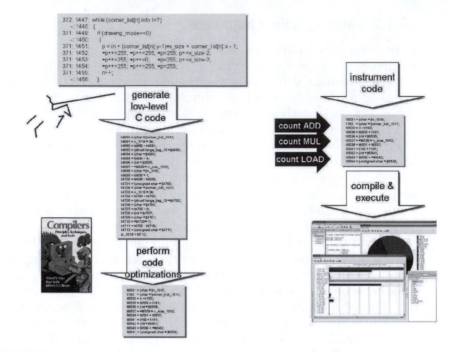

■ **FIGURE 16.7**

Processing of C source programs in the microprofiler (μP).

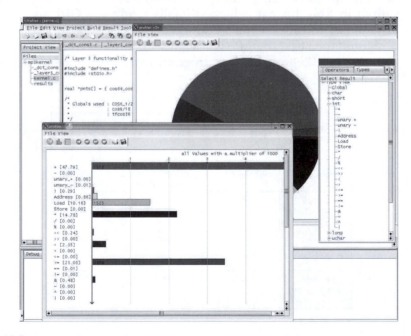

■ **FIGURE 16.8**

Microprofiler (μP) graphical user interface.

makefiles, making code instrumentation fully transparent for the user. The µP can be run with a variety of profiling options that can be configured through the GUI:

- *Operator execution frequencies*: Execution count for each C operator per C data type, in different functions and globally. This can help designers to decide which functional units should be included in the final architecture.

- *Occurrences, execution frequencies, and bit width of immediate constants*: This allows designers to decide the ideal bit width for integral immediate values and optimal instruction word length.

- *Conditional branch execution frequencies and average jump lengths*: This allows designers to take decisions in finalizing branch support HW (such as branch prediction schemes, length of jump addresses in instruction words, and zero-overhead loops).

- *Dynamic value ranges of integral C data types*: Helps designers to take decisions on data bus, register file, and functional unit bit widths. For example, if the dynamic value range of integers is between 17 and 3,031, then it is more efficient to have 16-bit, rather than 32-bit, register files and functional units.

Apart from these functionalities, the µP can also be used to obtain performance estimations (in terms of cycle count) and execution frequency information for C source lines to detect application hot spots.

Going back to the previously mentioned image processing example, the µP can first of all generate the same information as gcc, i.e., it identifies the program hot spot at C function granularity. In this example, the hot spot is a single function ("susan_corners") where almost 100% of the computation time is spent. The µP, however, allows for a more fine-grained analysis: Fig. 16.9 represents a "zoom" into the profile of the hot spot, showing the distribution of computations related to the different C data types. For instance, one can see there is a significant portion of pointer arithmetic, which may justify the inclusion of an address generation unit in the ASIP architecture. On the other hand, it becomes obvious that there are only very few floating point operations, so that an inclusion of an FPU would likely be useless for this type of application.

Fig. 16.10 shows a further "zoom" into the program hot spot, indicating the distribution of C operators for all integer variables inside the "susan_corners" hot spot. One can easily see that the largest fraction of operations is made up by ADD, SUB, and MUL, so that these deserve an efficient HW implementation in the ASIP ISA. On the other hand,

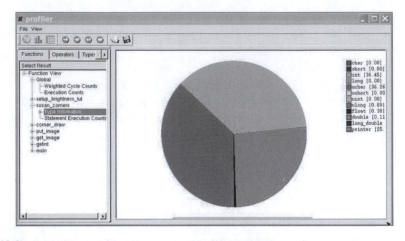

▪ **FIGURE 16.9**

Distribution of computations over C data types in program hot spot.

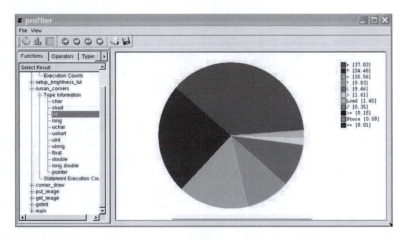

▪ **FIGURE 16.10**

Distribution of C operators over int variables in program hot spot.

there are other types of integer operations, e.g., comparisons, with a much lower dynamic execution frequency. Hence, it may be worthwhile to leave out such machine instructions and to emulate them by other instructions.

The μP approach permits us to obtain such types of information efficiently and without the need for a detailed ADL model. In this way, it can guide the ASIP designer in making early basic decisions on the optimal ASIP ISA. Additionally, as will be described in Section 16.2, the profiling data can be reused for determining optimal ISA extensions by means of complex, application-specific instructions.

16.1.3 Memory Access Microprofiling

In addition to guiding ASIP ISA design, the μP also provides comprehensive memory access profiling capabilities. To make sound decisions on architectural features related to the memory subsystem, a good knowledge of the memory access behavior of the application is necessary. If, for example, memory access profiling reveals that a single data object is accessed heavily throughout the whole program lifetime, then placing it into a scratch-pad memory will be a good choice. Another goal when profiling memory accesses is the estimation of the application's dynamic memory requirements. As the memory subsystem is one of the most expensive components of embedded systems, fine-tuning it (in terms of size) to the needs of the application can significantly reduce area and power consumption.

The main question in memory access profiling is what operations actually induce memory traffic. At the C source code (or IR) level, there are mainly three types of operations that can cause memory accesses:

- Accesses to global or static variables

- Accesses to locally declared composite variables (placed on the run-time stack, such as local arrays or structures)

- Accesses to (dynamically managed) heap memory

Accesses to local scalar variables usually do not cause traffic to main memory; as in a real RISC-like processor with a general-purpose register bank, scalars are held in registers. Possible, yet rare, memory accesses for local scalar variables can be considered as "noise" and are therefore left out from profiling. As outlined earlier, IR code instrumentation is used as the primary profiling vehicle. In the IR code, memory references are identified by pointer indirection operations. At the time of C to IR translation, a dedicated callback function is inserted before every memory access statement. During execution of the instrumented code, the profiler library searches its data structures and checks whether an accessed address lies in the memory range of either a global variable, a local composite, or an allocated heap memory chunk and updates corresponding read/write counters. In this way, by the end of program execution, all accesses to relevant memory locations are recorded.

16.1.4 Experimental Results

In order to evaluate the μP performance and capabilities, a number of experiments have been performed. The first group of experiments addresses the profiling speed. For fast design space exploration, the μP instrumented code needs to be at least as fast as the ISS of any arbitrary architecture. Preferably, it should be as fast as code generated by the underlying host compiler, such as gcc. Fig. 16.11 compares average speeds of instrumented code versus gcc generated code, and a fast compiled MIPS instruction accurate ISS (generated using CoWare LISATek) for the different configurations of the μP. As can be seen, the speed goals are achieved. The basic profiling options slow down instrumented code execution versus gcc by a factor of only 3. More advanced profiling options increase execution time significantly. However, even in the worst case, the instrumented code is almost an order of magnitude faster than ISS.

Next, we focused on the accuracy of predicting operator execution frequencies. Fig. 16.12 shows the operator count comparisons obtained from a MIPS ISS and instrumented code for the ADPCM benchmark. All the C operators are subdivided into five categories. As can be seen, the

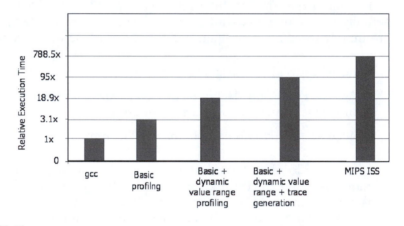

▪ **FIGURE 16.11**

Execution speed of compiled, instrumented, and simulated code.

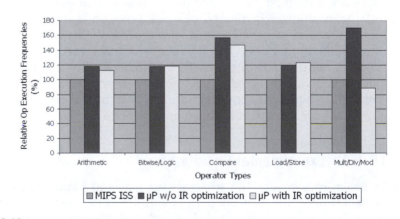

▪ **FIGURE 16.12**

Actual and μP estimated operator execution frequencies.

average deviation from the real values (as determined by ISS) is very reasonable. These data also emphasize the need to profile with optimized IR code: the average deviation is lower (23%) with IR optimizations enabled than without (36%).

Finally, Fig. 16.13 shows the accuracy of μP-based memory simulation for an ADPCM speech codec with regard to the LISATek on-the-fly memory simulator (integrated into the MIPS ISS). The memory hierarchy in consideration has only one cache level with associativity 1 and a block size of 4 bytes. The miss rates for different cache sizes have been plotted for both memory simulation strategies. As can be seen from the comparison, μP can almost accurately predict the miss rate for different cache sizes. This remains true as long as there is no or little overhead due to standard C library function calls. Since μP does not instrument library functions, the memory accesses inside binary functions remain unprofiled. This limitation can be overcome if the standard library source code is compiled using μP.

The μP has been applied, among others, in an MP3 decoder ASIP case study, where it has been used to identify promising optimizations of an initial RISC-like architecture in order to achieve an efficient implementation. In this scenario, the μP acts as a stand-alone tool for application code analysis and early ASIP architecture specification. Another typical use case is in connection with the ISA extension technology described in Section 16.2, which relies on the profiled operator execution frequencies in order to precisely determine data flow graphs of program hot spots.

Further information on the μP technology can be found in [6]. In [7], a more detailed description of the memory-profiling capabilities is given. Finally, [8] describes the integration of μP into a multiprocessor system-on-chip (MPSoC) design space exploration framework, where the μP performance estimation capabilities are exploited to accurately

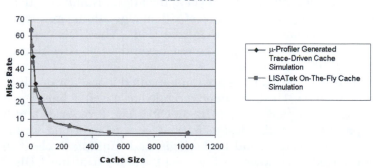

■ FIGURE 16.13

Miss-rate comparison between μP and ISS based cache simulation.

estimate cycle counts of tasks running on multiple processors, connected by a network on chip, at a high level.

16.2 SEMIAUTOMATIC ISA EXTENSION SYNTHESIS

While the μP tool described earlier supports a high-level analysis of the dynamic application characteristics, in this section we focus on the synthesis aspect of ASIPs. Related chapters (e.g., Chapters 6 and 7) in this book emphasize the same problem from different perspectives.

We assume a configurable processor as the underlying platform, i.e., a partially predefined processor template that can be customized toward a specific application by means of ISA extensions. Configurable processors have gained high popularity in the past years, since they take away a large part of the ASIP design risks and allow for reuse of existing IP, tools, and application SW. ISA extensions, or custom instructions, (CIs), are designed to speed up the application execution, at the expense of additional HW overhead, by covering (parts of) the program hot spots with application-specific, complex instruction patterns.

16.2.1 Sample Platform: MIPS CorExtend

The MIPS32 Pro Series CorExtend technology allows designers to tailor the performance of the MIPS32 CPU for specific applications or application domains while still maintaining the benefits of the industry-standard MIPS32 ISA. This is done by extending the MIPS32 instruction set with custom user-defined instructions (UDIs) with a highly specialized data path.

Traditionally, MIPS provides a complete tool chain consisting of a GNU-based development environment and a specific ISS for all of its processor cores. However, no design flow exists for deciding which instructions should be implemented as UDIs. The user is required to manually identify the UDIs via application profiling on the MIPS32 CPU using classical profiling tools. Critical algorithms are then considered to be implemented using specialized UDIs. The UDIs can enable a significant performance improvement beyond what is achievable with standard MIPS instructions.

These identified UDIs are implemented in a separate block, the CorExtend module. The CorExtend module has a well defined pin interface and is tightly coupled to the MIPS32 CPU. The Pro Series family CorExtend capability gives the designer full access to read and write the general-purpose registers, and both single and multiple cycle instructions are supported. As shown in Fig. 16.14 when executing UDIs, the MIPS32 CPU is signaling instruction opcodes to the CorExtend module via the pin interfaces, which are decoded subsequently. The module then signals back to the MIPS32 CPU which MIPS32 CPU resources (for

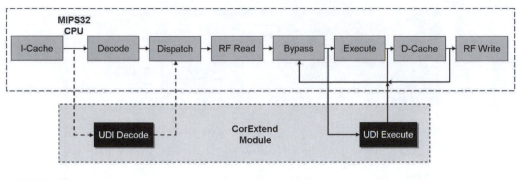

■ **FIGURE** 16.14

MIPS CorExtend Architecture (redrawn from [9]).

example, register or accumulators) are accessed by the UDI. The CorExtend module then has access to the requested resources while executing the UDI. At the end, the result is signaled back to the MIPS32 CPU.

Further examples of configurable and customizable processors include Tensilica Xtensa, ARCtangent, and CriticalBlue, which are described in more detail in Chapters 6, 8, and 10.

16.2.2   CoWare CorXpert Tool

The CorXpert system from CoWare (in its current version) specifically targets the MIPS CorExtend platform. It hides the complexity of the CorExtend RTL signal interface and the MIPSsim CorExtend simulation interface from the designer. The designer is just required to implement the instruction data path using ANSI C for the different pipeline stages of the identified CIs. It utilizes the LISATek processor design technology to automate the generation of the simulator and RTL implementation model. As shown in Fig. 16.15, the CorXpert GUI provides high-level abstraction to hide the complexity of processor specifications. The manually identified CIs are written in ANSI C in the data path editor. Furthermore, the designer can select an appropriate format for the instruction, defining which CPU registers or accumulators are read or written by the instructions. The generated model can then be checked and compiled with a single mouse click. The complete SW tool chain comprising the C compiler, assembler, linker, loader, and the ISS can also be generated by using this GUI.

16.2.3   ISA Extension Synthesis Problem

The problem of identifying ISA extensions is analogous to detecting the clusters of operations in an application that, if implemented as single

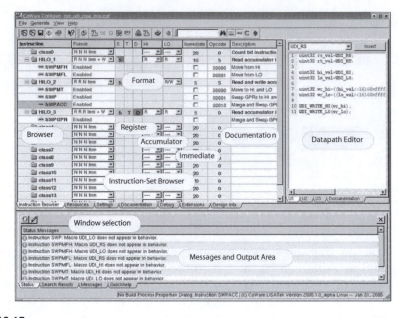

▪ FIGURE 16.15

CoWare CorXpert GUI.

complex instructions, maximize performance [10]. Such clusters must invariably satisfy some constraints; for instance, they must not have a critical path length greater than that supported by the clock length and the pipeline of the target processor used, and the combined silicon area of these clusters should not exceed the maximum permissible limit specified by the designer. Additionally, a high-quality extended instruction set generation approach needs to obtain results close to those achieved by the experienced designers, particularly for complex applications. So the problem of selecting operations for CIs is reduced to a problem of constrained partitioning of the data flow graph (DFG) of an application hot spot that maximizes execution performance. In other words, CIs are required to be optimally selected to maximize the performance such that all the given constraints are satisfied. The constraints on CIs can be broadly classified in the following categories:

▪ *Generic constraints*: These constraints are independent of the base processor architecture, but are still necessary for the proper functioning of the instruction set. One of the most important generic constraints is convexity. Convexity means that if any pair of nodes are in a CI, then all the nodes present on any path between the two nodes should also be in the same CI. This constraint is enforced by the fact that, for a CI to be architecturally feasible, all inputs of that CI should be available at the beginning of it and all results should be produced at the end of the same [10]. Thus any cluster of nodes can only be selected as a CI when the cluster is convex. For example, the CIs shown in Fig. 16.16(a) are not convex, since nodes n_a and n_c are present in the same CI ($CI1$), while the node n_b, which lies on the path between n_a and n_c, is in a different CI. The same holds true for the $CI2$. Thus $CI1$ and $CI2$ are both architecturally infeasible.

Schedulability is another important generic constraint required for legal CIs. Schedulability means that any pair of CIs in the instruction set should not have cyclic dependence between them. This generic constraint is needed because a particular CI or a subgraph

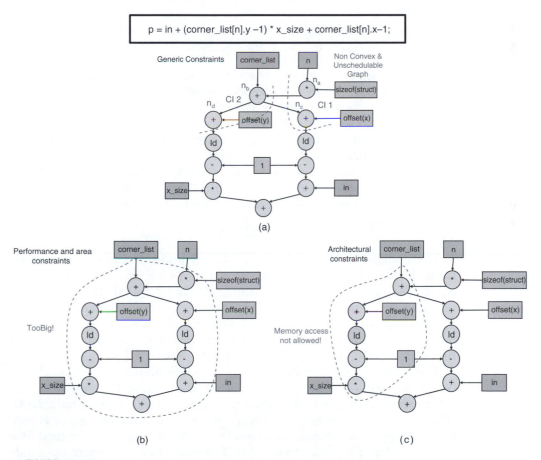

■ **FIGURE 16.16**

Examples of various constraints: (a) generic constraints, (b) performance constraints, (c) architectural constraints. The clusters of operations enclosed within dotted lines represent CIs.

can only be executed when all the predecessors of all the operators in this subgraph are already executed. This is not possible when the operators in different CIs have cyclic dependencies.

- *Performance constraints*: The performance matrices also introduce some constraints on the selection of operators for a CI. Usually embedded processors are required to deliver high performance under tight area and power consumption budgets. This in turn imposes constraints on the amount of HW resources that can be put inside a custom function unit (CFU). Additionally, the critical path length of the identified CIs may also be constrained by such tight budgets or by the architecture of the target processor [11]. Thus, the selection of operators for the CIs may also be constrained by the available area and permissible latencies of a CFU. Fig. 16.16(b) shows an example of the constraints imposed by such area restrictions. The CI obtained in this figure is valid except for the fact that it is too large to be implemented inside a CFU, which has strict area budget.

- *Architectural constraints*: As the name suggests, these constraints are imposed by the architecture itself. The maximum number of *general-purpose registers (GPRs)* that can be read and written from a CI forms an architectural constraint. Memory accesses from the customized instructions may be not allowed either, because having CIs that access memory creates CFUs with nondeterministic latencies, and the speedup becomes uncertain. This imposes an additional architectural constraint. This is shown in Fig. 16.16(c), where the CI is infeasible due to a memory access within the identified subgraph.

The complexity involved in the identification of CIs can be judged from the fact that, even for such a small graph with only architectural constraints in consideration, obtaining CIs is not trivial. The complexity becomes enormous with the additional consideration of generic and performance constraints during CI identification. Thus, the problem boils down to an optimization problem where the operators are selected to become a part of a CI based on the given constraints.

The proposed CI identification algorithm is applied to a basic block (possibly after performing if-conversion) of an application hot spot. The instructions within a basic block are typically represented as a DFG, $G = (N, E)$, where the nodes N represent operators and the edges E capture the data dependencies between them. A potential CI is defined as a cut C in the graph G such that $C \subseteq G$. The maximum number of operands of this CI, *from* or *to* the GPR file, are limited by the number of register file ports in the underlying core. We denote $IN(C)$ as the number of inputs from the GPR file to the cut C, and $OUT(C)$ as the number of outputs from cut C to the GPR file. Also, let $IN_{\max}$ be the maximum allowed inputs from GPR file and $OUT_{\max}$ be the maximum allowed outputs to the same. Additionally, the user can also specify area and the latency constraints for a CI. Now the problem of CI identification can be formally stated as follows:

Problem: *Given the DFG $G = (N, E)$ of a basic block and the maximum number of allowed CIs ($CI_{\max}$), find a set $S = \{C_1, C_2, \ldots, C_{CI_{\max}}\}$, that maximizes the speedup achieved for the complete application under the constraints*:

- $\forall C_k \in S$ the condition $C_k \subseteq G$ is satisfied, i.e., all elements of the set S are the subgraphs of the original graph G.

- $\forall C_k \in S$ the conditions $IN(C_k) \leq IN_{\max}$ and $OUT(C_k) \leq OUT_{\max}$ are satisfied, i.e., the number of GPR inputs/outputs to a CI does not exceed the maximum limit imposed by the architecture.

- $\forall C_k \in S$ the condition that C_k is *convex* is satisfied, i.e., there exists no path from node $a \in C_k$ to another node $b \in C_k$ through a node $w \notin C_k$.

- $\forall C_k, C_j \in S$ and $j \neq k$ the condition that C_k and C_j are mutually *schedulable* is satisfied, i.e., there exists no pair of nodes such that nodes $\{a, d\} \in C_k$ and nodes $\{b, c\} \in C_j$ where b is dependent on a and d is dependent on c.

- $\forall C_k \in S$ the condition $\sum_{k=1}^{CI_{\max}} AREA(C_k) \leq AREA_{\max}$ is satisfied, i.e., the combined silicon area of CIs does not exceed the maximum permissible value. Here $AREA(C_k)$ represents the area of a particular cut $C_k \in S$.

- $\forall C_k \in S$ the condition $CP(C_k) \leq LAT_{\max}$ is satisfied, i.e., the critical path, $CP(C_k)$, of any identified CI, C_k, is not greater than the maximum permissible HW latency.

The CIs identified through the solution of this problem are implemented in a CFU tightly attached to the base processor core. The execution of these CIs is usually dependent on the data accessed from the base processor core through GPRs implemented in the same. However, these GPR files usually have a limited number of access ports and thus impose strict constraints on the identification of CIs. Any CI that requires more communication ports than that supported by the GPR file is architecturally infeasible.

Our approach to overcome the restrictions imposed by the architectural constraints of GPR file access is to store the data needed for the execution of an identified CI as well as the outcome of this CI execution, in special registers, termed as internal registers. Such internal registers are assumed to be embedded in the CFU itself. For instance, let us assume that the execution of *CI*1 in Fig. 16.17 requires five GPR inputs, but the underlying architecture only supports two GPR read ports. This constraint can be overcome with the help of internal registers. A special instruction can be first used to move extra required inputs into these internal registers. Thus, the inputs of nodes n_j, n_r, and n_p can be moved from GPRs to internal registers using special move instructions. *CI*1 can then be executed by accessing five inputs—three from the internal registers and two from the GPRs. Similarly, internal registers can also be used to store the outcomes of CI execution, which cannot be written back to the GPRs directly due to the limited availability of GPR write ports. A special instruction can again be used to move this data back to the GPR in the next cycle. For example, the output of node n_j in Fig. 16.17 is moved from an internal register to the GPR with such a special instruction.

This approach of using internal registers converts an architectural constraint into a design parameter that is flexible and can be optimized based on the target application. More importantly, with this approach the problem stated earlier simplifies to two subproblems:

- Identify a cut under the all the remaining constraints of *convexity, schedulability, area, and latency*.

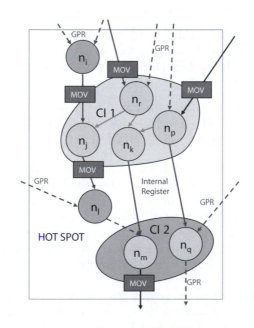

Communication via internal registers. Dotted lines represent GPRs, bold lines represent internal registers.

- Optimally use the combination of *internal*, as well as the *general-purpose, registers* for communication with the base processor core.

The first subproblem refers to identification of CIs independent of the architectural constraints imposed by the accessible GPR file ports. The second subproblem refers to the optimal use of the internal registers so that the overhead in terms of area and cycles wasted in moving values from and to these registers is minimized. The resulting instruction set customization problem can be solved in two independent steps as described next.

However, before describing the solutions to these mentioned subproblems, it is important to quantify the communication costs incurred due to the HW cycles wasted in moving the values to and from the internal registers. This can be better understood with the concept of HW and SW domains. All the nodes in clusters, which form the CIs, are implemented in HW in the CFU. As a result these nodes are said to be present in the *HW domain*. On the other hand, all the nodes of a DFG that are not part of any CI are executed in SW and thus are a part of the *SW domain*. There can be three possibilities for the use of these internal registers depending on the domains of communicating nodes. These three possibilities together with their communication costs are described here:

- A CI can use an internal register to store inputs from the SW domain. This situation is shown in Fig. 16.17, where the node n_r receives an input via internal register. The communication between node n_i and node n_j also falls under this category. In both cases, the loading of internal registers would incur communication cost. This cost depends on the number of GPR inputs allowed in any instruction.

- A CI can also use internal registers to provide outputs to the SW domain. This situation is shown in Fig. 16.17, where the node n_m produces an output via internal register. The communication between node n_j and node n_l also falls under this category. In both cases the storing back from internal registers would incur communication cost. As mentioned earlier, this cost depends on the number of GPR outputs allowed in any instruction.

- The third possibility of using internal registers is for communicating between two nodes, both of which are in different CIs.[1] This

---

[1] There exists another possibility where nodes within the same CI communicate. In such cases, both nodes are expected to be connected by a HW wire. Hence, this case is not treated here.

situation does not incur any communication cost, as the value is automatically stored in the communicating internal register after the execution of the producer CI and read by the consumer CI. The communication between nodes n_k and n_m and the nodes n_p and n_q illustrate this situation. Thus, communication between two CIs with *any kind of register (GPR or internal)* does not incur any communication overhead.

16.2.4   Synthesis Core Algorithm

We use the *integer linear programming (ILP)* technique to solve the optimization problem described in the previous section. This ILP technique tries to optimize an objective function so that none of the constraints are violated. If the constraints are contradictory and no solution can be obtained, it reports the problem as *infeasible*.

To keep the computation time within reasonable limits, as a first step all the constraints are considered, except the constraints on the number of inputs and outputs of a CI. However, the objective function in this step does take the penalty of using internal registers into account. The first step optimally decides whether a node remains in SW or forms a part of a CI that is implemented in HW. When this partition of the DFG in HW and SW domain is completed, the algorithm again uses another ILP model in the second step to decide about the means of communication between different nodes. As mentioned before, this communication is possible either through GPRs or through special internal registers. Fig. 16.18 presents the complete algorithm.

The algorithm starts with a DFG $G = (N, E)$ of a basic block, where all the nodes are in the SW domain. Each node and each edge is designated with an unique *identification (ID)* ranging from 1 to $|N|$ and 1 to $|E|$, respectively. Additionally, each node also has a *custom instruction identification (CID)*, that pertains to the CI that a particular node belongs to. All the nodes in the DFG are initialized with a CID of 0 or −1 (line 7 to 13 in Fig. 16.18). A value of −1 suggests that the node has the potential of becoming a part of a CI in some iteration. On the other hand, a CID value of 0 suggests that the node can never become a part of any CI. For example, the memory access (LOAD/STORE) nodes may come into this category. Such nodes never take part in any iteration. However, if the node accesses a local scratch-pad memory, it is possible to select this node to form a CI.

The first step of the instruction set customization algorithm is run iteratively over the graph, and in each iteration, exactly one CI is formed (lines 17 to 33). Each node with *ID* equal to j (written as n_j) is associated with a Boolean variable U_j in this step. An iteration counter is additionally used to keep track of the number of iterations (line 16). In each iteration the following sequence of steps is performed.

```
01    Procedure: InstSetCustomization()
02    Input:
03          DFG= (N, E )
04    Output:
05          Identified CIs with registers assigned
06    begin:
07          foreach node n ∈ N
08                if n cannot be part of CI//e.g n accesses memory
09                      Assign n with CID=0
10                else
11                      Assign n with CID=-1
12                endif
13          endfor
14
15    STEP 1:
16          iter = 1
17          while iter ≤ CImax
18                iter ++
19                // NILP is a set of nodes which are considered while constructing the ILP
20                NILP =φ
21                foreach node n ∈ N with CID < 0
22                      Add n to NILP
23                endfor
24                // Specify the objective function and the constraints
25                Develop an ILP problem with NILP
26                if (Infeasible)
27                      GOTO step2
28                endif
29                Solve the ILP problem
30                foreach node m ∈ NILP selected by the solver
31                      Assign CID of m as iter
32                endfor
33          endwhile
34
35    STEP 2:
36          // EILP is a set of edges which are considered while constructing the ILP
37          EILP= φ
38          foreach edge e ∈ E
39                if e is an input or output to CI
40                      Add e to EILP
41                endif
42          endfor
43          // Specify the objective function and the constraints
44          Develop an ILP problem with EILP
45          Solve the ILP problem
46          foreach edge b ∈ EILP
47                if b is selected by the solver
48                      Assign b as GPR
49                else
50                      Assign b as internal register
51                endif
52          endfor
53    end
```

■ **FIGURE 16.18**

Pseudo code for custom instruction generation.

- First, a set $NILP = \{U_1, U_2, \cdots, U_k\}$, such that the Boolean variable $U_i \in NILP$ if CID of the corresponding node n_i is less than 0, is created (lines 21 to 23).

- Then an objective function O_N where $O_N = f(U_1, U_2, \cdots, U_k)$, such that $U_i \in NILP$, is specified (line 25).

- Next, the constraints on the identification of clusters of nodes to form CIs are represented in form of mathematical inequalities (line 25). Let CN be such an inequality represented by $g(U_1, U_2, \cdots, U_k)$ OP $const$, where $g(U_1, U_2, \cdots, U_k)$ is a linear function of $U_i \in NILP$ and $OP \in \{<, >, \leq, \geq, =\}$.

- If feasible, the ILP solver then tries to maximize O_N, $\forall U_i \in NILP$ such that all the inequalities $CN_1, CN_2, \cdots, CN_{max}$ are satisfied (line 29). (The maximum number of inequalities depend on the structure of the DFG and the constraints considered.)

- When a variable U_i has a value of 1 in the solution, the corresponding node n_i is assigned with a CID value of the iteration counter. Otherwise, the initialized CID value of node n_i is not changed (lines 30 to 32).

The previously mentioned sequence is repeated until the solution becomes infeasible or the maximum bound of CI_{max} is reached. The role of the ILP in each iteration is to select a subgraph to form a CI such that the objective function is maximized under the constraints provided.

The working of step 1 is graphically represented in Fig. 16.19. Parts (b) to (f) of this figure shows the set of associated variables used for specifying the objective function, in iteration one and two, respectively. Moreover, in each iteration all the nodes whose associated variable U is solved for a value of 1 are assigned a CID value of iteration counter, as shown in part (c) and (f). All the other nodes whose associated variable U is solved for a value of 0 are not assigned any new CID but retain their initialized CID values. As a result, all the nodes selected in the same iteration are assigned the same CID, and they form parts of the same CI. Once selected in any iteration, the nodes do not compete with other nodes in subsequent iterations. This reduces the number of variables to be considered by the ILP solver in each iteration. Additionally, the iterative approach makes the complexity independent of the number of CIs to be formed.

In the second step, the ILP is used to optimally select the GPR inputs and outputs of the CIs. To develop an ILP optimization problem in this step each edge with an ID equal to k (written as e_k) is associated with a Boolean variable G_k. This step (lines 35 to 52 in Fig. 16.18) is only executed *once* at the end of the algorithm in the following sequence:

- First, a set $EILP = \{G_1, G_2, \cdots, G_k\}$, such that the Boolean variable $G_i \in EILP$ if the corresponding edge e_i is an input to/output from CI, is created (lines 38 to 42).

- Then an objective function O_G where $O_G = f(G_1, G_2, \cdots, G_k)$ such that $G_i \in EILP$ is specified (line 44).

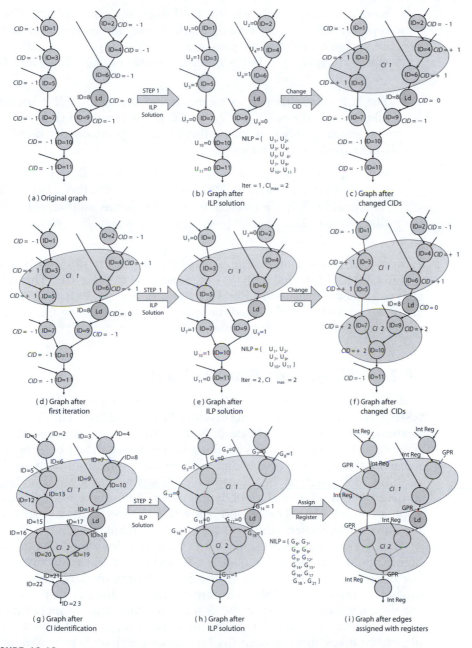

■ FIGURE 16.19

Instruction set customization algorithm: (a) original graph, (b), and (c) Step1 CI identification first iteration, (d), (e), and (f) Step1 CI identification second iteration, (g), (h), and (i) Step2 register assignment.

- Next, the constraints on the GPR inputs/outputs of the CIs are represented in form of mathematical inequalities (line 44). Let CG be such an inequality represented by $g(G_1, G_2, \cdots, G_k)$ OP $const$, where $g(G_1, G_2, \cdots, G_k)$ is a linear function of $G_i \in EILP$ and $OP \in \{<, >, \leq, \geq, =\}$.

- The ILP solver then tries to maximize O_G, $\forall G_i \in EILP$ such that all the inequalities $CG_1, CG_2, \cdots, CG_{max}$ are satisfied (line 45). (The maximum number of inequalities here depend on the number of CIs formed in the first step.)

- When a variable G_i has a value of 1 in the solution, the corresponding edge e_i forms a GPR input/output. On the other hand, if the variable G_i has a value of 0, the edge forms an internal register (lines 46 to 52).

Hence, the complete algorithm works in two steps to obtain CIs and to assign GPR and internal register inputs/outputs to these CIs. The second step is further illustrated in Fig. 16.19(g), (h), and (i). Part (h) shows the set of associated variables used for specifying the objective function and the constraints. The register assignment is shown in part (i).

16.2.5   ISA Synthesis–Based Design Flow

The previous ISA extension synthesis algorithm completes the envisioned ASIP design flow "on top" of the existing ADL-based flow. The design flow, as shown in Fig. 16.20, can be described as a three-phase process. The first phase consists of application profiling and hot spot detection. The designer is first required to profile and analyze the application with available profiling tools, such as the microprofiler described earlier in this chapter, to determine the computationally intensive code segment known as the application hot spot. The second phase then converts this hot spot to an intermediate representation using a compiler frontend.

This intermediate representation is then used by the CI identification algorithm to automatically identify CIs and to generate the corresponding instruction set extensions in the form of ADL models. The methodology is completely flexible to accommodate specific decisions taken by the designer. For example, if the designer is interested in including particular operators in CIs based on his experience, the CI identification algorithm takes into account (as far as possible) such priorities of the designer. These decisions of the designer are fed to the methodology using an interactive GUI. The third phase allows the automatic generation of complete SW toolkit (compiler, assembler, linker and loader, ISS) and the HW generation of the complete custom processor. This is specifically possible with the use of the ADL-based design flow described at the beginning of this chapter.

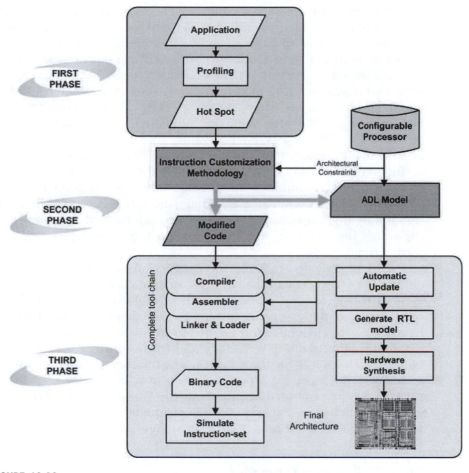

■ FIGURE 16.20

Complete design flow.

The instruction customization algorithm takes the intermediate representation of the C code (as generated by the LANCE compiler front-end [5]) as an input and identifies the CIs under the architectural, generic, and performance constraints. The architectural constraints are obtained from the base processor core, which needs to be specialized with the CFU(s). These constraints are fixed as far as the architecture of the base processor is fixed. The generic constraints are also fixed, as they need to be followed at any cost for the CIs to be physically possible. On the other hand, the performance constraints of chip area and the HW latency are variable and depend on the HW technology and the practical implications. For example, mobile devices have very tight area budgets, which might make the implementation of certain configurations of CFUs practically impossible, despite whatever the speedup achieved may be. Thus the final speedup attained is directly dependent on the amount of chip area permissible or the length of the critical path of the CIs.

The detailed instruction set customization algorithm is therefore run over a set of values with different combinations of area and critical path length to obtain a set of Pareto points. The various SW latency, HW latency, and the HW area values of individual operators are read from an XML file. All the HW latency values and the area values here are obtained with respect to a 32-bit multiplier. For example, a 32-bit (type = "int") adder would require only 12% of the area required for implementing a 32-bit multiplier. These values are calculated once using the Synopsys Design Compiler and remain the same as far as the simulation library does not change. The XML file can be further extended with the values for 16-bit or 8-bit operations as required by the application source code.

A GUI is provided to steer the instruction set customization algorithm. The GUI acts as a general SW integrated development environment (IDE). Therefore, it provides a file navigation facility, source code browsing, and parameter input facility. Based on the user inputs, the instruction set customization algorithm is run over different values of maximum permissible HW latency and maximum permissible area to obtain Pareto points for these values. The number of Pareto points generated directly depends on the input data provided by the user through the GUI. All the generated Pareto points can viewed in a two-dimensional graph in which the y-axis signifies the estimated speedup and the x-axis signifies the estimated area used in the potential CIs.

16.2.6 Speedup Estimation

A certain cost function is used to estimate the speedup of a potential CI. The speedup can be calculated as the ratio of number of cycles required by a cluster of nodes when executed in SW to that required when executed in HW. In other words, the speedup is the ratio of the SW latency to the HW latency of the potential CI.

$$Estimated\ Speedup = \frac{(SW\,Latency)_{Cluster}}{(HW\,Latency)_{Cluster}}$$

Calculation of the SW latency of the cluster is quite straightforward, as it is the summation of the SW latencies of all the nodes included in the cluster.

$$(SW\,Latency)_{Cluster} = \sum_{Nodes\,In\,Cluster} (SW\,Latency)_{node}$$

The calculation of the cycles needed, when an instruction is executed in HW, is much more complicated. Fig. 16.21 illustrates the approach needed for HW cycle estimation for a typical CI. As shown in the figure, the set $\{n_1, n_3, n_4, n_7, n_8, n_{11}\}$ forms $CI1$. Similarly, the sets $\{n_9\}$ and $\{n_{14}, n_{15}, n_6\}$ form parts of $CI2$ and $CI3$, respectively. On the other hand, the nodes $\{n_2, n_{10}, n_{13}, n_{12}, n_5\}$ are not included in any CI. The nodes that are a part of any CI are present in the HW domain, whereas the nodes that are not the part of any CI are present in the SW domain. Since the nodes in the HW domain can communicate via the internal registers, they do not incur any communication cost. But, when the nodes in HW communicate from (to) the nodes in SW, cycles are needed to load (store) the values. This communication cost can be better understood by considering the example of MIPS32 CorExtend configurable processor, which allows for only two GPR reads from an instruction in a single cycle (only two read ports available). If there are more than two inputs to an instruction like $CI1$, internal registers can be used to store the extra inputs. And any *two* operands can be moved into the internal registers in an extra cycle using the two available read ports, given such an architecture. The same processor only allows for a single output from an instruction in a single cycle (only one write port available). So any *one* operand can be stored back from the instruction using one extra cycle, given such a processor.

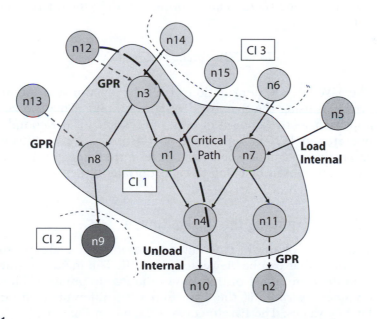

 FIGURE 16.21

Cost function.

Furthermore, the critical path determines the amount of cycles needed for the HW execution of the CI. The critical path is the longest path among the set of paths formed by the accumulated HW latencies of all the operators along that path. For instance in Fig. 16.21 a set of paths, $P = \{P_1, P_2, P_3, P_4\}$ can be created for $CI1$, where $P_1 = \{n_3, n_8\}$, $P_2 = \{n_3, n_1, n_4\}$, $P_3 = \{n_7, n_4\}$, and $P_4 = \{n_7, n_{11}\}$. Assuming that all the operators here individually take the same amount of cycles to be executed in HW, the critical path is P_2, since the cumulative HW latencies of nodes along this path are higher than the cumulative HW latencies of nodes along any other path.

If the cycle length of the clock used is greater than this critical path (the operators along the critical path can be executed in single cycle), it is guaranteed that all the other operators in other paths can also be executed within this cycle. This is possible by the execution of all the operators that are not the part of the critical path in parallel by providing appropriate HW resources. For example, in Fig. 16.21 nodes n_3 and n_7 are not dependent on any other node in $CI1$ and can be executed in parallel. After their execution, the nodes n_8, n_1, and n_{11} can be executed in parallel (assuming all operators have equal HW latencies). Therefore the HW cycles required for the execution of any CI depends on its critical path length.

The HW latency of the cluster can thus be given as the summation of cycles required to execute the operators in *critical path (CP)* and cycles used in loading (*LOAD*) and storing (*STORE*) the operands (using internal registers).

$$Total\ HW\ Latency = CP + LOAD + STORE$$

The HW latency information of all the operators is obtained by HW synthesis using the Synopsys Design Compiler. This is a coarse-grained approach, as the amount of cycles necessary for the completion of each operator in HW is dependent on the processor architecture and the respective instruction set. Nonetheless, the latency information can be easily configured by using the appropriate library.

16.2.7 Exploring the Design Space

The output of the instruction set customization algorithm is a set of solutions that can be viewed on the GUI. The major output of this GUI is in the form of a plot that shows the Pareto points of all the solutions obtained by running the algorithm with different maximum area and latency values. The Pareto curve is shown in Fig. 16.22.

The files of the selected configuration can be saved and the GUI can be used for further interaction. The GUI also displays the solution in the form of a DFG where all the nodes that belong to the same CI are

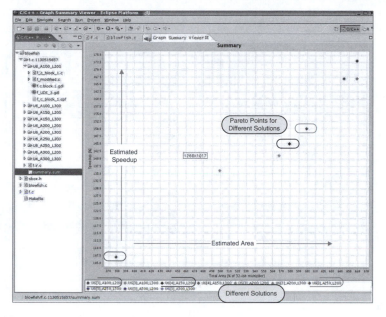

■ FIGURE 16.22

GUI: Pareto points.

marked with similar colors and boxes. This GUI facility is shown in Fig. 16.23. The complete output consist of the following files:

- *GDL file* contains the contains information on each node and edge of the DFG. This information is used by the GUI to display the graph so that the user can make out which node is a part of which CI.

- *Modified C source code file* contains the source code of the original application with the CIs inserted. This code can then be compiled with the updated SW tool chain.

- *ADL implementation file* contains the behavior code of the CIs in a format that can be used to generate the ADL model of the CFU.

After running the instruction set customization algorithm on the application DFG, each node is assigned a *CID* that designates the CI to which the node belongs. Additionally, the edges are assigned the register types (internal or GPR). Now for generating the modified C code with CI macros inserted in it, two more steps are to be performed:

- First it is required to *schedule* the nodes, taking into account the dependencies between the same. This step is important to maintain

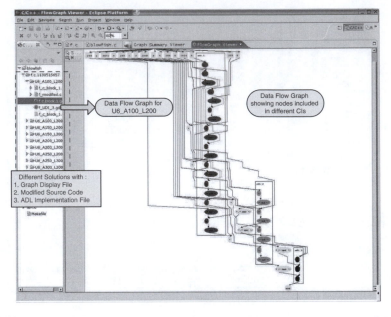

■ FIGURE 16.23

GUI: data flow graph of different solutions.

the correctness of the application code. We use a variant of the well known list scheduling algorithm for this purpose.

■ Another important link missing at this stage of the methodology is the allocation of internal registers. It is worth stating that in the instruction set customization algorithm described earlier, the edges were only assigned with the register types. Now it is required to allocate physical registers to these registers. In our approach, this is performed with a left-edge algorithm known from behavioral synthesis.

16.2.8 SW Tools Retargeting and Architecture Implementation

The final phase in the instruction set customization design flow is the automatic modification of the SW tools to accommodate new instructions in form of CIs. Additionally, the HW description of the complete CFU(s) that is tightly attached to the base processor core is generated in form of an ADL model. Fig. 16.24 illustrates this for the MIPS CorExtend customizable processor core. The region shown within the dotted line is the SW development environment used to design applications for MIPS processors. This environment is updated to accommodate the newly generated CIs using the ADL design flow. The idea is to use the ADL model generated in the second phase of the design process to automatically update the complete SW tool chain.

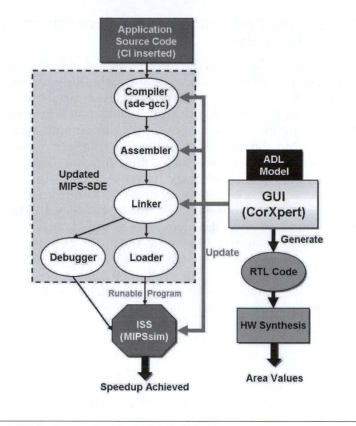

■ FIGURE 16.24

HW synthesis and SW tool generation for MIPS CorExtend.

To hide the complexity of the MIPS CorExtend RTL signal interface and the MIPSsim CorExtend simulation interface [9] from the designer, the CoWare CorXpert tool (see Section 16.2.2) is used. The instruction set customization methodology generates the instruction data path of the identified CIs in a form useable by this tool. CorXpert can then be used to update the SW tool chain and also to generate the RTL model description of the identified CIs. CorXpert utilizes the LISATek processor design technology to automate the generation of the simulator and RTL implementation model. It generates a LISA 2.0 description of the CorExtend module, which can be be exported from the tool for investigation or further modification using the LISATek Processor Designer.

The modified application can thus be compiled using the updated SW tool chain. The updated ISS (MIPSsim in this case) can then be used to obtain the values of real speedup achieved with the generated CIs. The RTL model generated from the ADL model is further synthesized, and the real values of area overhead are obtained. Thus the complete flow starting from the application hot spot down to HW synthesis of the identified CIs is achieved.

16.2.9   Case Study: Instruction Set Customization for Blowfish Encryption

This case study deals with the instruction set customization of a MIPS32 CorExtend configurable processor for a specific cryptographic application known as *Blowfish*. Blowfish [12] is a variable-length key, 64-bit block cipher. It is most suitable for applications where the key does not change often, like a communication link or an automatic file encryptor. The algorithm consists of two parts: data encryption and key expansion. The data encryption occurs via a simple function (termed as the *F* function), which is iterated 16 times. Each round consists of a key dependent *permutation*, and a key and data dependent *substitution*. All operations are XORs and additions on 32-bit words. The only additional operations are four indexed array data lookups per round. On the other hand, the key expansion converts a key of at most 448 bits (variable key-length) into several subkey arrays totaling 4,168 bytes. These keys must be precomputed before any data encryption or decryption. Blowfish uses four 32-bit key-dependent S-boxes, each of which has 256 entries.

The CIs identified for the Blowfish application are implemented in a CFU (CorExtend module), tightly coupled to the MIPS32 base processor core. This processor forces architectural constrains in the form of two inputs from the GPR file and one output to the same. Thus, all the identified CIs must use only two inputs and one output from and to the GPRs. Additionally, the CIs also use internal registers for communication between different CIs. Moreover, the MIPS32 CorExtend architecture does not allow main memory access from the CIs. This places another constraint on the identification of CIs. However, the CorExtend module does allow local scratch-pad memory accesses from the CIs. Later we will show how this facility can be exploited to achieve better speedup. First, the hot spot of Blowfish is determined by profiling. The instruction set customization methodology is then configured for the MIPS32 CorExtend processor core, which provides the architectural constraints for CI identification.

Automatic CI Identification for Blowfish

The instruction set customization algorithm is run over the hot spot function of Blowfish. In this case, three CIs are generated. Our methodology also automatically generates the modified source code of the application, with the CIs inserted in it. Fig. 16.25 shows the modified code of the *F* function after various CIs are identified in it. The modified code is a mixture of three address code and the CIs.

The automatically generated C behavior code of these CIs is further shown in Figs. 16.26 and 16.27. The term *UDI_INTERNAL* in this example refers to the internal register file that is implemented in the CFU. This internal register file is represented as an array, in the behavior code, where

```
/*****************************************
* Original Code of Function F
*****************************************/
extern unsigned long S[4][256];
unsigned long F(unsigned long x)
{
        unsigned short a;
        unsigned short b;
        unsigned short c;
        unsigned short d;
        unsigned long  y;

        d = x & 0x00FF;
        x >>= 8;
        c = x & 0x00FF;
        x >>= 8;
        b = x & 0x00FF;
        x >>= 8;
        a = x & 0x00FF;
        y = S[0][a] + S[1][b];
        y = y ^ S[2][c];
        y = y + S[3][d];

        return y;
}

/*****************************************
* Modified Code with CIs inserted
*****************************************/
extern unsigned long S[4][256];
unsigned long F (
    unsigned long x)
{
    int _dummy_RS = 1, _dummy_RT=1;
    unsigned long *t58, *t66, *t84, *t75;
    unsigned long t85, ldB29C1,ldB28C0;
    unsigned long ldB41C2,ldB53C3;
    //
    // C Code for Block 1
    //
    t58 = CI_1 ( x, S);
    t66=CI_Unload_Internal(_dummy_RS,_dummy_RT,3);

    ldB29C1 = */ * load */ t66;
    ldB28C0 = */ * load */ t58;

    t84=CI_2 (ldB28C0,ldB29C1);
    t75 = CI_Unload_Internal(_dummy_RS,_dummy_RT,5);

    ldB41C2 = */ * load */ t75;
    ldB53C3 = */ * load */ t84;
    t85 = CI_3(ldB53C3,ldB41C2);
    return t85;
}
```

■ FIGURE 16.25

Blowfish F function: original source code and modified code with CIs inserted.

the index represents the register identifications. The identifiers *UDI_RS* and *UDI_RT* and *UDI_WRITE_GPR*() are MIPS32 CorExtend specific macros and are used to designate the base processor GPRs. *UDI_RS* and *UDI_RT* designate the two input GPRs, and the *UDI_WRITE_GPR*() is used to designate the GPR that stores the result of the CI. The GPR allocation

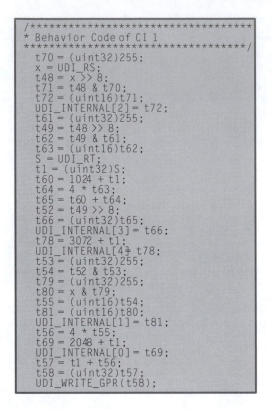

```
/**********************************
 * Behavior Code of CI 1
 **********************************/
t70 = (uint32)255;
x = UDI_RS;
t48 = x >> 8;
t71 = t48 & t70;
t72 = (uint16)t71;
UDI_INTERNAL[2] = t72;
t61 = (uint32)255;
t49 = t48 >> 8;
t62 = t49 & t61;
t63 = (uint16)t62;
S = UDI_RT;
t1 = (uint32)S;
t60 = 1024 + t1;
t64 = 4 * t63;
t65 = t60 + t64;
t52 = t49 >> 8;
t66 = (uint32)t65;
UDI_INTERNAL[3] = t66;
t78 = 3072 + t1;
UDI_INTERNAL[4] = t78;
t53 = (uint32)255;
t54 = t52 & t53;
t79 = (uint32)255;
t80 = x & t79;
t55 = (uint16)t54;
t81 = (uint16)t80;
UDI_INTERNAL[1] = t81;
t56 = 4 * t55;
t69 = 2048 + t1;
UDI_INTERNAL[0] = t69;
t57 = t1 + t56;
t58 = (uint32)t57;
UDI_WRITE_GPR(t58);
```

▪ **FIGURE 16.26**

Behavior code of CI 1 as implemented in CorXpert.

is the responsibility of the MIPS32 compiler. This behavior code is then used by the CorXpert tool to generate the HW description of the CIs.

It is worth stating that since MIPS32 has a LOAD/STORE architecture, efficient use of GPRs is very important for achieving optimum performance. The GPR allocation (done by the MIPS32 compiler) directly affects the amount of cycles required for the execution of any CI. Moreover, the most efficient GPR allocation is achieved when the CIs use all the three registers allowed (two for reading and one for writing the results). Hence we use *dummy_* variables in case the CI does not use all the three registers. The use of *dummy_* variables is shown in the example 16.25 in the instruction *CI_Unload_Internal*.

The modified code, with the CIs inserted, is then run on the MIPS ISS (termed as MIPSsim) to calculate the amount of speedup achieved with the identified CIs. The estimated and the simulated speedup for various configuration of the instruction set customization algorithm is illustrated in Fig. 16.28. A configuration corresponds to the area and latency constraints that the identified CIs need to follow. To get the trend in the speedup, the algorithm was run for various configurations.

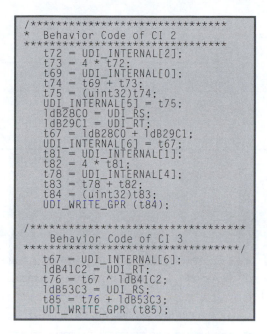

```
/*******************************
*  Behavior Code of CI 2
*******************************
    t72 = UDI_INTERNAL[2];
    t73 = 4 * t72;
    t69 = UDI_INTERNAL[0];
    t74 = t69 + t73;
    t75 = (uint32)t74;
    UDI_INTERNAL[5] = t75;
    1dB28C0 = UDI_RS;
    1dB29C1 = UDI_RT;
    t67 = 1dB28C0 + 1dB29C1;
    UDI_INTERNAL[6] = t67;
    t81 = UDI_INTERNAL[1];
    t82 = 4 * t81;
    t78 = UDI_INTERNAL[4];
    t83 = t78 + t82;
    t84 = (uint32)t83;
    UDI_WRITE_GPR (t84);


/***********************************
    Behavior Code of CI 3
***********************************/
    t67 = UDI_INTERNAL[6];
    1dB41C2 = UDI_RT;
    t76 = t67 ^ 1dB41C2;
    1dB53C3 = UDI_RS;
    t85 = t76 + 1dB53C3;
    UDI_WRITE_GPR (t85);
```

■ **FIGURE 16.27**

Behavior code of CIs 2 and 3 as implemented in CorXpert.

For each such configuration, the algorithm identifies a set of CIs. The speedup obtained by the identified CIs (under the area and latency constraints specified in the configuration) is shown in the figure. For example, $A < 200$ and $L = 2$ means that the amount of area *per* CI should not be greater than 200% (area with respect to 32-bit multiplier) and the CI should be executed within two cycles.[2] It can be seen that the speedup estimation in this case is quite accurate and clearly shows the trend in the amount of speedup achieved for various configuration. The highest simulated speedup achieved is 1.8 times.

The amount of area overhead incurred by the HW implementation of CIs is shown in Fig. 16.29. For obtaining these values, the CorXpert tool was first used to generate the RTL code in Verilog from the behavior code of the CIs. Synopsys Design Compiler was then used for the HW synthesis of the complete CFU.[3] The clock length was set at 5 ns, and the clock constraints were met in all cases. As seen in Fig. 16.29, the area estimation is not as accurate as the speedup estimation. The cost function for area calculation is quite coarsely grained, as it just provides the total area as the sum of individual area of all operators in CIs. The calculated area is more

---

[2] The CorExtend module supports a three-stage integer pipeline.

[3] It is assumed here that the constant operands (e.g., 7 is a constant in equation $t1 = x >> 7$) required as the inputs to some CIs are implemented in the CFU itself.

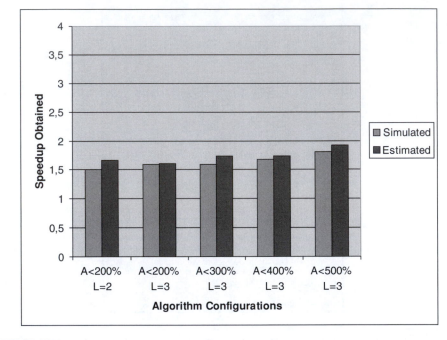

■ **FIGURE 16.28**

Speedup for Blowfish hot spot (*F*) for different algorithm configurations: simulated and estimated. '*A*' denotes maximum permissible area per CI, and '*L*' denotes maximum permissible HW latency for CI execution. Both '*A*' and '*L*' are specified by the user.

than the estimated area, since we do not consider the area used in the decoding logic in our cost function. However, the estimation does show the trend in the area overhead incurred with different configurations.

Use of Scratch-Pad Memories

Since the Blowfish hot spot contains four memory accesses, the speedup obtained with the identified CIs (without memory accesses) is not too high. To increase the speedup, the lookup tables accessed by the *F* function can be stored in the local scratch-pad memory and can be accessed from the CIs. These memories are local to the CFU and can be used to store tables and constants that do not require frequent updates. It is useful for the Blowfish application to store the *S-Boxes* in such scratch-pad memories, as they are initialized in the beginning and remain fixed throughout the encryption and decryption process.

Our tool framework provides a facility for the user to specify the scratch-pad memories from different CIs by marking the *LOAD* node as the *SCRATCH* accessing node. Each CI shown in the new solution is allowed single scratch-pad memory access (assuming the memory has

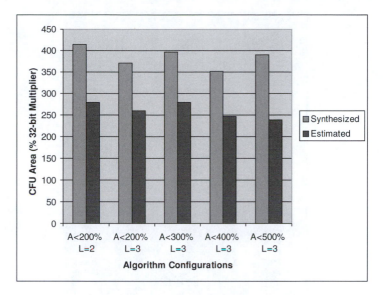

■ FIGURE 16.29

CFU area for different algorithm configurations: synthesized (with 5 ns clock length) and estimated. '*A*' denotes maximum permissible area per CI, and '*L*' denotes maximum permissible HW latency for CI execution. Both '*A*' and '*L*' are specified by the user.

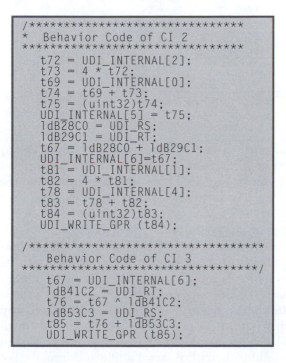

```
/********************************
 *  Behavior Code of CI 2
 ********************************
    t72 = UDI_INTERNAL[2];
    t73 = 4 * t72;
    t69 = UDI_INTERNAL[0];
    t74 = t69 + t73;
    t75 = (uint32)t74;
    UDI_INTERNAL[5] = t75;
    1dB28C0 = UDI_RS;
    1dB29C1 = UDI_RT;
    t67 = 1dB28C0 + 1dB29C1;
    UDI_INTERNAL[6]=t67;
    t81 = UDI_INTERNAL[1];
    t82 = 4 * t81;
    t78 = UDI_INTERNAL[4];
    t83 = t78 + t82;
    t84 = (uint32)t83;
    UDI_WRITE_GPR (t84);

/**********************************
    Behavior Code of CI 3
 **********************************/
    t67 = UDI_INTERNAL[6];
    1dB41C2 = UDI_RT;
    t76 = t67 ^ 1dB41C2;
    1dB53C3 = UDI_RS;
    t85 = t76 + 1dB53C3;
    UDI_WRITE_GPR (t85);
```

■ FIGURE 16.30

Blowfish *F* function: source code after CI identification (CIs use scratch-pad memories).

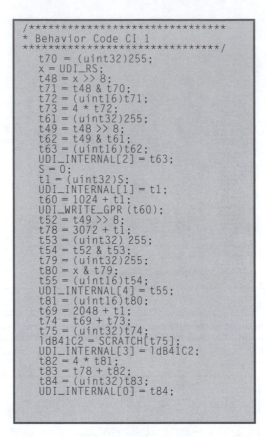

```
/******************************
 * Behavior Code CI 1
 ******************************/
     t70 = (uint32)255;
     x = UDI_RS;
     t48 = x >> 8;
     t71 = t48 & t70;
     t72 = (uint16)t71;
     t73 = 4 * t72;
     t61 = (uint32)255;
     t49 = t48 >> 8;
     t62 = t49 & t61;
     t63 = (uint16)t62;
     UDI_INTERNAL[2] = t63;
     S = 0;
     t1 = (uint32)S;
     UDI_INTERNAL[1] = t1;
     t60 = 1024 + t1;
     UDI_WRITE_GPR (t60);
     t52 = t49 >> 8;
     t78 = 3072 + t1;
     t53 = (uint32) 255;
     t54 = t52 & t53;
     t79 = (uint32)255;
     t80 = x & t79;
     t55 = (uint16)t54;
     UDI_INTERNAL[4] = t55;
     t81 = (uint16)t80;
     t69 = 2048 + t1;
     t74 = t69 + t73;
     t75 = (uint32)t74;
     ldB41C2 = SCRATCH[t75];
     UDI_INTERNAL[3] = ldB41C2;
     t82 = 4 * t81;
     t83 = t78 + t82;
     t84 = (uint32)t83;
     UDI_INTERNAL[0] = t84;
```

■ **FIGURE 16.31**

Behavior code of CI 1 using scratch-pad memories.

one read port). The modified source code of the application hot spot is shown in Fig. 16.30. As shown in this example, all the LOADs that were present in Fig. 16.25 are now present inside the CIs. The behavior code of the CIs formed is shown in Figs. 16.31 and 16.32. The variable *SCRATCH* shown in this example is the scratch-pad memory, implemented as an array in the behavior code.

The case study on the Blowfish application has demonstrated the importance of our multiphased methodology with user interaction. The user can typically start from the application source code (written in C) and obtain the description of the customized processor (with enriched instruction set) quickly. With the fully automated approach, the runtime of the instruction set customization algorithm forms the major part of the total time required by the methodology. Since our instruction set customization algorithm can handle moderately large DFGs (approximately 100 nodes) within at most a few CPU minutes, fully customized

```
/******************************
* Behavior Code of CI2
******************************/
    t84 = UDI_INTERNAL[0];
    1dB53C3 = SCRATCH[t84];
    UDI_INTERNAL[5] = 1dB53C3;
    t63 = UDI_INTERNAL[2];
    t64 = 4 * t63;
    t60 = UDI_RS;
    t65 = t60 + t64;
    t66 = (uint32)t65;
    UDI_INTERNAL[6] = t66;
    t55 = UDI_INTERNAL[4];
    t56 = 4 * t55;
    t1 = UDI_INTERNAL[1];
    t57 = t1 + t56;
    t58 = (uint32)t57;
    UDI_WRITE_GPR(t58);

/*******************************
* Behavior Code of CI 3
*******************************/
    t58 = UDI_RS;
    1dB28C0 = SCRATCH[t58];
    1dB29C1 = UDI_RT;
    t67 = 1dB28C0 + 1dB29C1;
    1dB41C2 = UDI_INTERNAL[3];
    t76 = t67^1dB41C2;
    1dB53C3 = UDI_INTERNAL[5];
    t85 = t76 + 1dB53C3;
    UDI_WRITE_GPR (t85);

/********************************
*Behavior Code of CI_4
********************************/
    t66 = UDI_INTERNAL[6];
    1dB29C1 = SCRATCH[t66];
    UDI_WRITE_GPR (1dB29C1);
```

■ **FIGURE 16.32**

Behavior code of CIs 2,3,4 using scratch-pad memories.

processor description for moderately sized application hot spots can be achieved within a short time. The case study has further proven the usefulness of user interaction, as the user identifies the scratch-pad memory access by marking the nodes in the DFG of the hot spot. Speedup of the application is considerably increased with the use of these scratch-pad memories.

Moreover, all the other less complex but tedious tasks of modified code generation and ADL model generation have been automated. As a result, the user can experiment with different configurations to obtain results as per the requirements and simultaneously meet the time-to market demand. It is worth mentioning here that the estimated speedup and area are provided for comparison of various solutions. These estimates in no case try to model the absolute values. For modeling the absolute parameters, advanced cost functions and libraries would be required.

16.3 SUMMARY AND OUTLOOK

The microprofiler and ISA synthesis tools described in this chapter cooperate to support automated ASIP design beyond today's ADL-based methodology. The microprofiler permits a much more fine-grained type of profiling than existing source-level profilers. Therefore, it meets the demands for optimizing ASIP architectures rather than optimizing source code. At the same time, it does not require a detailed ADL model of the target machine to help in early architecture decisions. The profiling data can be exported for the ISA extension synthesis methodology and tools proposed in this chapter. ISA extension synthesis provides a new level of automation in ASIP design, yet allows for incorporating the designer's expert knowledge through interaction.

We expect that the demand for such tools (and similar tools as described in various chapters of this book) will increase with the growing availability of *reconfigurable* processor architectures. The customizable portion of such reconfigurable ASIPs is stored in an embedded FPGA-like HW architecture, which allows for frequent reconfiguration.

For instance, Stretch [13] introduced the S5000 family of processor with a customizable FPGA-based coprocessor and a comprehensive suite of development tools that enable developers to automatically configure and optimize the processor using only their C/C++ code. The S5000 processor chip is powered by the Stretch S5 engine (Fig. 16.33), which incorporates the Tensilica Xtensa RISC processor core and the Stretch *instruction set extension fabric (ISEF)*. The ISEF is a software-configurable datapath based on proprietary programmable logic. Using

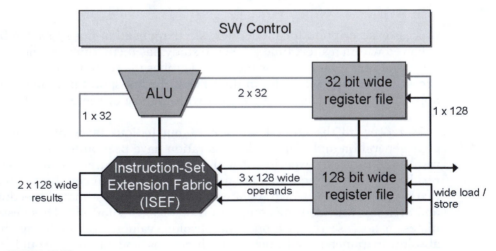

■ **FIGURE 16.33**

Stretch S5 engine (redrawn from [13]).

the ISEF, system, designers extend the processor instruction set and define the new instructions using only their C/C++ code.

Further semiconductor companies will shortly introduce architectures with similar post-fabrication reconfiguration capabilities. For the nonexpert user, use is enabled only via powerful ASIP design tools that (1) help the designer to efficiently and clearly identify *what* parts of an application to optimize, while (2) abstracting from tedious details on *how* to implement the necessary HW accelerator configurations.

ACKNOWLEDGEMENTS

The author gratefully acknowledges the contributions to the work described in this chapter made by K. Karuri, S. Kraemer, M. Al Faruque, and M. Pandey from the Institute for Integrated Signal Processing Systems at RWTH Aachen University, Germany. Research on the tools described in this chapter has been partially supported by CoWare, Inc., and Tokyo Electron Ltd.

CHAPTER 17

DESIGNING SOFT PROCESSORS FOR FPGAS

Göran Bilski, Sundararajarao Mohan, and Ralph Wittig

In this chapter we show how certain architectural elements, such as buses, muxes, ALUs, register files, and FIFOs, can be implemented on FPGAs, and the tradeoffs involved in selecting different size parameters for these architectural elements. The speed and area of these implementations dictate whether we choose a bus-based or mux-based processor implementation. The use of lookup tables (LUTs) instead of logic gates implies that some additional logic is free, but beyond a certain point there is a big variation in terms of the speed/area of an implementation. We propose to illustrate all these ideas with examples from the Micro-Blaze implementation. One such example is the FSL interface on Micro-Blaze: we have found that it is better to add a "function accelerator," with FSL-based inputs and outputs, rather than custom instructions to our soft processor.

17.1 OVERVIEW

This section provides an overview of FPGA architectures and the motivation for building soft processors on FPGAs. FPGAs have evolved from small arrays of programmable logic elements with a sparse mesh of programmable interconnect to large arrays consisting of tens of thousands of programmable logic elements, embedded in a very flexible programmable interconnect fabric. FPGA design and application-specific integrated circuit (ASIC) design have become very similar in terms of methodologies, flows, and the types of designs that can be implemented. FPGAs, however, have the advantage of lower costs and faster time to market, because there is no need to generate masks or wait for the IC to be fabricated. Additional details of FPGAs that distinguish them from ASIC or custom integrated circuits are described next. The rest of this section looks at the background and motivation for creating soft processors and processor accelerators on FPGAs, before introducing the MicroBlaze soft processor in the next section.

17.1.1 FPGA Architecture Overview

FPGAs have evolved from simple arrays of programmable logic elements connected through switchboxes to complex arrays containing LUTs, block random access memories (BRAMs), and other specialized blocks. For example, Fig. 17.1 shows a high-performance Virtex^{TM}-II Pro FPGA, from Xilinx, Inc., containing configurable logic blocks (CLBs) that contain several LUTs, BRAM, configurable input/output blocks, hard processor blocks, digital clock modules (DCMs) and multigigabit transceivers (MGTs).

While the high-performance PowerPC processor block is available in certain members of the Virtex family of FPGAs, it is possible to implement powerful 32-bit microprocessors using the other available resources, such as LUTs and BRAMs, on almost any FPGA. Typical FPGAs contain anywhere from a few thousand 4-input LUTs to a hundred thousand LUTs, and several hundred kilobits to a few megabits

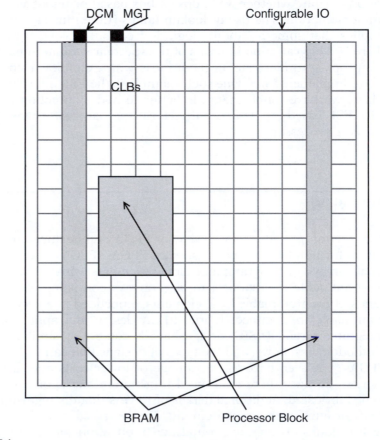

■ **FIGURE 17.1**

Typical FPGA with configurable logic blocks (CLBs), BRAMs, IO, and processor.

of BRAM [1], while a typical soft processor can be implemented with one thousand to two thousand LUTs. Designs implemented on FPGAs can take advantage of special-purpose blocks that are present on the FPGAs (in addition to the basic LUTs).

Fig. 17.2 shows the logic cell in a typical modern FPGA. A 4-input lookup table (4-LUT) can be programmed to implement any logical function of 4 or fewer inputs. The output signal can then be optionally latched or registered. Abundant routing resources are available to interconnect these lookup tables to form almost any desired logic circuit. In addition to the 4-LUT and the output register, various special-purpose circuits are added to the basic cell to speed up commonly used functions, such as multiplexing and addition/subtraction.

Early FPGAs exposed the details of all the routing resources, such as wire segments of various predefined lengths, switch boxes, and connection boxes, to the users so that users could make efficient use of scarce resources. Details of various classical FPGA architectures are described in [2]. Modern FPGAs have abundant routing resources to connect up the logic elements such as LUTs and muxes, and FPGA vendors do not document the interconnect details. Circuit designers typically assume that logic elements placed closer together can be connected together with less delay than logic elements that are placed further apart.

Special-purpose logic implemented on FPGAs from Xilinx or other companies typically includes BRAM (in chunks of several kilobits each), fast carry logic for arithmetic, special muxes, configurable flipflops, LUTs that can be treated as small (16- or 32-bit) RAMs or shift registers, and DSP logic in the form of multiply-accumulate blocks. Any circuit

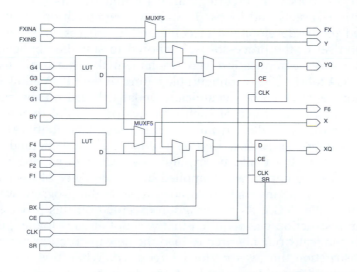

■ **FIGURE 17.2**

Simplified picture of a logic cell showing two 4-LUTs, 2 flipflops, and several muxes.

design that targets FPGAs can specifically target a portion of the circuit to these special-purpose blocks to improve speed or area. Synthesis tools that generate circuit implementations from high-level HDL descriptions typically incorporate several techniques to automatically map circuit fragments to these elements, but designers can still achieve significant improvements by manually targeting these elements in specific cases, as described in the subsection "Implementation Specific Details" in Section 17.3.2.

17.1.2 Soft Processors in FPGAs

Work on building soft processors in FPGAs has been motivated by two separate considerations. One of the motivations has been the fact that as FPGAs increased in size, it became possible for any FPGA user to define and build a soft processor. Another motivation was the fact that FPGAs were being used by researchers to build hardware accelerators for conventional processors [3]. It was natural to try to combine the processor and the accelerator on the same fabric by building soft processors. Yet another motivation for building soft processors was the desire to aggregate multiple functions on to the same chip. This was the same idea that led ASIC designers to incorporate processors into their designs. As a result of all these different motivations, there is no single application area for soft processors, and soft processors are used wherever FPGAs are used.

In the early 1990s, FPGAs became big enough to support the implementation of small 8-bit or 16-bit processors. At the same time, other FPGA users were creating hardware that was used to accelerate algorithm kernels and loops. The natural combination of these two ideas led to the development of configurable processors, where the instruction set of the processor was tailored to suit the application, and portions of the application were implemented in hardware using the same FPGA fabric. The reconfigurable nature of the FPGA allowed the instruction set itself to be dynamically updated based on the code to be executed. One example of such a processor was the dynamic instruction set computer (DISC) from Brigham Young University [4]. DISC had an instruction set where each instruction was implemented in one row of an FPGA that could be dynamically reconfigured one row at a time. The user program was compiled into a sequence of instructions, and a subset of the instructions needed to start the program was first loaded on to the FPGA. As the program executed on this processor, whenever an instruction was required that was not already loaded, the FPGA was dynamically reconfigured to load the new instruction, replacing another instruction that was no longer required. While this was an interesting concept, practical considerations such as the reconfiguration rates (microseconds to milliseconds), FPGA size limitations, and the lack of caches limited the use of these techniques.

While the potential for implementing custom processors using FPGAs was being explored by various researchers at conferences, such as the FPGA Custom Computing Machines Conference [5] and the International Conference on Field Programmable Logic (FPL) [6], other FPGA users saw the need for standard processors implemented on FPGAs, otherwise known as soft implementations of standard processors. CAST, Inc., described a soft implementation of the 8051 [7], Jan Gray reported his work on implementing a 16-bit processor on a Xilinx FPGA [8], and Ken Chapman reported a small 8/16 bit processor (now known as PicoBlaze) [9]. The key idea in all these implementations was that the processor itself was chosen to be small and simple in order to fit on the FPGAs of the day. Gray, for example, reported that his complete 16-bit processor system required 2 BRAMs, 257 4-LUTs, 71 flipflops, and 130 tristate buffers, and occupied just 16% of the LUT resources of a small SpartanII FPGA. While conventional processors were designed to be built using custom silicon or ASIC fabrication technologies, some of these new soft processors were built for efficient implementation on FPGA fabrics. The KCPSM (now known as PicoBlaze [9]) processor was designed for implementing simple state machines, programmed in assembly language, on FPGAs, while the NIOS family of 16-bit and 32-bit soft processors was designed as a general-purpose processor family [10]. The introduction of the NIOS soft processor family was quickly followed by the introduction of the MicroBlaze soft processor, which is described in detail in the rest of this chapter. Both NIOS and MicroBlaze were general-purpose RISC architectures designed to be used with general-purpose compiler tools and programmed in high-level languages such as C.

17.1.3 Overview of Processor Acceleration

While processors (soft or hard) are good at performing general-purpose tasks, many applications contain special tasks that general-purpose processors cannot perform at the required speeds. Classic examples of such tasks are vector computations, floating point calculations, graphics, and multimedia processing. Examples of such applications are the floating point coprocessors that accompanied the early Intel processors, the auxiliary processing unit on the IBM PowerPC [11], and the MMX instructions on the Intel Pentium processors [12]. Coprocessing typically involves the use of a special-purpose interconnect between the main processor and the coprocessor and requires the main processor to recognize certain instructions as coprocessor instructions that are handed off to the auxiliary processor. The complexity of the coprocessor instructions is similar to, or slightly more than, the complexity of the main processor instructions—for example, floating-point-add instead of integer-add.

When researchers started using FPGAs for acceleration of external hard processors, the FPGA-based circuits were much slower than the hardwired processors. Even today, typical soft logic implemented on FPGAs might run at 100 MHz or so, compared to the 400-MHz to 3-GHz clock rates for external processors. To accelerate code running on such a processor using soft logic implemented on an FPGA, the soft logic has to compute the equivalent of more than 4 to 30 processor instructions in 1 clock cycle. As a result, it becomes more attractive to accelerate larger functions or procedures rather than single instructions, so that the communication overhead between the processor and the accelerator is minimized. Starting from the late 1980s researchers have been using FPGAs for function acceleration. A well known early example of FPGA-based function acceleration is the PAM project [3]. Function acceleration is typically implemented over general-purpose buses, shared memories, coprocessor interfaces, or special point-to-point connections such as FSLs (described in more detail in Section 17.3).

Another method to improve application run times is to profile the application and create special-purpose instructions to speed up commonly occurring kernels [13]. While this method has been used successfully for customizable hard processors such as the Tensilica processor, it has not been widely applied to soft processors. While some soft processors such as NIOS allow users to create custom instructions, other soft processors such as MicroBlaze rely on function acceleration, as described in the subsection "Custom Instructions and Function Acceleration" in Section 17.3.2.

17.2   MICROBLAZE SOFT PROCESSOR ARCHITECTURE

17.2.1   Short Description of MicroBlaze

MicroBlaze is a 32-bit RISC processor with a Harvard architecture. It has 32 general-purpose registers, and the datapath is 32 bits wide. It supports the IBM standard onchip peripheral bus (OPB). It supports optional caches for data and instructions and multiple FIFO data links known as fast simplex links (FSLs). MicroBlaze has several additional specialized buses or connections, such as the local memory bus (LMB) or the Xilinx cache link (XCL) channels.

Fig. 17.3 shows the block diagram of MicroBlaze V4.0. The OPB and LMB bus interfaces are shown, along with multiple FSL and XCL interfaces. The optional XCL interfaces are special high-speed interfaces to memory controllers. The other optional elements of MicroBlaze are the barrel shifer, the divider, and the floating point unit (FPU).

MicroBlaze supports a three-operand instruction format. The register file has two read ports and one write port.

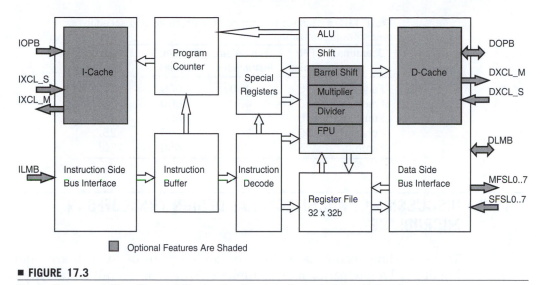

Optional Features Are Shaded

■ **FIGURE 17.3**

Block diagram of MicroBlaze v4.0, showing bus interfaces and function units.

17.2.2 Highlights of Architectural Features

MicroBlaze 4.00 has a three-stage pipeline, with the instruction fetch, operand fetch, and the execution stages. At most, one instruction can be in the execute stage at any time. The register file is read during the operand fetch stage and written during the execute stage. While most instructions execute in one cycle, some instructions, such as the hardware multiply, divide, and floating point instruction take multiple cycles to execute. MicroBlaze contains a small instruction prefetch buffer to allow up to four instructions to be prefetched, so the IF stage can be active even when the OF stage is stalled, waiting for the previous instruction to complete execution.

The FPU requires approximately 1,000 LUTs, or almost the area of the basic MicroBlaze processor. The FPU is tightly coupled to the processor and is not a coprocessor or auxiliary processor that resides outside the base processor, and the FPU uses the same general-purpose registers as the other function units, resulting in low latency. The FPU and MicroBlaze run at the same clock frequency. The fact that MicroBlaze is a soft processor allows users to customize the processor, trading off area for performance according to their needs by choosing options such as the FPU. When implemented on a Virtex 4 device, the FPU and MicroBlaze can run at 200 MHz, to obtain a performance of 33 MFLOPs. The FPU offers a speedup of almost 100 times more than software emulation of floating point instructions, at the cost of doubling the area. Table 17.1 compares the speeds of several floating point instructions on the FPU and in software emulation.

TABLE 17.1 ▪ Sample floating point instructions and speedup over software emulation

Operation	FPU Cycles	Emulation Instructions	Emulation Cycles	Speedup
Fadd	6	450	600	100
Frsub	6	450	600	100
Fmul	6	1200	1600	266
Fdiv	30	600	750	25
Fcmp.lt	3	350	450	150

17.3   DISCUSSION OF ARCHITECTURAL DESIGN TRADEOFFS IN MICROBLAZE

The preceding section described the architecture of MicroBlaze, and highlighted a few performance numbers. While the overall architecture is similar to the basic RISC machines described in textbooks, users can take advantage of the "soft" nature of the processor to create a custom configuration to meet their performance and cost requirements. The tools that allow the user to study these tradeoffs are beyond the scope of this chapter, but it is worth pointing out that standard compiler tools such as gcc are easily adapted to support configurable soft processors.

17.3.1   Architectural Building-Blocks and Their FPGA Implementation

FPGA Logic and Memory Structures

The basic logic structure in FPGAs is a 4-input LUT that can be programmed to implement any function of 4 or fewer inputs. However, there are several additional structures connected to the LUTs to optimize the implementation of commonly used functions. Some of the notable special structures are:

- Dedicated 2-input multiplexers

    - The outputs of two adjacent LUTs can be multiplexed to implement any 5-input function, or to implement a 4:1 multiplexer, or several other functions with 9 or fewer inputs. These implementations are typically faster and more area efficient than an implementation that uses only LUTs.

    - A second 2-input MUX allows the output of the functions created in the previous step to be further expanded to implement 8:1 multiplexers, 6-input LUTs, or other functions.

- Fast local memory
  - The 4-input LUT is really a 16-bit RAM whose address lines are the same as the LUT inputs. Adjacent LUTs are combined to form dual-ported memories.
  - The 4-input LUT can also be treated as a cascadable 16-bit shift register.
- Dedicated carry logic to implement adderspt
  - Fast carry chain: A chain of MUXCY elements controlled by the LUT output to either propagate the carry from the previous stage or place a new "generated" carry signal up the chain.
  - Special XOR gate to compute SUM function: An n-bit adder is implemented using a column of n LUTs controlling n MUXCY elements to generate the carry bits, and n XOR gates to compute the 1-bit sum from the two corresponding input bits and the carry output bit from the previous stage MUXCY.

The carry chain has multiple additional uses for implementing wide functions. Fig. 17.4 shows a wide (12-input) AND function implemented using LUTs and the carry chain. Each 4-input LUT implements a 4-input AND function on 4 of the 12 inputs. Instead of using another LUT to implement a 3-input AND function to build a complete 12-input AND, the output of each LUT is fed to the control input of a 2-input mux that is part of the carry chain. When the output of a LUT is 0, it makes the mux output 0; otherwise, it propagates the value on the right-hand input of the MUXCY element. In this case, the carry chain implements a 3-input AND function without incurring the extra routing and logic delay of an extra LUT stage.

In addition to these logic structures, some larger structures such as dedicated large multipliers and BRAMs are very useful in FPGA applications.

A major difference between ASIC implementations and FPGA implementations of processors is that the area cost in FPGAs is expressed in terms of the number of LUTs, while the area cost in ASICs is expressed in terms of 2-input AND gate implementations. In an ASIC, a 2-input AND gate is much cheaper than a 4-input XOR, which in turn is much cheaper than a 16-bit shift register. In an FPGA, all of these use exactly one LUT and hence cost the same.

Another major difference between ASIC and FPGA implementations is the way the function delay is computed. In ASICs, the delay of a function is dependent on the number of levels of logic in the function, and routing delay plays a negligible role. In FPGAs, the delay of a function is dependent on the number of LUT levels (unless the special structures described earlier are used), and routing delays between LUTs can be a significant part of the total delay.

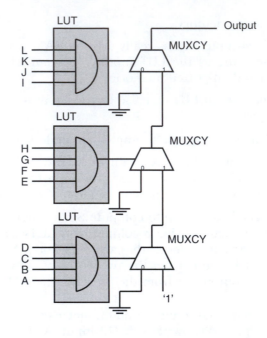

■ **FIGURE 17.4**

Implementation of a 12-input AND gate using LUTs and carry logic.

17.3.2   Examples of Architectural Decisions in MicroBlaze

Instruction Set Design

The essential property of a processor is the instruction set that is supported. When the processor is implemented on an FPGA, the instruction set must match the building blocks of the FPGA logic for efficient implementation. Each LUT performs one bit of a 32-bit operation. MicroBlaze has four logic instructions. The choice of this number is influenced by the fact that a 4-input LUT can process at most four inputs, two of which come from two operands, and two of which select the instruction type. Any fewer than four instructions would result in wasted inputs, and any more than four instructions would require an additional LUT, leading to a doubling of the size.

Datapath Design

In silicon-based processor design, the number of pipeline stages is determined by performance and area requirements, and the area or delay cost of a mux is very small compared to the cost of an ALU or other function unit. In FPGA-based soft processor design, the cost of a forwarding or bypass multiplexer is one LUT per bit for a 2-input multiplexer, and this is the same cost as the ALU described in the previous section.

If more pipeline stages are added, the number, the size, and the delay of the required multiplexers increases. The optimal pipeline depth for speed is between three and five, and a deeper pipeline will run slower, negating the benefits of additional pipe stages. The optimal pipeline depth for area is three to four. MicroBlaze 4.00 is optimized for a combination of speed and area with three pipeline stages.

Implementation-Specific Details

This section describes some specific implementation details of processor modules to take advantage of the FPGA fabric. As a general rule, all timing-critical functions are implemented using the carry chain to reduce logic delays. Wherever possible, the shift register mode of the LUT is used to implement small FIFOs such as FSLs and instruction fetch buffers. Additional set/reset pins on the basic FPGA latch/flip-flop

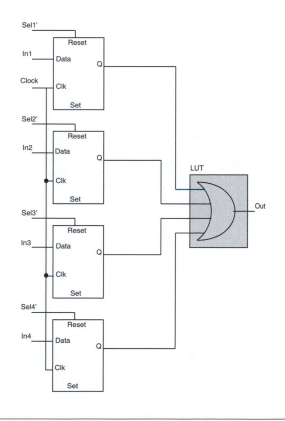

■ **FIGURE 17.5**

Implementation of a 4:1 mux with fully decoded select signals using latches and a LUT.

are used to implement additional combinational function inputs, as shown in Fig. 17.5. This figure shows the implementation of a 4-input MUX with fully decoded select inputs. There are four data input signals and four select input signals. Implementing this 8-input function would require more than two 4-LUTs. However, if there are additional unused latches available on the FPGA, the same function can be implemented using four latches and a single 4-LUT as shown. The latch/flipflop elements are configured as transparent latches with the clock signal connected to logic 1. The output of the latch is then equal to the input, unless the reset signal goes to logic 1, causing the output to become logic 0. The inverted form of the select signal is used so each latch implements the AND function of its data input and select input, and at any time exactly one of the four inverted select signals is at logic 0, so that exactly one data input is selected. The latch outputs are combined with an OR function implemented in the 4-LUT to create the 4:1 mux function.

A good example of implementation-based optimization in MicroBlaze is the jump detection logic, where the delay is reduced by two orders of

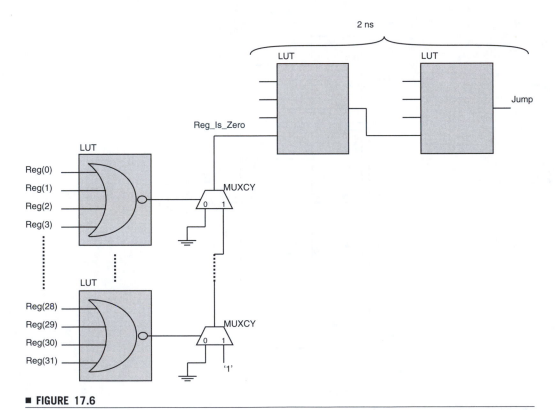

▪ **FIGURE 17.6**

Standard implementation of jump detection logic using LUTs.

magnitude, from 2 ns, to 0.02 ns, by using carry chains to implement logic.

Fig. 17.6 shows the jump detection logic implemented using two levels of LUTs, with a total delay of 2 ns. The jump signal goes high when a particular 32-bit register contains the value 0 and the opcode corresponds to a jump instruction. The register bits are compared to 0 using LUTs implementing 4-input NOR functions whose outputs are 1 if all four inputs are 0, and 0 otherwise. The carry chain muxes are controlled by these LUT outputs to pass a 1 to the output of the carry chain only if all the LUT outputs are 1. The carry chain output is thus 1 when the register bits correspond to the value 0, and 0 otherwise. Two additional LUTs are used to combine this output called "Reg_is_Zero" with the opcode bits to produce the jump output. The combined delay of the last two cascaded LUTs is 2 ns.

Fig. 17.7 shows how the same circuit is implemented using LUTs and carry logic to reduce the total additional delay to 0.02 ns. The function

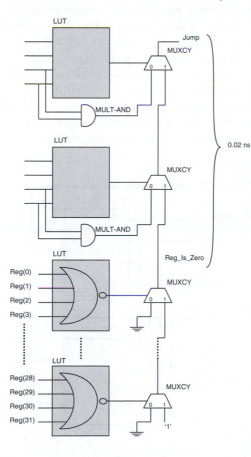

■ **FIGURE 17.7**

Jump detection logic implemented using LUTs and carry logic to reduce delay.

of the opcode bits and the Reg_is_Zero signal is refactored so that each LUT and the corresponding MULT-AND gate produce an output that is independent of the Reg_is_Zero signal. The first MUXCY mux is used to conditionally select the Reg_is_Zero value or the MULT-AND value, and this result is conditionally propagated through the next MUXCY. Since the LUT outputs are computed independently of Reg_is_Zero, and the LUT inputs are available well before the Reg_is_Zero signal is available, the delay from the time the Reg_is_Zero signal is available to the time the jump signal becomes available is just the propagation time of the carry chain, which is very small (0.02 ns).

Fig. 17.8 shows a 2-read-port, 1-write-port, 16×1 bit register file implemented using the dual-ported RAM building block. Two adjacent 4-LUTs can be configured as a dual-ported 16×1 RAM block called the RAM16X1D. The RAM16X1D block takes two 4-bit addresses labeled as W_ADDR and RD_ADDR. The DPO and SPO outputs correspond to the data stored at the "read address" and "write address," respectively. The register file uses just the DPO output and ignores the SPO output corresponding to the data stored at the "write address." The Din input is the data that is written to the register. This building block by itself is not sufficient to implement the actual register file, because three addresses are required, two for read and one for write, and these addresses are all valid in the same clock cycle. Combining two of these blocks with a common write address and write data, and separate read addresses results in a 16×1-bit register with two read ports and one write port that can be accessed simultaneously as shown.

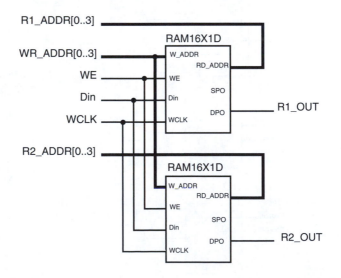

■ **FIGURE 17.8**

16×1-bit Register file with two read ports and one write port implemented using dual-ported LUT-RAM building blocks.

In the absence of the 16-bit LUT-RAM blocks, a register file could be implemented using the larger BRAM present on FPGAs. In either case, the number of read/write ports on the register file is limited by the capabilities of the dual-port RAM blocks present on the FPGA. Adding additional ports would require additional muxing and duplication of scarce RAM resources and would slow down the register file itself.

Performance of Soft Processors

Data sheets published by Xilinx and Altera provide performance numbers for soft processors from these companies. Typically the most advanced, speed-optimized versions of the soft processors run at 100 to 200 MHz on the fastest FPGA devices. Low-cost and low-performance versions of the soft processors run at 50 to 100 MHz on low-cost and low-performance devices from the FPGA manufacturers. These speeds compare with speeds of 1 to 4 GHz for commercial processors from Intel and other vendors, and speeds of 300 to 400 MHz for the hard PowerPC processor in Xilinx FPGAs. The desire to improve the performance of the soft processors leads to the development of custom instructions and function accelerators.

Custom Instructions and Function Acceleration

When the processor is implemented as soft logic on an FPGA, and the processor itself can be customized to improve area and performance as described earlier, it is natural to think about adding custom instructions to the processor. Typically, a custom instruction reads and writes the processor registers just like the regular instructions but performs some special function of the data it sees. However, the MicroBlaze implementation has a fixed floor plan to improve performance, and custom instructions disturb this optimal floor plan because it is not possible to allocate a priori, special areas for these functions. Instead of custom instructions, MicroBlaze provides the fast simplex link (FSL) connections and special instructions to read and write FSLs. Micro-Blaze can have up to 32 FSL connections, each consisting of a 32-bit data bus, along with a few control signals to connect function accelerator modules. A function accelerator module can connect to as many FSLs as required. The FSL read/write instructions move data to/from any of the 32 general-purpose registers of MicroBlaze from/to any of the FSL wires. The FSL instructions and wires are predefined so they do not change the area/speed of the processor itself. FSL instructions come in two flavors, blocking and nonblocking. A blocking fsl_read instruction causes the processor to wait until the read data is available, while a nonblocking fsl_read returns immediately. Each FSL connection has a FIFO that allows the function accelerator logic to run at its own speed and perform internal pipelining when possible. The following simple example illustrates how FSLs can be used in function acceleration.

Consider the following code fragment that calls three functions to perform a two-dimensional discrete cosine transform (2D-DCT) function on a two-dimensional array of numbers as follows. The first function call is to a function that performs a one-dimensional discrete cosine transform (1D-DCT) on each row of the two-dimensional array. The second call is to a function that performs a matrix transpose operation, and the third call is to the same 1D-DCT function on each row of the transposed matrix.

```
/* BEFORE – ALL SOFTWARE*/
  int dct2d (int * array2d) {
    dct1d(array2d);
    transpose(array2d);
    dct1d(array2d);
}
```

This function can be accelerated by creating a "soft logic" hardware block that performs the entire 2D-DCT function using data that is sent from the processor to the hardware block using the FSL connections on MicroBlaze. The function body for 2D-DCT is replaced by a series of "fsl_write" instructions that transfer data from MicroBlaze registers to the FSL connection called FSL1, followed by a series of "fsl_read" instructions that read the result data back from the hardware using the FSL connection called FSL2.

```
/* AFTER – HARDWARE ACCELERATED OVER FSL CONNECTION */
  int dct2d (int * array2d) {
    /* write out the array to FSL */
    for (i = 0 to n-1)
      fsl_write(FSL1, array2d[i]);
    for (i = 0 to n-1)
      fsl_read(FSL2, array2d[i]);
  }
```

The acceleration that can be obtained using this scheme depends on factors such as the relative clock rates of the processor and the accelerator, the number of clock cycles required to send the data to the accelerator, the number of clock cycles used for computing the function in hardware, the number of clock cycles used for computing the function in software, and so on.

While there has been extensive research on generating custom instructions for processors implemented as ASICs [13], there has been some recent interest in custom instructions and function acceleration for soft processors. Jason Cong et al. [14] focus on architectural extensions for soft processors to overcome some of the data bandwidth limitations specific to the FPGA fabric. They describe how the act of placing a custom instruction within the datapath of soft processor can affect the speed of the processor itself, while adding a special link such as the FSL is very useful but is not sufficient by itself to provide a complete solution. While they describe a good solution in terms of shadow registers, this is still an open area for research.

17.3.3 Tool Support

FPGA vendors such as Xilinx [1] and Altera [10] provide tools such as "Xilinx Platform Studio" and "SOPC Builder" to simplify the task of building a processor-based system on FPGAs. The focus of these tools is on describing a processor system at a high level of abstraction, but the tools provide some support for integrating custom instructions and function accelerators with the rest of the hardware. However, the task of identifying and actually implementing the custom instruction or accelerator is left to the system designers. The custom instructions are typically used by programmers writing assembly code or using functional wrappers that invoke assembly code. Designers use standard off-the-shelf compilers, debuggers, and profiling tools. Profiling tools help to identify sections of code that can be replaced with a custom instruction or accelerator.

17.4 CONCLUSIONS

In this chapter we described how a soft processor is implemented in an FPGA. The architecture of the soft processor is constrained by the FPGA implementation fabric. For example, the number of pipe stages is limited to three in our example, because adding more pipe stages increases the number and the size of multiplexers in the processor, and the relatively high cost of these multiplexers (relative to ASIC implementations) reduces the possible speed advantage. However, these relative costs can change as the FPGA fabric evolves, and new optimizations might be necessary. Several implementation techniques specific to FPGAs were presented to allow designers of soft processors to overcome speed and area limitations. Finally, we pointed out some interesting ideas in extending the processor instruction set and accelerating the soft processor to point out future research areas.

ACKNOWLEDGEMENTS

The authors wish to thank Professor Tobias Noll, who reviewed the original version of the chapter and provided valuable feedback to improve our presentation.

CHAPTER REFERENCES

Chapter 1

[1] R. Kumar, D. M. Tullsen, N. P. Jouppi, and P. Ranganathan. Heterogeneous chip multiprocessors. *Computer*, 38(11):32–38, Nov. 2005.

[2] M. J. Bass and C. M. Christensen. The future of the microprocessor business. *IEEE Spectrum*, 39(4):34–39, Feb. 2002.

Chapter 2

[1] J. L. Hennessy and D. A. Patterson. *Computer Architecture—A Quantitative Approach*, 3rd ed. Morgan Kaufmann, 2003.

[2] H. Blume, H. Hubert, H. T. Feldkamper, and T. G. Noll. Model-based exploration of the design space for heterogeneous systems on chip. In *Proceedings of the 13th International Conference on Application-Specific Systems, Architectures and Processors*, 2002, pp. 29–40.

[3] K. Keutzer, M. Gries, G. Martin, and H. Meyr. *Designing ASIP: The MESCAL Methodology*. Kluwer, 2005.

[4] H. Meyr, M. Moeneclaey, and S. Fechtel. *Digital Communication Receivers—Synchronization, Channel Estimation and Signal Processing*. John Wiley & Sons, 1998.

[5] H. James and G. Harrison. Implementation of a 3GPP turbo decoder on a programmable DSP core. In *Proceedings of the Communications Design Conference*, 2001.

[6] H. Blume, T. Gemmeke, and T. G. Noll. Platform-specific turbo decoder implementations. In *Proceedings of the DSP Design Workshop*, 2003.

[7] A. La Rosa, L. Lavagno, and C. Passerone. Implementation of a UMTS turbo-decoder on a dynamically reconfigurable platform. *IEEE Transactions on Computer-Aided Design of Integrated Circuits and Systems*, 24(1):100–106, Jan. 2005.

[8] F. Gilbert, M. J. Thul, and N. Wehn. Communication-centric architectures for turbo-decoding on embedded multiprocessors. In *Proceedings of the Design Automation and Test in Europe Conference and Exhibition*, 2003, pp. 356–361.

[9] P. C. Tseng and L. G. Chen. Perspectives of multimedia SoC. In *Proceedings of the IEEE Workshop on Signal Processing Systems*, 2003, p. 3.

[10] O. Lüthje. A methodology for automated test generation for LISA processor models. In *Proceedings of the 12th Workshop on Synthesis and System Integration of Mixed Information Technologies*, 2004, pp. 263–273.

[11] P. Yu and T. Mitra. Scalable custom instructions identification for instruction set extensible processors. In *Proceedings of the International Conference on Compilers, Architectures, and Synthesis for Embedded Systems*, 2004, pp. 69–78.

[12] F. Sun, S. Ravi, A. Raghunathan, and N. Jha. Synthesis of custom processors based on extensible platforms. In *Proceedings of the International Conference on Computer Aided Design*, 2002, pp. 641–648.

[13] M. Arnold and H. Corporaal. Designing domain specific processors. In *Proceedings of the 9th International Workshop on Hardware/Software Codesign*, 2001, pp. 61–66.

[14] N. T. Clark, H. Zhong, and S. A. Mahlke. Processor acceleration through automated instruction set customization. In *Proceedings of the 36th International Symposium on Microarchitecture*, 2003, pp. 129–140.

[15] M. Ravasi. *An Automatic C-code Instrumentation Framework for High Level Algorithmic Complexity Analysis and System Design*. PhD dissertation, Ecole Polytechnique Fédérale de Lausanne, Lausanne, Switzerland, 2003.

[16] L. Cai, A. Gerstlauer, and D. Gajski. *Retargetable Profiling for Rapid, Early System-Level Design Space Exploration*. Technical report Center for Embedded Computer Systems, University of California, Irvine, California, 2004.

[17] M. Ravasi and M. Mattavelli. High-level algorithmic complexity evaluation for system design. *Journal of Systems Architecture*, 48(13–15):403–427, May 2003.

[18] K. Atasu, L. Pozzi, and P. Ienne. Automatic application-specific instruction-set extensions under microarchitectural constraints. In *Proceedings of the 40th Design Automation Conference*, 2003, pp. 256–261.

[19] P. Biswas, V. Choudhary, K. Atasu, L. Pozzi, P. Ienne, and N. Dutt. Introduction of local memory elements in instruction set extensions. In *Proceedings of the 41st Design Automation Conference*, 2004, pp. 729–734.

[20] K. Karuri, M. Al Faruque, S. Kraemer, R. Leupers, G. Ascheid, and H. Meyr. Fine-grained application source code profiling for ASIP design. In *Proceedings of the 42nd Design Automation Conference*, Anaheim, 2005, pp. 329–334.

[21] I. Verbauwhede, P. Schaumont, and H. Kuo. Design and performance testing of a 2.29-GB/s Rijndael processor. *IEEE Journal of Solid-State Circuits*, 38(3):569–572, Mar. 2003.

[22] O. Schliebusch, A. Chattopadhyay, D. Kammler, R. Leupers, G. Ascheid, H. Meyr, and T. Kogel. A framework for automated and optimized ASIP implementation supporting multiple hardware description languages. In *Proceedings of the Asia and South Pacific Design Automation Conference*, 2005, pp. 280–285.

[23] O. Schliebusch, A. Chattopadhyay, E. M. Witte, D. Kammler, G. Ascheid, R. Leupers, and H. Meyr. Optimization techniques for ADL-driven RTL processor synthesis. In *Proceedings of the 16th IEEE International Workshop on Rapid System Prototyping*, 2005, pp. 165–171.

[24] E. M. Witte, A. Chattopadhyay, O. Schliebusch, D. Kammler, R. Leupers, G. Ascheid, H. Meyr, and T. Kogel. Applying resource-sharing algorithms to ADL-driven automatic ASIP implementation. In *Proceedings of the International Conference of Computer Design*, 2005, pp. 193–199.

[25] L. Su. Digital media—the new frontier for supercomputing. Opening Keynote, *MPSoC*, 2005.

[26] T. Glökler and H. Meyr. *Design of Energy-Efficient Application-Specific Instruction Set Processors (ASIPs)*. Kluwer, 2004.

[27] J. H. Lee, J. H. Moon, K. L. Heo, M. H. Sunwoo, S. K. Oh, and I. H. Kim. Implementation of application-specific DSP for OFDM systems. In *Proceedings of the 2004 International Symposium on Circuits and Systems*, 2004, vol. 3, pp. 665–668.

Chapter 3

[1] J. Fisher, P. Faraboschi, and C. Young. *Embedded Computing: A VLIW Approach to Architecture, Compilers and Tools*. Morgan Kaufmann, 2004.

[2] P. Faraboschi, J. Fisher, G. Brown, G. Desoli, and F. Homewood. Lx: A technology platform for customizable VLIW embedded processing. In *Proceedings of the 27th Annual International Symposium on Computer Architecture*, 2000, pp. 203–213.

[3] P. Faraboschi, G. Desoli, and J. Fisher. *Clustered Instruction-Level Parallel Processors*. Technical Report HPL-98-204. HP Labs, Cambridge, Massachusetts, Dec. 1998.

[4] J. A. Fisher, P. Faraboschi, and G. Desoli. Custom-fit processors: letting applications define architectures. In *Proceedings of the 29th Annual International Symposium on Microarchitecture*, 1996, pp. 324–335.

[5] P. Faraboschi, G. Desoli and J. Fisher. VLIW architectures for DSP and multimedia applications. *IEEE Signal Processing*, 15(2):59–85, Mar. 1998.

[6] B. Rau and M. Schlansker. *Embedded Computing: New Directions in Architecture and Automation*. Technical Report HPL-2000-115, HP Labs, Palo Alto, California, Sep. 2000.

[7] R. Schreiber, S. Aditya, B. Rau, V. Kathail, S. Mahlke, S. Abraham, and G. Snider. *High-Level Synthesis of Nonprogrammable Hardware Accelerators*. Technical Report HPL-2000-31, HP Labs, Palo Alto, California, May 2000.

[8] V. K. S. Aditya, B. R. Rau, and V. Kathail. Automatic architectural synthesis of VLIW and EPIC processors. In *Proceedings of the 12th International Symposium on System Synthesis*, 1999, pp. 107–113.

[9] A. Wang, E. Killian, D. Maydan, and C. Rowen. Hardware/software instruction set configurability for system-on-chip processors. In *Proceedings of the Design Automation Conference*, 2001, pp. 184–190.

[10] Tensilica, Inc. Xtensa processor. http://www.tensilica.com.

[11] R. E. Gonzalez. Xtensa: a configurable and extensible processor. *IEEE Micro*, 20(2):60–70, Mar.–Apr. 2000.

[12] ARC International. ARCtangent processor. http://www.arc.com.

[13] Improv Systems, Inc. Jazz DSP. http://www.improvsys.com.

[14] R. J. Cloutier and D. E. Thomas. Synthesis of pipelined instruction set processors. In *Proceedings of the 30th Design Automation Conference*, 1993, pp. 583–588.

[15] M. Gschwind. Instruction set selection for ASIP design. In *Proceedings of the International Symposium HW/SW Codesign*, 1999, pp. 7–11.

[16] I.-J. Huang and A. M. Despain. Generating instruction sets and microarchitectures from applications. In *Proceedings of the International Conference on Computer-Aided Design*, 1994, pp. 391–396.

[17] V. S. Lapinski. *Algorithms for Compiler-Assisted Design Space Exploration of Clustered VLIW ASIP Datapaths*. PhD dissertation, University of Texas, Austin, Texas, May 2001.

[18] T. Sherwood, M. Oskin and B. Calder. Balancing design options with Sherpa. In *Proceedings of the International Conference on Compilers, Architecture, and Synthesis for Embedded Systems*, 2004, pp. 57–68.

[19] Y. Arai, T. Agui, and M. Nakajima. A fast DCT-SQ scheme for images. *Transactions of the IEICE*, E-71(11):1095–1097, Nov. 1988.

[20] E. Feig and S. Winograd. Fast algorithms for the discrete cosine transform. *IEEE Transactions on Signal Processing*, 40(9):2174–2193, Sept. 1992.

[21] A. Ligtenberg and M. Vetterli. A discrete fourier/cosine transform chip. *IEEE Journal on Selected Areas in Communications*, 4(1):49–61, Jan. 1986.

[22] G. Desoli, N. Mateev, E. Duesterwald, P. Faraboschi, and J. Fisher. DELI: a new run-time control point. *Proceedings of the 35th Annual International Symposium on Microarchitecture*, 2002, pp. 257–258.

[23] A. Klaiber. *The Technology behind Crusoe Processors*, Technical report, Transmeta Corp., Santa Clara, California, Jan. 2000. http://www.transmeta.com/pdfs/paper_aklaiber_19jan00.pdf.

[24] A. Linden and J. Fenn. Understanding Gartner's hype cycles. *Gartner Research Report*, May 2003.

Chapter 4

[1] P. C. Clements. A survey of architecture description languages. In *Proceedings of International Workshop on Software Specification and Design*, 1996, pp. 16–25.

[2] M. R. Barbacci. Instruction set processor specifications (ISPS): the notation and its applications. *IEEE Transactions on Computers*, C-30(1):24–40, Jan. 1981.

[3] W. Qin and S. Malik. Architecture description languages for retargetable compilation. In *The Compiler Design Handbook: Optimizations & Machine Code Generation*. CRC, 2002.

[4] H. Tomiyama, A. Halambi, P. Grun, N. Dutt, and A. Nicolau. Architecture description languages for systems-on-chip design. In *Proceedings of Asia Pacific Conference on Chip Design Language*, 1999, pp. 109–116.

[5] R. Leupers and P. Marwedel. Retargetable code generation based on structural processor descriptions. *Design Automation for Embedded Systems*, 3(1):75–108, 1998.

[6] M. Freericks. *The nML Machine Description Formalism*. Technical report TR SM-IMP/DIST/08, Computer Science Department, TU Berlin, Berlin, 1993.

[7] G. Hadjiyiannis, S. Hanono, and S. Devadas. ISDL: an instruction set description language for retargetability. In *Proceedings of the 34th Design Automation Conference*, 1997, pp. 299–302.

[8] A. Fauth and A. Knoll. Automatic generation of DSP program development tools. In *Proceedings of International Conference of Acoustics, Speech and Signal Processing*, 1993, pp. 457–460.

[9] D. Lanneer, J. Praet, A. Kifli, K. Schoofs, W. Geurts, F. Thoen, and G. Goossens. CHESS: retargetable code generation for embedded DSP

processors. In P. Marwedel and G. Goossens, editors *Code Generation for Embedded Processors*, pp. 85–102. Kluwer, 1995.

[10] F. Lohr, A. Fauth, and M. Freericks. *Sigh/sim: an Environment for Retargetable Instruction Set Simulation*. Technical report 1993/43, Department of Computer Science, Tu Berlin, Berlin, 1993.

[11] Target Compiler Technologies N. V. http://www.retarget.com.

[12] J. Paakki. Attribute grammar paradigms—a high level target methodology in language implementation. *ACM Computing Surveys*, 27(2):196–256, June 1995.

[13] S. Hanono and S. Devadas. Instruction selection, resource allocation, and scheduling in the AVIV retargetable code generator. In *Proceedings of the 35th Design Automation Conference*, 1998, pp. 510–515.

[14] G. Hadjiyiannis, P. Russo, and S. Devadas. A methodology for accurate performance evaluation in architecture exploration. In *Proceedings of the 36th Design Automation Conference*, 1999, pp. 927–932.

[15] J. Gyllenhaal, B. Rau, and W. Hwu. *HMDES Version 2.0 Specification*. Technical report IMPACT-96-3, IMPACT Research Group, University of Illinois, Urbana, Illinois, 1996.

[16] Trimaran Consortium. *The MDES User Manual*. http://www.trimaran.org. 1997.

[17] P. Grun, A. Halambi, N. Dutt, and A. Nicolau. RTGEN: an algorithm for automatic generation of reservation tables from architectural descriptions. *IEEE Transactions on Very Large Scale Integration (VLSI) Systems*, 11(4):731–737, August 2003.

[18] J. L. Hennessy and D. A. Patterson. *Computer Architecture—A Quantitative Approach*, 3rd ed. Morgan Kaufmann, 2003.

[19] P. Mishra, M. Mamidipaka, and N. Dutt. Processor-memory co-exploration using an architecture description language. *ACM Transactions on Embedded Computing Systems (TECS)*, 3(1):140–162, Feb. 2004.

[20] V. Zivojnovic, S. Pees, and H. Meyr. LISA—machine description language and generic machine model for HW/SW co-design. In *Proceedings of the IEEE Workshop on VLSI Signal Processing*, 1996, pp. 127–136.

[21] S. Pees, A. Hoffmann, and H. Meyr. Retargetable compiled simulation of embedded processors using a machine description language. *ACM Transactions on Design Automation of Electronic Systems (TODAES)* 5(4):815–834, Oct. 2000.

[22] M. Hohenauer, O. Wahlen, K. Karuri, H. Scharwächter, T. Kogel, R. Leupers, G. Ascheid, H. Meyr, G. Braun, and H. Van Someren. A methodology and tool suite for C compiler generation from ADL processor models. In *Proceedings of Design, Automation and Test in Europe*, Conference and Exhibition, 2004, pp. 1276–1283.

[23] W. Qin, S. Rajagopalan, and S. Malik. A formal concurrency model based architecture description language for synthesis of software development tools. In *Proceedings of ACM Conference on Languages, Compilers, and Tools for Embedded Systems (LCTES)*, 2004, pp. 47–56.

[24] A. Halambi, P. Grun, V. Ganesh, A. Khare, N. Dutt, and A. Nicolau. EXPRESSION: a language for architecture exploration through compiler/simulator retargetability. In *Proceedings of Design Automation and Test in Europe Conference and Exhibition*, 1999, pp. 485–490.

[25] A. Nohl, G. Braun, O. Schliebusch, R. Leupers, H. Meyr, and A. Hoffmann. A universal technique for fast and flexible instruction-set architecture simulation. In *Proceedings of the 39th Design Automation Conference*, 2002, pp. 22–27.

[26] M. Reshadi, P. Mishra, and N. Dutt. Instruction set compiled simulation: A technique for fast and flexible instruction set simulation. In *Proceedings of the 40th Design Automation Conference*, 2003, pp. 758–763.

[27] M. Itoh, Y. Takeuchi, M. Imai, and A. Shiomi. Synthesizable HDL generation for pipelined processors from a micro-operation description. *IEICE Transactions on Fundamentals of Electronics, Communications and Computer Sciences*, E83-A(3):394–400, Mar. 2000.

[28] O. Schliebusch, A. Chattopadhyay, M. Steinert, G. Braun, A. Nohl, R. Leupers, G. Ascheid, and H. Meyr. RTL processor synthesis for architecture exploration and implementation. In *Proceedings of the Design Automation and Test in Europe Conference and Exhibition*, 2004, pp. 156–160.

[29] P. Mishra, A. Kejariwal, and N. Dutt. Synthesis-driven exploration of pipelined embedded processors. In *Proceedings of International Conference on VLSI Design*, 2004, p. 921.

[30] P. Mishra and N. Dutt. Graph-based functional test program generation for pipelined processors. In *Proceedings of the Design, Automation and Test in Europe Conference and Exhibition*, 2004, pp. 182–187.

[31] P. Mishra and N. Dutt. Functional coverage driven test generation for validation of pipelined processors. In *Proceedings of Design, Automation and Test in Europe Conference and Exhibition*, 2005, pp. 678–683.

[32] H. -M. Koo and P. Mishra. Functional test generation using property decompositions for validation of pipelined processors. In *Proceedings of Design, Automation and Test in Europe Conference and Exhibition*, 2006, pp. 1240–1245.

[33] O. Lüthje. A methodology for automated test generation for LISA processor models. In *Proceedings of the 12th Workshop on Synthesis and System Integration of Mixed Technologies*, 2004, pp. 266–273.

[34] P. Mishra and N. Dutt. *Functional Verification of Programmable Embedded Architectures: A Top-Down Approach*. Springer, 2005.

[35] P. Mishra and N. Dutt. Automatic modeling and validation of pipeline specifications. *ACM Transactions on Embedded Computing Systems (TECS)*, 3(1):114–139, Feb. 2004.

[36] P. Mishra, N. Dutt, N. Krishnamurthy, and M. Abadir. A top-down methodology for validation of microprocessors. *IEEE Design & Test of Computers*, 21(2):122–131, Mar. 2004.

Chapter 5

[1] A. V. Aho, R. Sethi, and J. D. Ullman. *Compilers—Principles, Techniques, and Tools*. Addison-Wesley, 1986.

[2] A. W. Appel. *Modern Compiler Implementation in C*. Cambridge University Press, 1998.

[3] S. S. Muchnik. *Advanced Compiler Design & Implementation*. Morgan Kaufmann, 1997.

[4] R. Wilhelm and D. Maurer. *Compiler Design*, Addison-Wesley, 1995.

[5] T. Mason and D. Brown. *lex & yacc*, O'Reilly & Associates, 1991.

[6] K. M. Bischoff. *Design, Implementation, Use, and Evaluation of Ox: an Attribute-Grammar Compiling System Based on Yacc, Lex, and C*. Technical report 92-31, Department of Computer Science, Iowa State University, Ames, Iowa, 1992.

[7] R. Leupers, O. Wahlen, M. Hohenauer, T. Kogel, and P. Marwedel. An executable intermediate representation for retargetable compilation and high-level code optimization. In *Proceedings of the International Workshop on Systems, Architectures, Modeling, and Simulation*, 2003, pp. 120–125.

[8] A. Aho, S. Johnson, and J. Ullman. Code generation for expressions with common subexpressions. *Journal of the ACM*, 24(1), Jan. 1977.

[9] C. W. Fraser, D. R. Hanson, and T. A. Proebsting. Engineering a simple, efficient code generator. *ACM Letters on Programming Languages and Systems*, 1(3):216–226, Sept. 1992.

[10] S. Bashford and R. Leupers. Constraint driven code selection for fixed-point DSPs. In *Proceedings of the 36th Design Automation Conference*, 1999, pp. 817–822.

[11] M. A. Ertl. Optimal code selection in DAGs. In *Proceedings of the ACM Symposium on Principles of Programming Languages*, 1999, pp. 242–249.

[12] F. Chow and J. Hennessy. Register allocation by priority-based coloring. *ACM SIGPLAN Notices*, 19(6):222–232, Jun. 1984.

[13] P. Briggs. *Register allocation via graph coloring*. PhD dissertation, Department of Computer Science, Rice University, Houston, Texas, 1992.

[14] S. Liao, S. Devadas, K. Keutzer, S. Tjiang, and A. Wang. Storage assignment to decrease code size. In *Proceedings of the ACM SIGPLAN Conference on Programming Language Design and Implementation*, 1995, pp. 186–195.

[15] M. Lam. Software pipelining: an effective scheduling technique for VLIW machines. In *Proceedings of the ACM SIGPLAN Conference on Programming Language Design and Implementation*, 1988, pp. 318–328.

[16] J. Wagner and R. Leupers. C compiler design for a network processor. *IEEE Transactions on Computer-Aided Design of Integrated Circuits and Systems*, 20(11):1302–1308, Nov. 2001.

[17] T. Wilson, G. Grewal, B. Halley, and D. Banerji. An integrated approach to retargetable code generation. In *Proceedings of the 7th International Symposium on High-Level Synthesis*, 1994, pp. 70–75.

[18] R. Leupers and P. Marwedel. *Retargetable Compiler Technology for Embedded Systems—Tools and Applications*. Kluwer, 2001.

[19] P. Marwedel and L. Nowak. Verification of hardware descriptions by retargetable code generation. In *Proceedings of the 26th Design Automation Conference*, 1989, pp. 441–447.

[20] R. Leupers and P. Marwedel. Retargetable code generation based on structural processor descriptions. *Design Automation for Embedded Systems*, 3(1):75–108, Jan. 1998.

[21] R. Leupers and P. Marwedel. A BDD-based frontend for retargetable compilers. In *Proceedings of the European Design & Test Conference*, 1995, p. 239.

[22] J. Van Praet, D. Lanneer, G. Goossens, W. Geurts, and H. De Man. A graph based processor model for retargetable code generation. In *Proceedings of the European Design and Test Conference*, 1996, p. 102.

[23] Target Compiler Technologies N.V. http://www.retarget.com.

[24]   P. Mishra, N. Dutt, and A. Nicolau. Functional abstraction driven design space exploration of heterogenous programmable architectures. In *Proceedings of the 4th International Symposium on System Synthesis*, 2001, p. 256–261.

[25]   Free Software Foundation, Inc. EGCS. http://gcc.gnu.org.

[26]   C. Fraser and D. Hanson. *A Retargetable C Compiler: Design and Implementation*. Addison-Wesley, 1995.

[27]   C. Fraser and D. Hanson. LCC. http://www.cs.princeton. edu/software/lcc.

[28]   ACE Associated Compiler Experts. http://www.ace.nl.

[29]   A. Oraioglu and A. Veidenbaum. Application specific microprocessors (guest editors introduction). *IEEE Design & Test Magazine*, 20(1):6–7, Jan. 2003.

[30]   Tensilica, Inc. http://www.tensilica.com.

[31]   Stretch, Inc. http://www.stretchinc.com.

[32]   A. Hoffmann, H. Meyr, and R. Leupers. *Architecture Exploration for Embedded Processors with LISA*. Kluwer, 2002.

[33]   CoWare, Inc. http://www.coware.com.

[34]   H. Scharwächter, D. Kammler, A. Wieferink, M. Hohenauer, K. Karuri, J. Ceng, R. Leupers, G. Ascheid, and H. Meyr. ASIP architecture exploration for efficient IPSec encryption: a case study. In *Proceedings of the International Workshop on Software and Compilers for Embedded Systems*, 2004.

[35]   A. Nohl, G. Braun, O. Schliebusch, R. Leupers, H. Meyr and A. Hoffmann. A universal technique for fast and flexible instruction set simulation. In *Proceedings of 39th Design Automation Conference*, 2002, pp. 22–27.

[36]   O. Wahlen, M. Hohenauer, G. Braun, R. Leupers, G. Ascheid, H. Meyr, and X. Nie. Extraction of efficient instruction schedulers from cycle-true processor models. In *Proceedings of the 7th International Workshop on Software and Compilers for Embedded Systems*, 2003, pp. 167–181.

[37]   M. Hohenauer, O. Wahlen, K. Karuri, H. Scharwächter, T. Kogel, R. Leupers, G. Ascheid, H. Meyr, G. Braun, and H. van Someren. A methodology and tool suite for C compiler generation from ADL processor models. In *Proceedings of the Design, Automation & Test in Europe Conference and Exhibition*, 2004, pp. 1276–1283.

[38]   J. Ceng, W. Sheng, M. Hohenauer, R. Leupers, G. Ascheid, H. Meyr, and G. Braun. Modeling instruction semantics in ADL processor descriptions for C compiler retargeting. In *Proceedings of the International Workshop on Systems, Architectures, Modeling, and Simulation*, 2004, pp. 463–473.

[39]   G. Braun, A. Nohl, W. Sheng, J. Ceng, M. Hohenauer, H. Scharwächter, R. Leupers, and H. Meyr. A novel approach for flexible and consistent ADL-driven ASIP design. In *Proceedings of the 41st Design Automation Conference*, 2004, pp. 717–722.

[40]   M. Lee, V. Tiwari, S. Malik, and M. Fujita. Power analysis and minimization techniques for embedded DSP software. *IEEE Transactions on Very Large Scale Integration (VLSI) Systems*, 5(2):123–135, Mar. 1997.

[41]   S. Steinke, M. Knauer, L. Wehmeyer, and P. Marwedel. An accurate and fine grain instruction-level energy model supporting software optimizations. In *Proceedings of the 11th International Workshop on Power and Timing Modeling, Optimization, and Simulation*, 2001.

[42]   S. Steinke, N. Grunwald, L. Wehmeyer, R. Banakar, M. Balakrishnan, and P. Marwedel. Reducing energy consumption by dynamic copying of instructions onto onchip memory. In *Proceedings of the 15th International Symposium on System Synthesis*, 2002, pp. 213–218.

[43]   M. Kandemir, M. J. Irwin, G. Chen, and I. Kolcu. Banked scratch-pad memory management for reducing leakage energy consumption. In *Proceedings of the International Conference on Computer-Aided Design*, 2004, pp. 120–124.

[44]   H. Falk and P. Marwedel. Control flow driven splitting of loop nests at the source code level. In *Proceedings of the Design, Automation and Test in Europe Conference and Exhibition*, 2003, pp. 410–415.

[45]   C. Liem, P. Paulin, and A. Jerraya. Address calculation for retargetable compilation and exploration of instruction-set architectures. In *Proceedings of the 33rd Design Automation Conference*, 1996, pp. 597–600.

[46]   B. Franke and M. O'Boyle. Compiler transformation of pointers to explicit array accesses in DSP applications. In *Proceedings of the International Conference on Compiler Construction*, 2001, pp. 69–85.

[47]   PowerEscape, Inc. http://www.powerescape.com.

[48]   F. Sun, S. Ravi, A. Raghunathan, and N. Jha. Synthesis of custom processors based on extensible platforms. In *Proceedings of the International Conference on Computer-Aided Design*, 2002, pp. 641–648.

[49]   D. Goodwin and D. Petkov. Automatic generation of application specific processors. In *Proceedings of the International Conference on Compilers, Architectures, and Synthesis for Embedded Systems*, 2003, pp. 137–147.

[50]   K. Atasu, L. Pozzi, and P. Ienne. Automatic application-specific instruction-set extensions under microarchitectural constraints. In *Proceedings of the 40th Design Automation Conference*, 2003, pp. 256–261.

Chapter 6

[1]   C. Rowen and S. Leibson. *Engineering the Complex SoC. Fast, Flexible Design with Configurable Processors*. Prentice Hall, 2004.

[2]   S. Leibson and J. Kim. Configurable processors: a new era in chip design. *Computer*, 38(7):51–59, Jul. 2005.

[3]   M. Gries, K. Keutzer, H. Meyr, and G. Martin. *Building ASIPs: The MESCAL Methodology*. Springer, 2005.

[4]   D. Jani, C. Benson, A. Dixit, and G. Martin. Functional verification of configurable embedded processors. In *The Functional Verification of Electronic Systems: An Overview from Various Points of View*, B. Bailey, editor IEC, 2005.

[5]   J. Wei and C. Rowen. Implementing low-power configurable processors—practical options and tradeoffs. In *Proceedings of the 42nd Design Automation Conference*, 2005, pp. 706–711.

[6]   D. Goodwin and D. Petkov. Automatic generation of application specific processors. In *Proceedings of the International Conference on Compilers, Architectures, and Synthesis for Embedded Systems*, 2003 pp. 137–147.

[7]   D. Goodwin. The end of ISA design: power tools for optimal processor generation. In *Proceedings of the Hot Chips 16 Symposium*, 2004.

Chapter 7

[1]   L. Pozzi, K. Atasu, and P. Ienne. Optimal and approximate algorithms for the extension of embedded processor instruction sets. *IEEE Transactions on Computer-Aided Design of Integrated Circuits and Systems*, 25(7):1209–1229, Jul. 2006.

[2]   K. Atasu, L. Pozzi, and P. Ienne. Automatic application-specific instruction-set extensions under microarchitectural constraints. In *Proceedings of the 40th Design Automation Conference*, 2003, pp. 256–261.

[3]   P. Yu and T. Mitra. Scalable custom instructions identification for instruction set extensible processors. In *Proceedings of the International Conference on Compilers, Architectures, and Synthesis for Embedded Systems*, 2004, pp. 69–78.

[4]   K. Atasu, G. Dündar, and C. Özturan. An integer linear programming approach for identifying instruction-set extensions. In *Proceedings of the International Conference on Hardware/Software Codesign and System Synthesis*, 2005, pp. 172–177.

[5]   C. Lee, M. Potkonjak, and W. H. Mangione-Smith. Media-Bench: a tool for evaluating and synthesizing multimedia and communicatons systems. In *Proceedings of the 30th Annual International Symposium on Microarchitecture*, 1997, pp. 330–335.

[6]   P. Biswas, V. Choudhary, K. Atasu, L. Pozzi, P. Ienne, and N. Dutt. Introduction of local memory elements in instruction set extensions. In *Proceedings of the 41st Design Automation Conference*, 2004, pp. 729–734.

[7]   P. Biswas, N. Dutt, P. Ienne, and L. Pozzi. Automatic identification of application-specific functional units with architecturally visible storage. In *Proceedings of the Design, Automation and Test in Europe Conference and Exhibition*, 2006, pp. 212–217.

[8]   T. R. Halfhill. EEMBC releases first benchmarks. *Microprocessor Report*, 1 May 2000.

[9]   M. Baleani, F. Gennari, Y. Jiang, Y. Patel, R. K. Brayton, and A. Sangiovanni-Vincentelli. HW/SW partitioning and code generation of embedded control applications on a reconfigurable architecture platform. In *Proceedings of the 10th International Workshop on Hardware/Software Codesign*, 2002, pp. 151–156.

[10] C. Alippi, W. Fornaciari, L. Pozzi, and M. Sami. A DAG-based design approach for reconfigurable VLIW processors. In *Proceedings of the Design, Automation and Test in Europe Conference and Exhibition*, 1999, pp. 778–779.

[11] P. Biswas, S. Banerjee, N. Dutt, L. Pozzi, and P. Ienne. ISEGEN: generation of high-quality instruction set extensions by iterative improvement. In *Proceedings of the Design, Automation and Test in Europe Conference and Exhibition*, 2005, pp. 1246–1251.

[12] N. T. Clark, H. Zhong, and S. A. Mahlke. Processor acceleration through automated instruction set customisation. In *Proceedings of the 36th Annual International Symposium on Microarchitecture*, 2003, pp. 129–140.

[13] R. Razdan and M. D. Smith. A high-performance microarchitecture with hardware-programmable functional units. In *Proceedings of the 27th International Symposium on Microarchitecture*, 1994, pp. 172–180.

[14] S. Hauck, T. W. Fry, M. M. Hosler, and J. P. Kao. The Chimaera reconfigurable functional unit. In *Proceedings of the 5th IEEE Symposium on Field-Programmable Custom Computing Machines*, 1997, pp. 87–96.

[15] N. T. Clark, H. Zhong, and S. A. Mahlke. Automated custom instruction generation for domain-specific processor acceleration. *IEEE Transactions on Computers*, 54(10):1258–1270, Oct. 2005.

[16] T. Austin, E. Larson, and D. Ernst. SimpleScalar: An infrastructure for computer system modeling. *Computer*, 35(2):59–67, February 2002.

[17] L. Pozzi and P. Ienne. Exploiting pipelining to relax register-file port constraints of instruction-set extensions. In *Proceedings of the International Conference on Compilers, Architectures, and Synthesis for Embedded Systems*, 2005, pp. 2–10.

[18] P. Y. Calland, A. Mignotte, O. Peyran, Y. Robert, and F. Vivien. Retiming DAGs. *IEEE Transactions on Computer-Aided Design of Integrated Circuits and Systems*, 17(12):1319–1325, Dec. 1998.

Chapter 8

[1] J. A. Fisher, P. Faraboschi, and G. Desoli. Custom-fit processors: letting applications define architectures. In *Proceedings of the 29th Annual International Symposium on Microarchitecture*, 1996, pp. 324–335.

[2] A. C. Cheng and G. S. Tyson. An energy efficient instruction set synthesis framework for low power embedded system designs. *IEEE Transactions on Computers*, 54(6):698–713, June 2005.

[3] T. Karnik, Y. Ye, J. Tschanz, L. Wei, S. Burns, V. Govindarajulu, V. De, and S. Borkar. Total power optimization by simultaneous dual-Vt allocation and device sizing in high performance microprocessors. In *Proceedings of the 39th Design Automation Conference*, 2002, pp. 486–491.

[4] J. Lee, K. Choi, and N. D. Dutt. Energy-efficient instruction set synthesis for application-specific processors. In *Proceedings of the 2003 International Symposium on Low Power Electronics and Design*, 2003, pp. 330–333.

[5] N. Clark, M. Kudlur, H. Park, S. Mahlke, and K. Flautner. Application-specific processing on a general-purpose core via transparent instruction set customization. In *Proceedings of the 37th Annual International Symposium on Microarchitecture*, 2004, pp. 30–40.

[6] K. Atasu, L. Pozzi, and P. Ienne. Automatic application-specific instruction-set extensions under microarchitectural constraints. In *Proceedings of the 40th Design Automation Conference*, 2003, pp. 256–261.

[7] P. Brisk, A. Kaplan, and M. Sarrafzadeh. Area-efficient instruction set synthesis for reconfigurable system-on-chip designs. In *Proceedings of the 41st Design Automation Conference*, 2004, pp. 395–400.

[8] ARC International. *ARCompact*™. *ISA Programmer's Reference Manual*. http://www.arc.com/documentation/productbriefs/.

[9] T. R. Halfhill. ARC 700 aims higher. *Microprocessor Report*, 8 Mar. 2004.

[10] T. R. Halfhill. ARC shows SIMD extensions. *Microprocessor Report*, 21 Nov. 2005.

[11] Taiwan Semiconductor Manufacturing Company, Inc. *TSMC 0.13-micron Technology Platform*.

[12] R. P. Weicker. Dhrystone benchmark: rationale for version 2 and measurement rules. *ACM SIGPLAN Notices.*, 23(8):49–62, Aug. 1988.

[13] J. K. F. Lee and A. J. Smith. Branch prediction strategies and branch target buffer design. In *Instruction-Level Parallel Processors*, IEEE Computer Society, 1995, pp. 83–99.

[14] S. McFarling. *Combining Branch Predictors*. Technical report TN-36, Digital Equipment Corporation, Palo Alto, California, Jun. 1993.

[15] S.-H. Wang, W.-H. Peng, Y. He, G.-Y. Lin, C.-Y. Lin, S.-C. Chang, C.-N. Wang, and T. Chiang. A software-hardware co-implementation of MPEG-4 advanced video coding (AVC) decoder with block level pipelining. *Journal of VLSI Signal Processing*, 41(1):93–110, Aug. 2005.

[16] N. Ahmed, T. Natarajan, and K. Rao. Discrete cosine transform. *IEEE Transactions on Computers*, 23(1):90–93, Jan. 1974.

[17] D. Hankerson, P. D. Johnson, and G. A. Harris. *Introduction to Information Theory and Data Compression*. CRC, 1998.

[18] D. A. Lelewer and D. S. Hirschberg. Data compression. *ACM Computing Surveys*, 19(3):261–296, Sep. 1987.

[19] H. C. Hunter and J. H. Moreno. A new look at exploiting data parallelism in embedded systems. In *CASES '03: Proceedings of the 2003 International Conference on Compilers, Architecture and Synthesis for Embedded Systems*, New York, NY, ACM Press, 2003, pp. 159–169.

[20] C. Kozyrakis and D. Patterson. Vector vs. superscalar and VLIW architectures for embedded multimedia benchmarks. In *Proceedings of the 35th Annual International Symposium on Microarchitecture*, 2002, pp. 283–293.

[21] H. Liao and A. Wolfe. Available paralellism in video applications. In *Proceedings of the 30th Annual International Symposium on Microarchitecture*, 1997, pp. 321–329.

[22] M. Lorenz, P. Marwedel, T. Dräger, G. Fettweis, and R. Leupers. Compiler based exploration of DSP energy savings by SIMD operations. In *Proceedings of the Asia South Pacific Design Automation Conference*, 2004, pp. 838–841.

[23] G. M. Amdahl. On the validity of the single-processor approach to achieving large-scale computing capability. In *Proceedings of the AFIPS Conference*, 1967, pp. 483–485.

[24] M. Fowler, K. Beck, J. Brant, W. Opdyke, and D. Roberts. *Refactoring: Improving the Design of Existing Code*. Addison-Wesley, 1999.

Chapter 9

[1] Intel Corporation, *IA-32 Intel Architecture Optimization Reference Manual*, 2005.

[2] Advanced Micro Devices, Inc., *AMD64 Architecture Programmer's Manual*, 2005.

[3] M. Gries, K. Keutzer, H. Meyr, and G. Martin. *Building ASIPs: The MESCAL Methodology*. Springer, 2005.

[4] D. Sima, T. Fountain, and P. Kacsuk. *Advanced Computer Architectures: A Design Space Approach*. Addison-Wesley, 1997.

[5] I. Park, M. D. Powell, and T. N. Vijaykumar. Reducing register ports for higher speed and lower energy. In *Proceedings of the 35th Annual Symposium on Microarchitecture*, 2002, pp. 171–182.

[6] H. Corporaal, *Microprocessor Architectures: From VLIW to TTA*. John Wiley and Sons, 1997.

[7] J. Fisher, P. Faraboschi, and C. Young. *Embedded Computing: A VLIW Approach to Architecture, Compilers, and Tools*. Morgan Kaufmann, 2004.

[8] R. Morelos-Zaragoza. Encoder/decoder for binary BCH codes in C (version 3.1), 1997. http://www.eccpage.com/bch3.c.

Chapter 10

[1] P. Brisk, A. Kaplan, and M. Sarrafzadeh. Area-efficient instruction set synthesis for reconfigurable system-on-chip designs. In *Proceedings of the 41st Design Automation Conference*, 2004, pp. 395–400.

[2] N. T. Clark, H. Zhong, and S. A. Mahlke. Processor acceleration through automated instruction set customization. In *Proceedings of the 36th Annual International Symposium on Microarchitecture*, 2003, pp. 129–140.

[3] D. Herrmann and R. Ernst. Improved interconnect sharing by identity operation insertion. In *Proceedings of the International Conference on Computer-Aided Design*, 1999, pp. 489–493.

[4] H. Bunke, C. Guidobaldi, and M. Vento. Weighted minimum common supergraph for cluster representation. In *Proceedings of the IEEE International Conference on Image Processing*, 2003, pp. II-25–II-28.

[5] M. R. Garey and D. S. Johnson. *Computers and Intractability: A Guide to the Theory of NP-Completeness*. W. H. Freeman and Co., 1979.

[6] W. J. Masek and M. S. Peterson. A faster algorithm computing string edit distances. *Journal of Computer System Sciences*, 20:18–31, 1980.

[7] E. Ukkonen. Online construction of suffix trees. *Algorithmica*, 14:249–260, 1995.

[8] C. Lee, M. Potkonjak, and W. H. Mangione-Smith. Media-Bench: a tool for evaluating and synthesizing multimedia and communications systems. In *Proceedings of the 30th Annual International Symposium on Microarchitecture*, 1997, pp. 330–335.

[9] D. Pearson and V. Vazirani. Efficient sequential and parallel algorithms for maximal bipartite sets. *Journal of Algorithms*, 14:171–179, 1993.

[10] F. J. Kurdahi and A. C. Parker. REAL: A program for REgister ALlocation. In *Proceedings of the 24th Design Automation Conference*, 1987, pp. 210–215.

[11] R. Kastner, A. Kaplan, S. Memik, and E. Bozorgzadeh. Instruction generation for hybrid-reconfigurable systems. *ACM Transactions on Design Automation of Embedded Systems (TODAES)*, pp. 605–627, Oct. 2002.

[12] K. Bazargan and M. Sarrafzadeh. Fast online placement for reconfigurable computing systems. In *Proceedings of the IEEE Symposium on FPGAs for Custom Computing Machines*, 1999, p. 300.

[13] S. O. Memik, E. Bozorgzadeh, R. Kastner, and M. Sarrafzadeh. A superscheduler for embedded reconfigurable systems. In *Proceedings of the International Conference on Computer-Aided Design*, 2001, pp. 391–394.

[14] G. De Micheli. *Synthesis and Optimization of Digital Circuits*, McGraw-Hill, 1994.

Chapter 11

[1] M. R. Garey and D. S. Johnson. *Computers and Intractability: A Guide to the Theory of NP-Completeness*. W. H. Freeman & Co., 1979.

[2]   R. B. Bryant. Graph-based algorithms for Boolean function manipulation. *IEEE Transactions on Computers*. 35(8):677–691, Aug. 1986.

[3]   R. Rudell. Dynamic variable ordering for ordered binary decision diagrams. In *Proceedings of the International Conference on Computer Aided Design*, 1993, pp. 42–47.

[4]   J. C. Madre and J. P. Billion. Proving circuit correctness using formal comparison between expected and extracted behaviour. In *Proceedings of the 25th Design Automation Conference*, 1988, pp. 205–210.

[5]   J. C. Madre, O. Coudert, and J. P. Billion. Automating the diagnosis and the rectification of design errors with PRIAM. In *Proceedings of the International Conference on Computer Aided Design*, 1989, pp. 30–33.

[6]   N. Cheung, S. Parameswaran, and J. Henkel. Inside: Instruction selection/identification & design exploration for extensible processors. In *Proceedings of the International Conference on Computer-Aided Design*, 2003, pp. 291–297.

[7]   R. Cytron, J. Ferrante, B. Rosen, et al. An efficient method of computing static single assignment form. In *Proceedings of the 16th ACM SIGPLAN-SIGACT Symposium on Principles of Programming Languages*, 1989, pp. 25–35.

[8]   R. Brayton, G. Hachtel, A. Sangiovanni-Vincentelli, et al. VIS: a system for verification and synthesis. In *Proceedings of the 8th International Conference in Computer Aided Verfication*, 1996, pp. 428–432.

[9]   S. Cheng, R. Brayton, G. York, et al. Compiling Verilog into timed finite state machines. In *Proceedings of the Verilog HDL Conference*, 1995, p. 32.

[10] Tensilica, Inc. Xtensa processor, http://www.tensilica.com.

[11] C. Lee, M. Potkonjak, and W. H. Mangione-Smith. Media-Bench: a tool for evaluating and synthesizing multimedia and communications systems. In *Proceedings of the 30th Annual International Symposium on Microarchitecture*, 1997, pp. 330–335.

[12] N. Cheung, J. Henkel, and S. Parameswaran. Rapid configuration & instruction selection for an ASIP: a case study. In *Proceedings of the Design, Automation and Test in Europe Conference and Exhibition*, 2003, pp. 802–807.

[13] V. K. S. Aditya, B. R. Rau, and V. Kathail. Automatic architectural synthesis of VLIW and EPIC processors. In *Proceedings of the 12th International Symposium on System Synthesis*, 1999, pp. 107–113.

[14] A. Peymandoust, L. Pozzi, P. Ienne, and G. Micheli. Automatic instruction-set extension and utilization for embedded processors. In *Proceedings of the International Conference on Application Specific Array Processors*, 2003, pp. 103–114.

[15] H. Zima and B. Chapman. *Supercompilers for Parallel and Vector Computers*. ACM, 1990.

[16] Y. Wand and R. Weber. An ontological model of an information system. *IEEE Transactions on Software Engineering*, 16(11):1282–1292, Nov. 1990.

[17] Insightful Corporation. S-PLUS. http://www.insightful.com/products/splus.

Chapter 12

[1]   B. Bailey, G. Martin, and T. Anderson, eds. *Taxonomies for the Development and Verification of Digital Systems*. Springer, 2005.

[2]   R. E. Bryant. Graph-based algorithms for Boolean function manipulation. *IEEE Transactions on Computers*, 35(8):677–691, Aug. 1986.

[3]   R. S. French, M. S. Lam, J. R. Levitt, and K. Olukotun. A general method for compiling event-driven simulations. In *Proceedings of the 22nd Design Automation Conference*, 1995, pp. 151–156.

[4]   H. Foster, A. Krolnik, and D. Lacey. *Assertion-Based Design*. Kluwer, 2003.

[5]   R. E. Bryant. Symbolic simulation—techniques and applications. In *Proceedings of the 27th Design Automation Conference*, 1990, pp. 517–521.

[6]   M. W. Moskewicz, C. F. Madigan, Y. Zhao, L. Zhang, and S. Malik. Chaff: engineering an efficient SAT solver. In *Proceedings of the 38th Design Automation Conference*, 2001, pp. 530–535.

[7]   R. Drechsler and D. Große. System level validation using formal techniques. In *IEE Proceedings—Computer & Digital Techniques*, 152(3):393–406, May 2005.

[8]   J. R. Burch, E. M. Clarke, K. L. McMillan, and D. L. Dill. Sequential circuit verification using symbolic model checking. In *Proceedings of the 27th Design Automation Conference*, 1990, pp. 46–51.

[9]   A. Biere, A. Cimatti, E. M. Clarke, M. Fujita, and Y. Zhu. Symbolic model checking using SAT procedures instead of BDDs. In *Proceedings of the 36th Design Automation Conference*, 1999, pp. 317–320.

[10] H. Chockler, O. Kupferman, and M. Y. Vardi. Coverage metrics for formal verification. In *Proceedings of the 12th Advanced Research Working Conference on Correct Hardware Design and Verfication Methods*, 2003, pp. 111–125.

[11] K. Winkelmann, H.-J. Trylus, D. Stoffel, and G. Fey. A cost-efficient block verification for a UMTS up-link chip-rate coprocessor. In *Proceedings of the Design, Automation and Test in Europe Conference and Exhibition*, 2004, pp. 162–167.

[12] Accellera Organization, Inc. *Property Specification Language—Reference Manual*, version 1.1. June 2005. http://www.eda.org/vFv/docs/PSL-vl.l.pdf/.

[13] P. Bjesse and K. Claessen. SAT-based verification without state space traversal. In *Formal Methods in Computer-Aided Design*, pp. 372–389, Nov. 2000.

[14] M. Sheeran, S. Singh, and G. Stålmarck. Checking safety properties using induction and a SAT-solver. In *Proceedings of the Third International Conference on Formal Methods in Computer-Aided Design*, 2000, pp. 108–125.

[15] B. Becker, R. Drechsler, and P. Molitor. *Technische Informatik—Eine Einführung*. Pearson Education Deutschland, 2005.

[16] Open SystemC Initiative. *Functional Specification for SystemC 2.0.* http://www.systemc.org.

[17] T. Grötker, S. Liao, G. Martin, and S. Swan. *System Design with SystemC*. Kluwer, 2002.

[18] D. Große, U. Kühne, C. Genz, F. Schmiedle, B. Becker, R. Drechsler, and P. Molitor. Modellierung eines Mikroprozessors in SystemC. In *GI/ITG/GMM-Workshop, Methoden und Beschreibungssprachen zur Modellierung und Verifikation von Schaltungen und Systemen*, 2005.

[19] D. Große and R. Drechsler. *CheckSyC:* An efficient property checker for RTL SystemC designs. In *Proceedings of the IEEE International Symposium on Circuits and Systems*, 2005, pp. 4167–4170.

[20]   R. Drechsler, G. Fey, C. Genz, and D. Große. SyCE: an integrated environment for system design in SystemC. In *Proceedings of the 16th IEEE International Workshop on Rapid System Prototyping*, 2005, pp. 258–260.

[21]   L. M. Chirica and D. F. Martin. Toward compiler implementation correctness proofs. *ACM Transactions on Programming Languages and Systems (TOPLAS)*, 8(2):185–214, Apr. 1986.

[22]   A. Heberle, T. Gaul, W. Goerigk, G. Goos, and W. Zimmermann. Construction of verified compiler front-ends with program-checking. In *Proceedings of the Third International Andrei Ershov Memorial Conference on Perspectives of System Informatics*, LNCS-1755, 1999, pp. 481–492.

[23]   S. Glesner. Using program checking to ensure the correctness of compiler implementations. *Journal of Universal Computer Science*, 9(3):191–222, Mar. 2003.

[24]   Synopsys Inc. VCS documentation. *SystemVerilog Assertion Library*, 2004.

Chapter 13

[1]   N. Shah. *Understanding Network Processors*. Master's thesis, University of California, Berkeley, California, 2001.

[2]   E. Kohler, R. Morris, B. Chen, J. Jannotti, and M. F. Kaashoek. The Click modular router. *ACM Transactions on Computer Systems (TOCS)*, 18(3):263–297, Aug. 2000.

[3]   B. Chen and R. Morris. Flexible control of parallelism in a multiprocessor PC router. In *Proceedings of the 2002 USENIX Annual Technical Conference*, 2002, pp. 333–346.

[4]   N. Shah, W. Plishker, K. Ravindran, and K. Keutzer. NP-Click: A productive software development approach for network processors. *IEEE Micro*, 24(5):45–54, Sep.–Oct. 2004.

[5]   C. Sauer, M. Gries, and S. Sonntag. Modular domain-specific implementation and exploration framework for embedded software platforms. In *Proceedings of the Design Automation Conference*, 2005, pp. 254–259.

[6]   P. Paulin, C. Pilkington, E. Bensoudane, M. Langevin, and D. Lyonnard. Application of a multi-processor SoC platform to high-speed packet forwarding. In *Proceedings of the Design, Automation and Test in Europe Conference and Exhibition (Designer Forum)*, 2004, vol. 3, pp. 58–63.

[7]   K. Ravindran, N. Satish, Y. Jin, and K. Keutzer. An FPGA-based soft multiprocessor for IPv4 packet forwarding. In *Proceedings of the 15th International Conference on Field Programmable Logic and Applications*, 2005, pp. 487–492.

[8]   C. Kulkarni, G. Brebner, and G. Schelle. Mapping a domain specific language to a platform FPGA. In *Proceedings of the 41st Design Automation Conference*, 2004, pp. 924–927.

[9]   Tensilica, Inc. Tensilica Xtensa LX processor tops EEMBC networking 2.0 benchmarks, May 2005. http://www.tensilica.com/html/pr_2005_05_16.html.

[10]   I. Sourdis, D. Pnevmatikatos, and K. Vlachos. An efficient and low-cost input/output subsystem for network processors. In *Proceedings of the 1st Workshop on Application Specific Processors*, 2002, pp. 56–64.

[11]   G. Hadjiyiannis, S. Hanono, and S. Devadas. ISDL: An instruction set description language for retargetability. In *Proceedings of the 34th Design Automation Conference*, 1997, pp. 299–302.

[12] A. Fauth, J. Van Praet, and M. Freericks. Describing instruction set processors using nML. In *Proceedings of the European Design and Test Conference*, 1995, pp. 503–507.

[13] A. Hoffman, H. Meyr, and R. Leupers. *Architecture Exploration for Embedded Processors with LISA*. Kluwer, 2002.

[14] A. Halambi, P. Grun, V. Ganesh, A. Khare, N. Dutt, and A. Nicolau. EXPRESSION: a language for architecture exploration through compiler/simulator retargetability. In *Proceedings of Design, Automation and Test in Europe Conference and Exhibition*, 1999, pp. 485–490.

[15] R. Leupers and P. Marwedel. Retargetable code generation based on structural processor descriptions. In *Design Automation for Embedded Systems*, 3(1):75–108, Jan. 1998.

[16] M. Gries, K. Keutzer, H. Meyr, and G. Martin. *Building ASIPs: The MESCAL Methodology*. Springer, 2005.

[17] S. Weber and K. Keutzer. Using minimal minterms to represent programmability. In *Proceedings of the International Conference on Hardware/Software Codesign and System Synthesis*, 2005, pp. 63–66.

[18] Free Software Foundation, Inc. GNU Multiple Precision Arithmetic Library. http://www.swox.com/gmp.

[19] S. Weber, M. Moskewicz, M. Gries, C. Sauer, and K. Keutzer. Fast cycle-accurate simulation and instruction set generation for constraint-based descriptions of programmable architectures. In *Proceedings of the International Conference on Hardware/Software Codesign and System Synthesis*, 2004, pp. 18–23.

[20] E. Lee. Embedded software. In *Advances in Computers*, M. Zelkowitz, editor, Academic, 2002, pp. 56-99.

[21] E. Willink, J. Eker, and J. Janneck. Programming specifications in CAL. In *Proceedings of the OOPSLA Workshop on Generative Techniques in the Context of Model Driven Architecture*, 2002.

[22] Y. Jin, N. Satish, K. Ravindran, and K. Keutzer. An automated exploration framework for FPGA-based soft multiprocessor systems. In *Proceedings of the International Conference on Hardware/Software Codesign and System Synthesis*, 2005, pp. 273–278.

[23] T. Henriksson and I. Verbauwhede. Fast IP address lookup engine for SoC integration. In *Proceedings of the IEEE Design and Diagnostics of Electronic Circuits and Systems Workshop*, 2002, pp. 200–210.

[24] D. Taylor, J. Lockwood, T. Sproull, J. Turner, and D. Parlour. Scalable IP lookup for internet routers. *IEEE Journal on Selected Areas in Communications*, 21:522–534, May 2003.

[25] M. Degermark, A. Brodnik, S. Carlsson, and S. Pink. Small forwarding tables for fast routing lookups. In *Proceedings of the ACM SIGCOMM '97 Conference on Applications Technologies, Architectures, and Protocols for Computer Communication*, 1997, pp. 3–14.

[26] T. Henriksson, U. Nordqvist, and D. Liu. Embedded protocol processor for fast and efficient packet reception. In *Proceedings of the IEEE International Conference on Computer Design*, 2002, pp. 414–419.

[27] A. DeHon, J. Adams, M. DeLorimier, N. Kapre, Y. Matsuda, H. Naeimi, M. Vanier, and M. Wrighton. Design patterns for reconfigurable computing. In *Proceedings of the IEEE Symposium on Field-Programmable Custom Computing Machines*, 2004, pp. 13–23.

[28]   D. Meng, R. Gunturi, and M. Castelino. IXP2800 Intel network processor IP forwarding benchmark full disclosure report for OC192-POS. In *Intel Corp. Technical Report for the Network Processing Forum (NPF)*, 2003.

Chapter 14

[1]   The Global Mobile Suppliers Association. 3G/WCDMA Fact Sheet. Aug. 2005, http://www.gsacom.com.

[2]   Third Generation Partnership Project (3GPP). 3GPP Technical Specification 25.913, Requirements for Evolved UTRA (E-UTRA) and Evolved UTRAN (E-UTRAN). V7.0.0, Jun. 2005.

[3]   H. Holma and A. Toskala, editors *WCDMA for UMTS: Radio Access for Third Generation Mobile Communications*. John Wiley & Sons, 2000, p. 315.

[4]   Third Generation Partnership Project. 3GPP Technical Specification 25.213, Spreading and modulation (FDD). V7.0.0, Mar. 2006.

[5]   T. Ristimäki and J. Nurmi. Reconfigurable IP blocks: a survey. In *Proceedings of the International Symposium on System-on-Chip*, 2004, pp. 117–122.

[6]   Tuulos, K. *3GPP FDD Baseband Test Data Generator*. Master's thesis, University of Turku, Turku, Finland, February 2002, p. 74.

[7]   S. Goto, T. Yamada, N. Takayama, and H. Yasuura. A design for a low-power digital matched filter applicable to W-CDMA. In *Proceedings of the Euromicro Symposium on Digital System Design*, 2002, pp. 210–217.

[8]   L. Chi-Fang et al. A low-power ASIC design for cell search in the W-CDMA system. *IEEE Journal of Solid State Circuits*, 39(5): 852–857, May 2004.

Chapter 15

[1]   C. E. Shannon. A mathematical theory of communication. *Bell System Technical Journal*, 27:379–423 and 623–656, Jul. and Oct. 1948.

[2]   C. Berrou, A. Glavieux, and P. Thitimajshima. Near Shannon limit error-correcting coding and decoding: Turbo-Codes. In *Proceedings of the International Conference on Communications*, 1993, pp. 1064–1070.

[3]   Consultative Committee for Space Data Systems. www.ccsds.org.

[4]   A. Worm. *Implementation Issues of Turbo-Decoders*. PhD dissertation, Institute of Microelectronic Systems, Department of Electrical Engineering and Information Technology, University of Kaiserslautern, Kaiserslautern, Germany, 2001.

[5]   Third Generation Partnership Project (3GPP). www.3gpp.org.

[6]   J. Vogt, K. Koora, A. Finger, and G. Fettweis. Comparison of different turbo decoder realizations for IMT-2000. In *Proceedings of the Global Telecommunications Conference*, 1999, vol.5, pp. 2704–2708.

[7]   F. Gilbert, F. Kienle, and N. Wehn. Low complexity stopping criteria for UMTS turbo-decoders. In *Proceedings of the 2003-Spring Vehicular Technology Conference*, 2003, pp. 2376–2380.

[8]   P. Robertson, P. Hoeher, and E. Villebrun. Optimal and sub-optimal maximum a posteriori algorithms suitable for turbo decoding. *European Transactions on Telecommunications*, 8(2):119–125, Mar.–Apr. 1997.

[9] M. J. Thul, F. Gilbert, T. Vogt, G. Kreiselmaier, and N. Wehn. A scalable system architecture for high-throughput turbo-decoders. In *Proceedings of the 2002 Workshop on Signal Processing Systems*, 2002, pp. 152–158.

[10] H. Dawid, G. Gehnen, and H. Meyr. MAP channel decoding: algorithm and VLSI architecture. In *Proceedings of the Sixth Workshop on VLSI Signal Processing VI*, 1993, pp. 141–149.

[11] T. Ngo and I. Verbauwhede. Turbo codes on the fixed point DSP TMS320C55x. In *Proceedings of the 2000 Workshop on Signal Processing Systems*, 2000, pp. 255–264.

[12] ST Microelectronics N.V. http://www.st.com.

[13] StarCore LLC. http://www.starcore-dsp.com.

[14] Analog Devices, Inc. http://www.analog.com.

[15] Tensilica, Inc. http://www.tensilica.com.

[16] B. Ackland and C. Nicol. High performance DSPs—What's hot and what's not? In *Proceedings of the International Symposium on Low Power Electronics and Design*, 1998, pp. 1–6.

[17] CoWare, Inc. http://www.coware.com.

[18] R. E. Gonzalez. Xtensa: a configurable and extensible processor. *IEEE Micro*, 20(2):60–70, Mar.–Apr. 2000.

[19] H. Michel. *Implementation of Turbo-Decoders on Programmable Architectures*. PhD dissertation, Microelectronic System Design Reseach Group, Department of Electrical Engineering and Information Technology, University of Kaiserslautern, Kaiserslautern, Germany, 2002.

[20] C. Rowen and S. Leibson. *Engineering the Complex SoC. Fast, Flexible Design with Configurable Processors*. Prentice Hall, 2004.

[21] M. J. Thul, C. Neeb, and N. Wehn. Network-on-chip-centric approach to interleaving in high throughput channel decoders. In *Proceedings of the 2005 IEEE International Symposium on Circuits and Systems*, 2005, pp. 1766–1769.

[22] F. Gilbert. *Optimized, Highly Parallel Architectures for Iterative Decoding Algorithms*. PhD dissertation, Microelectronic Systems Design Research Group, Department of Electrical Engineering and Information Technology, University of Kaiserslautern, Kaiserslautern, Germany, May 2003.

[23] R. G. Gallager. *Low-Density Parity-Check Codes*. M.I.T. Press, 1963.

[24] D. MacKay and R. Neal. Near Shannon limit performance of low-density parity-check codes. *Electronic Letters*, 32(18):1645–1646, Mar. 1996.

Chapter 16

[1] CoWare, Inc. LISATek tools. http://www.coware.com.

[2] A. Hoffmann, H. Meyr, and R. Leupers. *Architecture Exploration for Embedded Processors with LISA*. Kluwer, 2002.

[3] G. Braun, A. Nohl, O. Schliebusch, A. Hoffmann, R. Leupers, and H. Meyr. A universal technique for fast and flexible instruction-set architecture simulation. In *IEEE Transactions on Computer-Aided Design of Integrated Circuits and Systems*, 23(12):1625–1639, Dec. 2004.

[4] M. Hohenauer, O. Wahlen, K. Karuri, H. Scharwächter, T. Kogel, R. Leupers, G. Ascheid, H. Meyr, G. Braun, and H. van Someren. A methodology and tool suite for C compiler generation from ADL

processor models. In *Proceedings of the Design, Automation and Test in Europe Conference and Exhibition*, 2004, pp. 1276–1283.

[5] RWTH Aachen University. LANCE Retargetable C compiler. http://www.lancecompiler.com.

[6] K. Karuri, M. Al Faruque, S. Kraemer, R. Leupers, G. Ascheid, and H. Meyr. Fine-grained application source code profiling for ASIP design. In *Proceedings of the 42nd Design Automation Conference*, Anaheim, 2005, pp. 329–334.

[7] K. Karuri, C. Huben, R. Leupers, G. Ascheid, and H. Meyr. Memory access micro-profiling for ASIP design. In *3rd IEEE Int. Workshop on Electronic Design, Test, & Applications (DELTA)*, Kuala Lumpur, Malaysia, 2006.

[8] T. Kempf, K. Karuri, R. Leupers, G. Ascheid, and H. Meyr. A SW performance estimation framework for early system level design using fine-grained instrumentation. In *Proceedings of the Design, Automation and Test in Europe Conference and Exhibition*, 2006, pp. 468–473.

[9] MIPS Technologies, Inc. http://www.mips.com.

[10] K. Atasu, L. Pozzi, and P. Ienne. Automatic application-specific instruction-set extensions under microarchitectural constraints. In *Proceedings of the 40th Design Automation Conference*, 2003, pp. 256–261.

[11] N. T. Clark, H. Zhong, and S. A. Mahlke. Processor acceleration through automated instruction set customization. *Proceedings of the 36th Annual International Symposium on Microarchitecture*, 2003, pp. 129–140.

[12] B. Schneier. Description of a new variable-length key, 64-bit block cipher (Blowfish). In *Proceedings of the Cambridge Security Workshop on Fast Software Encryption*, LNCS-809, 1994, pp. 191–204.

[13] Stretch, Inc. http://www.stretchinc.com.

Chapter 17

[1] Xilinx, Inc. Virtex4 and other FPGA families. http://www.xilinx.com, 2005.

[2] S. Trimberger, editor. *Field Programmable Gate Array Technology*. Kluwer 1994.

[3] J. Vuillemin, P. Bertin, D. Roncin, M. Shand, H. Touati, and P. Boucard. Programmable active memories: reconfigurable systems come of age. *IEEE Transactions on Very Large Scale Integration (VLSI) Systems*, 4(1):56–69, Mar. 1996.

[4] M. J. Wirthlin and B. L. Hutchings. A dynamic instruction set computer. In *Proceedings of the IEEE Workshop on FPGAs for Custom Computing Machines*, 1995, pp. 99–107.

[5] IEEE Symposium on Field-Programmable Custom Computing Machines. http://www.fccm.org.

[6] International Conference on Field Progammable Logic and Applications. http://www.fpl.org.

[7] CAST, Inc. Details of 8051 processor implemented on various FPGA families. http://www.cast-inc.com/cores/c8051.

[8] J. Gray. Building a RISC system in an FPGA—Part 1: Tools, instruction set, and datapath. *Circuit Cellar Magazine*, (116):26, Mar. 2000.

[9] Xilinx, Inc. PicoBlaze soft processor. http://www.origin.xilinx.com/bvdocs/appnotes/xapp213.pdf, Sep. 2000.

[10] Altera Corporation. Details of NIOS Soft CPU. http://www.altera.com/
      literature/ds/ds_nios_cpu.pdf.

[11] International Business Machines Corporation. *The IBM PowerPC 405
      Core*. 1998.

[12] Intel Corporation. Details of Intel MMX technology. http://www.intel.com/
      design/intarch/mmx/mmx.htm.

[13] K. Atasu, L. Pozzi, and P. Ienne. Automatic application-specific
      instruction-set extensions under microarchitectural constraints. In *Pro-
      ceedings of the 40th Design Automation Conference*, 2003, pp. 256–261.

[14] J. Cong, Y. Fan, G. Han, A. Jagannathan, G. Reinman, and Z. Zhang.
      Instruction set extension with shadow registers for configurable proces-
      sors. In *Proceedings of the 13th International Symposium on Field Pro-
      grammable Gate Arrays*, 2005, pp. 99–106.

BIBLIOGRAPHY

Accellera Organization, Inc. *Property Specification Language—Reference Manual*, version 1.1, June 2005. http://www.eda.org/vfv/docs/PSL-v1.1.pdf.

B. Ackland and C. Nicol. High performance DSPs—what's hot and what's not? In *Proceedings of the International Symposium on Low Power Electronics and Design*, 1998, pp. 1–6.

V. K. S. Aditya, B. R. Rau, and V. Kathail. Automatic architectural synthesis of VLIW and EPIC processors. In *Proceedings of the 12th International Symposium on System Synthesis*, 1999, pp. 107–113.

N. Ahmed, T. Natarajan, and K. Rao. Discrete cosine transform. *IEEE Transactions on Computers*, 23(1):90–93, Jan. 1974.

A. Aho, S. Johnson, and J. Ullman. Code generation for expressions with common subexpressions. *Journal of the ACM*, 24(1), Jan. 1977.

A. V. Aho, R. Sethi, and J. D. Ullman. *Compilers—Principles, Techniques, and Tools*, Addison-Wesley, 1986.

C. Alippi, W. Fornaciari, L. Pozzi, and M. Sami. A DAG-based design approach for reconfigurable VLIW processors. In *Proceedings of the Design, Automation and Test in Europe Conference and Exhibition*, 1999, pp. 778–779.

Altera Corporation. Details of NIOS Soft CPU. http://www.altera.com/literature/ds/ds_nios_cpu.pdf.

Advanced Micro Devices, Inc. *AMD64 Architecture Programmer's Manual*. 2005.

G. M. Amdahl. On the validity of the single-processor approach to achieving large-scale computing capability. In *Proceedings of the AFIPS Conference*, 1967, pp. 483–485.

Analog Devices, Inc. http://www.analog.com.

A. W. Appel. *Modern Compiler Implementation in C*. Cambridge University Press, 1998.

ARC International. *ARCompact™ ISA Programmer's Reference Manual*. http://www.arc.com/documentation/productbriefs/.

——. ARCtangent processor. http://www.arc.com.

Y. Arai, T. Agui, and M. Nakajima. A fast DCT-SQ scheme for images. *Transactions of the IEICE*, E-71(11):1095–1097, Nov. 1988.

M. Arnold and H. Corporaal. Designing domain specific processors. In *Proceedings of the 9th International Workshop on Hardware/Software Codesign*, 2001, pp. 61–66.

ACE Associated Compiler Experts. http://www.ace.nl.

K. Atasu, L. Pozzi, and P. Ienne. Automatic application-specific instruction-set extensions under microarchitectural constraints. In *Proceedings of the 40th Design Automation Conference*, 2003, pp. 256–261.

K. Atasu, G. Dündar, and C. Özturan. An integer linear programming approach for identifying instruction-set extensions. In *Proceedings of the International Conference on Hardware/Software Codesign and System Synthesis*, 2005, pp. 172–177.

T. Austin, E. Larson, and D. Ernst. SimpleScalar: an infrastructure for computer system modeling. *Computer*, 35(2):59–67, Feb. 2002.

B. Bailey, G. Martin, and T. Anderson, eds. *Taxonomies for the Development and Verification of Digital Systems*. Springer, 2005.

M. Baleani, F. Gennari, Y. Jiang, Y. Patel, R. K. Brayton, and A. Sangiovanni-Vincentelli. HW/SW partitioning and code generation of embedded control applications on a reconfigurable architecture platform. In *Proceedings of the 10th International Workshop on Hardware/Software Codesign*, 2002, pp. 151–156.

M. R. Barbacci. Instruction set processor specifications (ISPS): the notation and its applications. *IEEE Transactions on Computers*, C-30(1):24–40, Jan. 1981.

S. Bashford and R. Leupers. Constraint driven code selection for fixed-point DSPs. In *Proceedings of the 36th Design Automation Conference*, 1999. pp. 817–822.

M. J. Bass and C. M. Christensen. The future of the microprocessor business. *IEEE Spectrum*, 39(4):34–39, Feb. 2002.

K. Bazargan and M. Sarrafzadeh. Fast online placement for reconfigurable computing systems. In *Proceedings of the IEEE Symposium on FPGAs for Custom Computing Machines*, 1999, p. 300.

B. Becker, R. Drechsler, and P. Molitor. *Technische Informatik–Eine Einführung*. Pearson Education Deutschland, 2005.

C. Berrou, A. Glavieux, and P. Thitimajshima. Near Shannon limit error-correcting coding and decoding: Turbo-Codes. In *Proceedings of the International Conference on Communications*, 1993, pp. 1064–1070.

A. Biere, A. Cimatti, E. M. Clarke, M. Fujita, and Y. Zhu. Symbolic model checking using SAT procedures instead of BDDs. In *Proceedings of the 36th Design Automation Conference*, 1999, pp. 317–320.

K. M. Bischoff. *Design, Implementation, Use, and Evaluation of Ox: an Attribute-Grammar Compiling System Based on Yacc, Lex, and C*. Technical report 92–31, Department of Computer Science, Iowa State University, Ames, Iowa, 1992.

P. Biswas, S. Banerjee, N. Dutt, L. Pozzi, and P. Ienne. ISEGEN: generation of high-quality instruction set extensions by iterative improvement. In *Proceedings of the Design, Automation and Test in Europe Conference and Exhibition*, 2005, pp. 1246–1251.

P. Biswas, V. Choudhary, K. Atasu, L. Pozzi, P. Ienne, and N. Dutt. Introduction of local memory elements in instruction set extensions. In *Proceedings of the 41st Design Automation Conference*, 2004, pp. 729–734.

P. Biswas, N. Dutt, P. Ienne, and L. Pozzi. Automatic identification of application-specific functional units with architecturally visible storage. In *Proceedings of the Design, Automation and Test in Europe Conference and Exhibition*, 2006, pp. 212–217.

P. Bjesse and K. Claessen. SAT-based verification without state space traversal. In *Formal Methods in Computer-Aided Design*, pp. 372–389, Nov. 2000.

H. Blume, H. Hubert, H. T. Feldkamper, and T. G. Noll. Model-based exploration of the design space for heterogeneous systems on chip. In *Proceedings of the 13th International Conference on Application-Specific Systems, Architectures and Processors*, 2002, pp. 29–40.

H. Blume, T. Gemmeke, and T. G. Noll. Platform-specific turbo decoder implementations. In *Proceedings of the DSP Design Workshop*, 2003.

G. Braun, A. Nohl, O. Schliebusch, A. Hoffmann, R. Leupers, and H. Meyr. A universal technique for fast and flexible instruction-set architecture simulation. *IEEE Transactions on Computer-Aided Design of Integrated Circuits and Systems*, 23(12):1625–1639, Dec. 2004.

G. Braun, A. Nohl, W. Sheng, J. Ceng, M. Hohenauer, H. Scharwaechter, R. Leupers, and H. Meyr. A novel approach for flexible and consistent ADL-driven ASIP design. In *Proceedings of the 41st Design Automation Conference*, 2004. pp. 717–722.

R. Brayton, G. Hachtel, A. Sangiovanni-Vincentelli, et al. VIS: a system for verification and synthesis. In *Proceedings of the 8th International Conference in Computer Aided Verification*, 1996, pp. 428–432.

P. Briggs. *Register Allocation via Graph Coloring*. PhD dissertation, Department of Computer Science, Rice University, Houston, Texas, 1992.

P. Brisk, A. Kaplan, and M. Sarrafzadeh. Area-efficient instruction set synthesis for reconfigurable system-on-chip designs. In *Proceedings of the 41st Design Automation Conference*, 2004, pp. 395–400.

R. E. Bryant. Graph-based algorithms for Boolean function manipulation. *IEEE Transactions on Computers*, 35(8):677–691, Aug. 1986.

——. Symbolic simulation—techniques and applications. In *Proceedings of the 27th Design Automation Conference*, 1990, pp. 517–521.

H. Bunke, C. Guidobaldi, and M. Vento. Weighted minimum common supergraph for cluster representation. In *Proceedings of the IEEE International Conference on Image Processing*, 2003, pp. II-25–II-28.

J. R. Burch, E. M. Clarke, K. L. McMillan, and D. L. Dill. Sequential circuit verification using symbolic model checking. In *Proceedings of the 27th Design Automation Conference*, 1990, pp. 46–51.

L. Cai, A. Gerstlauer, and D. Gajski. *RetargetableProfiling for Rapid, Early System-Level Design Space Exploration*. Technical report. Center for Embedded Computer Systems, University of California, Irvine, California, 2004.

P. Y. Calland, A. Mignotte, O. Peyran, Y. Robert, and F. Vivien. Retiming DAGs. *IEEE Transactions on Computer-Aided Design of Integrated Circuits and Systems*, 17(12):1319–1325, Dec. 1998.

CAST, Inc. Details of the 8051 processor implemented on various FPGA families. http://www.cast-inc.com/cores/c8051.

J. Ceng, W. Sheng, M. Hohenauer, R. Leupers, G. Ascheid, H. Meyr, and G. Braun. Modeling instruction semantics in ADL processor descriptions for C compiler retargeting. In *Proceedings of the International Workshop on Systems, Architectures, Modeling, and Simulation*, 2004, pp. 463–473.

B. Chen and R. Morris. Flexible control of parallelism in a multiprocessor PC router. In *Proceedings of the 2002 USENIX Annual Technical Conference*, 2002, pp. 333–346.

S. Cheng, R. Brayton, G. York, et al. Compiling Verilog into timed finite state machines. In *Proceedings of the Verilog HDL Conference*, 1995, p. 32.

H. Chockler, O. Kupferman, and M. Y. Vardi. Coverage metrics for formal verification. In *Proceedings of the 12th Advanced Research Working Conference on Correct Hardware Design and Verification Methods*, 2003, pp. 111–125.

A. C. Cheng and G. S. Tyson. An energy efficient instruction set synthesis framework for low power embedded system designs. *IEEE Transactions on Computers*, 54(6):698–713, June 2005.

N. Cheung, J. Henkel, and S. Parameswaran. Rapid configuration & instruction selection for an ASIP: a case study. In *Proceedings of the Design, Automation and Test in Europe Conference and Exhibition*, 2003, pp. 802–807.

N. Cheung, S. Parameswaran, and J. Henkel. Inside: Instruction selection/identification & design exploration for extensible processors. In *Proceedings of the International Conference on Computer-Aided Design*, 2003, pp. 291–297.

L. Chi-Fang et al. A low-power ASIC design for cell search in the W-CDMA system. *IEEE Journal of Solid State Circuits*, 39(5): 852–857, May 2004.

L. M. Chirica and D. F. Martin. Toward compiler implementation correctness proofs. *ACM Transactions on Programming Languages and Systems (TOPLAS)*, 8(2):185–214, Apr. 1986.

F. Chow and J. Hennessy. Register allocation by priority-based coloring. *ACM SIGPLAN Notices*, 19(6):222–232, Jun.1984.

N. Clark, M. Kudlur, H. Park, S. Mahlke, and K. Flautner. Application-specific processing on a general-purpose core via transparent instruction set customization. In *Proceedings of the 37th Annual International Symposium on Microarchitecture*, 2004, pp. 30–40.

N. T. Clark, H. Zhong, and S. A. Mahlke. Processor acceleration through automated instruction set customization. In *Proceedings of the 36th Annual International Symposium on Microarchitecture*, 2003, pp. 129–140.

——. Automated custom instruction generation for domain-specific processor acceleration. *IEEE Transactions on Computers*, 54(10):1258–1270, Oct. 2005.

P. C. Clements. A survey of architecture description languages. In *Proceedings of International Workshop on Software Specification and Design*, 1996, pp. 16–25.

R. J. Cloutier and D. E. Thomas. Synthesis of pipelined instruction set processors. In *Proceedings of the 30th Design Automation Conference*, 1993, pp. 583–588.

J. Cong, Y. Fan, G. Han, A. Jagannathan, G. Reinman, and Z. Zhang. Instruction set extension with shadow registers for configurable processors. In *Proceedings of the 13th International Symposium on Field Programmable Gate Arrays*, 2005, pp. 99–106.

Consultative Committee for Space Data Systems. http://www.ccsds.org.

H. Corporaal. *Microprocessor Architectures: From VLIW to TTA*. John Wiley and Sons, 1997.

CoWare, Inc. http://www.coware.com.

——. LISATek tools. http://www.coware.com.

R. Cytron, J. Ferrante, B. Rosen, et al. An efficient method of computing static single assignment form. In *Proceedings of the 16th ACM SIGPLAN-SIGACT Symposium on Principles of Programming Languages*, 1989, pp. 25–35.

H. Dawid, G. Gehnen, and H. Meyr. MAP channel decoding: algorithm and VLSI architecture. In *Proceeding of the Sixth Workshop on VLSI Signal Processing*, 1993, pp. 141–149.

M. Degermark, A. Brodnik, S. Carlsson, and S. Pink. Small forwarding tables for fast routing lookups. In *Proceedings of the ACM SIGCOMM '97 Conference on Applications Technologies, Architectures, and Protocols for Computer Communication*, 1997, pp. 3–14.

A. DeHon, J. Adams, M. DeLorimier, N. Kapre, Y. Matsuda, H. Naeimi, M. Vanier, and M. Wrighton. Design patterns for reconfigurable computing. In *Proceedings of the IEEE Symposium on Field-Programmable Custom Computing Machines*, 2004, pp. 13–23.

G. De Micheli. *Synthesis and Optimization of Digital Circuits.*, McGraw-Hill, 1994.

G. Desoli, N. Mateev, E. Duesterwald, P. Faraboschi, and J. Fisher. DELI: a new run-time control point. *Proceedings of the 35th Annual International Symposium on Microarchitecture*, 2002, pp. 257–268.

R. Drechsler and D. Große. System level validation using formal techniques. In *IEE Proceedings—Computer and Digital Techniques*, 152(3):393–406, May 2005.

R. Drechsler, G. Fey, C. Genz, and D. Große. SyCE: an integrated environment for system design in SystemC. In *Proceedings of the 16th IEEE International Workshop on Rapid System Prototyping*, 2005, pp. 258–260.

M. A. Ertl. Optimal code selection in DAGs. In *Proceedings of the ACM Symposium on Principles of Programming Languages*, 1999, pp. 242–249.

H. Falk and P. Marwedel. Control flow driven splitting of loop nests at the source code level. In *Proceedings of the Design Automation and Test in Europe Conference and Exhibition*, 2003, pp. 410–415.

P. Faraboschi, J. Fisher, G. Brown, G. Desoli, and F. Homewood. Lx: A technology platform for customizable VLIW embedded processing. In *Proceedings of the 27th Annual International Symposium on Computer Architecture*, 2000, pp. 203–213.

P. Faraboschi, G. Desoli, and J. Fisher. VLIW architectures for DSP and multimedia applications. *IEEE Signal Processing*, 15(2):59–85, Mar. 1998.

——. *Clustered Instruction-Level Parallel Processors*. Technical report HPL-98–204. HP Labs, Cambridge, Massachussets, Dec. 1998.

A. Fauth and A. Knoll. Automatic generation of DSP program development tools. In *Proceedings of International Conference of Acoustics, Speech and Signal Processing*, 1993, pp. 457–460.

A. Fauth, J. Van Praet, and M. Freericks. Describing instruction set processors using nML. In *Proceedings of the European Design and Test Conference*, 1995, pp. 503–507.

E. Feig and S. Winograd. Fast algorithms for the discrete cosine transform. *IEEE Transactions on Signal Processing*, 40(9):2174–2193, Sept. 1992.

J. A. Fisher, P. Faraboschi, and G. Desoli. Custom-fit processors: letting applications define architectures. In *Proceedings of the 29th Annual International Symposium on Microarchitecture*, 1996, pp. 324–335.

J. Fisher, P. Faraboschi, and C. Young. *Embedded Computing: A VLIW Approach to Architecture, Compilers and Tools*. Morgan Kaufmann, 2004.

H. Foster, A. Krolnik, and D. Lacey. *Assertion-Based Design*. Kluwer, 2003.

M. Fowler, K. Beck, J. Brant, W. Opdyke, and D. Roberts. *Refactoring: Improving the Design of Existing Code*. Addison-Wesley, 1999.

B. Franke and M. O'Boyle. Compiler transformation of pointers to explicit array accesses in DSP applications. In *Proceedings of the International Conference on Compiler Construction*, 2001, pp. 69–85.

C. W. Fraser, D. R. Hanson, and T. A. Proebsting. Engineering a simple, efficient code generator generator. *ACM Letters on Programming Languages and Systems*, 1(3):216–226, Sept. 1992.

C. Fraser and D. Hanson. *A Retargetable C Compiler: Design and Implementation*. Addison-Wesley, 1995.

——. LCC. http://www.cs.princeton.edu/software/lcc.

Free Software Foundation, Inc. EGCS. http://gcc.gnu.org.

——. GNU Multiple Precision Arithmetic Library. http://www.swox.com/gmp.

M. Freericks. *The nML Machine Description Formalism*. Technical report TR SM-IMP/DIST/08, Computer Science Department, TU Berlin, Berlin, 1993.

R. S. French, M. S. Lam, J. R. Levitt, and K. Olukotun. A general method for compiling event-driven simulations. In *Proceedings of the 32nd Design Automation Conference*, 1995, pp. 151–156.

R. G. Gallager. *Low-Density Parity-Check Codes*. M.I.T. Press, 1963.

M. R. Garey and D. S. Johnson. *Computers and Intractability: A Guide to the Theory of NP-Completeness*. W. H. Freeman and Co., 1979.

F. Gilbert. *Optimized, Highly Parallel Architectures for Iterative Decoding Algorithms*. PhD dissertation, Microelectronic Systems Design Research Group, Department of Electrical Engineering and Information Technology, University of Kaiserslautern, Germany, May 2003.

F. Gilbert, F. Kienle, and N. Wehn. Low complexity stopping criteria for UMTS turbo-decoders. In *Proceedings of the Spring Vehicular Technology Conference*, 2003, pp. 2376–2380.

F. Gilbert, M. J. Thul, and N. Wehn. Communication-centric architectures for turbo-decoding on embedded multiprocessors. In *Proceedings of the Design Automation and Test in Europe Conference and Exhibition*, 2003, pp. 356–361.

S. Glesner. Using program checking to ensure the correctness of compiler implementations. *Journal of Universal Computer Science*, 9(3):191–222, Mar. 2003.

The Global Mobile Suppliers Association. 3G/WCDMA Fact Sheet. Aug. 2005, http://www.gsacom.com.

T. Glökler and H. Meyr. *Design of Energy-Efficient Application-Specific Instruction Set Processors (ASIPs)*. Kluwer, 2004.

D. Goodwin and D. Petkov. Automatic generation of application specific processors. In *Proceedings of the International Conference on Compilers, Architectures, and Synthesis forEmbedded Systems*, 2003, pp. 137–147.

D. Goodwin. The end of ISA design: power tools for optimal processor generation. In *Proceedings of the Hot Chips 16 Symposium*, 2004.

R. E. Gonzalez. Xtensa: a configurable and extensible processor. *IEEE Micro*, 20(2):60–70, Mar.-Apr. 2000.

S. Goto, T. Yamada, N. Takayama, and H. Yasuura. A design for a low-power digital matched filter applicable to W-CDMA. In *Proceedings of the Euromicro Symposium on Digital System Design*, 2002, pp. 210–217.

J. Gray. Building a RISC system in an FPGA—Part 1: Tools, instruction set, and datapath. *Circuit Cellar Magazine*, (116):26, Mar. 2000.

M. Gries, K. Keutzer, H. Meyr, and G. Martin. *Building ASIPs: The MESCAL Methodology*. Springer, 2005.

D. Große, U. Kühne, C. Genz, F. Schmiedle, B. Becker, R. Drechsler, and P. Molitor. Modellierung eines Mikroprozessors in SystemC. In *Proceedings of the GI/ITG/GMM-Workshop, Methoden und Beschreibungssprachen zur Modellierung und Verifikation von Schaltungen und Systemen*, 2005.

D. Große and R. Drechsler. *CheckSyC:* An efficient property checker for RTL SystemC designs. In *Proceedings of the IEEE International Symposium on Circuits and Systems*, 2005, pp. 4167–4170.

T. Grötker, S. Liao, G. Martin, and S. Swan. *System Design with SystemC*. Kluwer, 2002.

P. Grun, A. Halambi, N. Dutt, and A. Nicolau. RTGEN: an algorithm for automatic generation of reservation tables from architectural descriptions. *IEEE Transactions on Very Large Scale Integration (VLSI) Systems*, 11(4):731–737, Aug. 2003.

M. Gschwind. Instruction set selection for ASIP design. In *Proceedings of the International Symposium HW/SW Codesign*, 1999, pp. 7–11.

J. Gyllenhaal, B. Rau, and W. Hwu. *HMDES Version 2.0 Specification*. Technical report IMPACT-96–3. IMPACT Research Group, University of Illinois, Urbana, Illinois, 1996.

G. Hadjiyiannis, S. Hanono, and S. Devadas. ISDL: an instruction set description language for retargetability. In *Proceedings of the 34th Design Automation Conference*, 1997, pp. 299–302.

G. Hadjiyiannis, P. Russo, and S. Devadas. A methodology for accurate performance evaluation in architecture exploration. In *Proceedings of the 36th Design Automation Conference*, 1999, pp. 927–932.

A. Halambi, P. Grun, V. Ganesh, A. Khare, N. Dutt, and A. Nicolau. EXPRESSION: a language for architecture exploration through compiler/simulator retargetability. In *Proceedings of the Design Automation and Test in Europe Conference and Exhibition*, 1999, pp. 485–490.

T. R. Halfhill. EEMBC releases first benchmarks. *Microprocessor Report*, 1 May 2000.

——. ARC 700 aims higher. *Microprocessor Report*, 8 Mar. 2004.

——. ARC shows SIMD extensions. *Microprocessor Report*, 21 Nov. 2005.

D. Hankerson, P. D. Johnson, and G. A. Harris. *Introduction to Information Theory and Data Compression*. CRC, 1998.

S. Hanono and S. Devadas. Instruction selection, resource allocation, and scheduling in the AVIV retargetable code generator. In *Proceedings of the 35th Design Automation Conference*, 1998, pp. 510–515.

S. Hauck, T. W. Fry, M. M. Hosler, and J. P. Kao. The Chimaera reconfigurable functional unit. In *Proceedings of the 5th IEEE*

Symposium on Field-Programmable Custom Computing Machines, 1997, pp. 87–96.

A. Heberle, T. Gaul, W. Goerigk, G. Goos, and W. Zimmermann. Construction of verified compiler front-ends with program-checking. In *Proceedings of the Third International Andrei Ershov Memorial Conference on Perspectives of System Informatics,* LNCS-1755, 1999, pp. 481–492.

J. L. Hennessy and D. A. Patterson. *Computer Architecture—A Quantitative Approach,* 3rd ed. Morgan Kaufmann, 2003.

T. Henriksson, U. Nordqvist, and D. Liu. Embedded protocol processor for fast and efficient packet reception. In *Proceedings of the IEEE International Conference on Computer Design,* 2002, pp. 414–419.

T. Henriksson and I. Verbauwhede. Fast IP address lookup engine for SoC integration. In *Proceedings of the IEEE Design and Diagnostics of Electronic Circuits and Systems Workshop,* 2002, pp. 200–210.

D. Herrmann and R. Ernst. Improved interconnect sharing by identity operation insertion. In *Proceedings of the International Conference on Computer-Aided Design,* 1999, pp. 489–493.

A. Hoffmann, H. Meyr, and R. Leupers. *Architecture Exploration for Embedded Processors with LISA.* Kluwer, 2002.

M. Hohenauer, O. Wahlen, K. Karuri, H. Scharwächter, T. Kogel, R. Leupers, G. Ascheid, H. Meyr, G. Braun, and H. van Someren. A methodology and tool suite for C compiler generation from ADL processor models. In *Proceedings of the Design, Automation and Test in Europe Conference and Exhibition,* 2004, pp. 1276–1283.

H. Holma and A. Toskala, editors. *WCDMA for UMTS: Radio Access for Third Generation Mobile Communications.* John Wiley & Sons, 2000, p. 315.

I.-J. Huang and A. M. Despain. Generating instruction sets and microarchitectures from applications. In *Proceedings of the International Conference on Computer-Aided Design,* 1994, pp. 391–396.

H. C. Hunter and J. H. Moreno. A new look at exploiting data parallelism in embedded systems. In *Proceedings of the 2003 International Conference on Compilers, Architecture and Synthesis for Embedded Systems,* 2003, pp. 159–169.

IEEE Symposium on Field-Programmable Custom Computing Machines. http://www.fccm.org.

Insightful Corporation. S-PLUS. http://www.insightful.com/products/splus.

Intel Corporation. *IA-32 Intel Architecture Optimization Reference Manual.* 2005.

——. Details of Intel MMX technology. http://www.intel.com/design/intarch/mmx/mmx.htm.

International Business Machines Corporation. *The IBM PowerPC 405 Core.* 1998.

International Conference on Field Programmable Logic and Applications. http://www.fpl.org.

M. Itoh, Y. Takeuchi, M. Imai, and A. Shiomi. Synthesizable HDL generation for pipelined processors from a micro-operation description. *IEICE Transactions on Fundamentals of Electronics, Communications and Computer Sciences*, E83-A(3):394–400, Mar. 2000.

H. James and G. Harrison. Implementation of a 3GPP turbo decoder on a programmable DSP core. In *Proceedings of the Communications Design Conference*, 2001.

D. Jani, C. Benson, A. Dixit, and G. Martin. Functional verification of configurable embedded processors. In *The Functional Verification of Electronic Systems: An Overview from Various Points of View*, B. Bailey, editor. IEC, 2005

Improv Systems, Inc. Jazz DSP. http://www.improvsys.com.

Y. Jin, N. Satish, K. Ravindran, and K. Keutzer. An automated exploration framework for FPGA-based soft multiprocessor systems. In *Proceedings of the International Conference on Hardware/Software Codesign and System Synthesis*, 2005, pp. 273–278.

M. Kandemir, M. J. Irwin, G. Chen, and I. Kolcu. Banked scratch-pad memory management for reducing leakage energy consumption. In *Proceedings of the International Conference on Computer-Aided Design*, 2004, pp. 120–124.

T. Karnik, Y. Ye, J. Tschanz, L. Wei, S. Burns, V. Govindarajulu, V. De, and S. Borkar. Total power optimization by simultaneous dual-Vt allocation and device sizing in high performance microprocessors. In *Proceedings of the 39th Design Automation Conference*, 2002, pp. 486–491.

K. Karuri, C. Huben, R. Leupers, G. Ascheid, and H. Meyr. Memory access micro-profiling for ASIP design. In *Proceedings of the 3rd IEEE International Workshop on Electronic Design, Test, & Applications*, 2006, pp. 255–262.

K. Karuri, M. Al Faruque, S. Kraemer, R. Leupers, G. Ascheid, and H. Meyr. Fine-grained application source code profiling for ASIP design. In *Proceedings of the 42nd Design Automation Conference*, Anaheim, 2005, pp. 329–334.

R. Kastner, A. Kaplan, S. Memik, and E. Bozorgzadeh. Instruction generation for hybrid-reconfigurable systems. *ACM Transactions on Design Automation of Electronic Systems (TODAES)*, pp. 605–627, Oct. 2002.

T. Kempf, K. Karuri, R. Leupers, G. Ascheid, and H. Meyr. A SW performance estimation framework for early system level design using fine-grained instrumentation. In *Proceedings of the Design, Automation and Test in Europe Conference and Exhibition*, 2006, pp. 468–473.

K. Keutzer, M. Gries, G. Martin, and H. Meyr. *Designing ASIP: The MESCAL Methodology*. Kluwer, 2005.

A. Klaiber. *The Technology behind Crusoe Processors*. Technical report, Transmeta Corp., Santa Clara, California. Jan. 2000. http://www.transmeta.com/pdfs/paper_aklaiber_19jan00.pdf.

E. Kohler, R. Morris, B. Chen, J. Jannotti, and M. F. Kaashoek. The Click modular router. *ACM Transactions on Computer Systems (TOCS)*, 18(3):263–297, Aug. 2000.

H.-M. Koo and P. Mishra. Functional test generation using property decompositions for validation of pipelined processors. In *Proceedings of the Design, Automation and Test in Europe Conference and Exhibition*, 2006, pp. 1240–1245.

C. Kozyrakis and D. Patterson. Vector vs. superscalar and VLIW architectures for embedded multimedia benchmarks. In *Proceedings of the 35th Annual International Symposium on Microarchitecture*, 2002, pp. 283–293.

C. Kulkarni, G. Brebner, and G. Schelle. Mapping a domain specific language to a platform FPGA. In *Proceedings of the 41st Design Automation Conference*, 2004, pp. 924–927.

R. Kumar, D. M. Tullsen, N. P. Jouppi, and P. Ranganathan. Heterogeneous chip multiprocessors. *Computer*, 38(11):32–38, Nov. 2005.

F. J. Kurdahi and A. C. Parker. REAL: a program for REgister ALlocation. In *Proceedings of the 24th Design Automation Conference*, 1987, pp. 210–215.

M. Lam. Software pipelining: an effective scheduling technique for VLIW machines. In *Proceedings of the ACM SIGPLAN Conference on Programming Language Design and Implementation*, 1988, pp. 318–328.

D. Lanneer, J. Praet, A. Kifli, K. Schoofs, W. Geurts, F. Thoen, and G. Goossens. CHESS: retargetable code generation for embedded DSP processors. In P. Marwedel and G. Goossens, editors, *Code Generation for Embedded Processors*, pp. 85–102. Kluwer, 1995.

V. S. Lapinski. *Algorithms for Compiler-Assisted Design Space Exploration of Clustered VLIW ASIP Datapaths*. PhD dissertation, University of Texas, Austin, Texas, May 2001.

A. La Rosa, L. Lavagno, and C. Passerone. Implementation of a UMTS turbo-decoder on a dynamically reconfigurable platform. *IEEE Transactions on Computer-Aided Design of Integrated Circuits and Systems*, 24(1):100–106, Jan.2005.

C. Lee, M. Potkonjak, and W. H. Mangione-Smith. Media-Bench: a tool for evaluating and synthesizing multimedia and communications systems. In *Proceedings of the 30th Annual International Symposium on Microarchitecture*, 1997, pp. 330–335.

E. Lee. Embedded software. In *Advances in Computers*, M. Zelkowitz, editor, Academic, 2002, pp. 56–99.

J. Lee, K. Choi, and N. D. Dutt. Energy-efficient instruction set synthesis for application-specific processors. In *Proceedings of the 2003*

International Symposium on Low Power Electronics and Design, 2003, pp. 330–333.

J. H. Lee, J. H. Moon, K. L. Heo, M. H. Sunwoo, S. K. Oh, and I. H. Kim. Implementation of application-specific DSP for OFDM systems. In *Proceedings of the 2004 International Symposium on Circuits and Systems*, 2004, vol. 3, pp. 665–668.

J. K. F. Lee and A. J. Smith. Branch prediction strategies and branch target buffer design. In *Instruction-Level Parallel Processors*, IEEE Computer Society, 1995, pp. 83–99.

M. Lee, V. Tiwari, S. Malik, and M. Fujita. Power analysis and minimization techniques for embedded DSP software. *IEEE Transactions on Very Large Scale Integration (VLSI) Systems*, 5(2)123–135, Mar. 1997.

S. Leibson and J. Kim. Configurable processors: a new era in chip design. *Computer*, 38(7):51–59, Jul. 2005.

D. A. Lelewer and D. S. Hirschberg. Data compression. *ACM Computing Surveys*, 19(3):261–296, Sep. 1987.

R. Leupers and P. Marwedel. A BDD-based frontend for retargetable compilers. In *Proceedings of the European Design & Test Conference*, 1995, p. 239.

——. Retargetable code generation based on structural processor descriptions. *Design Automation for Embedded Systems*, 3(1):75–108, Jan. 1998.

——. *Retargetable Compiler Technology for Embedded Systems—Tools and Applications*. Kluwer, 2001.

R. Leupers, O. Wahlen, M. Hohenauer, T. Kogel, and P. Marwedel. An executable intermediate representation for retargetable compilation and high-level code optimization. In *Proceedings of the International Workshop on Systems, Architectures, Modeling, and Simulation*, 2003, pp. 120–125.

H. Liao and A. Wolfe. Available paralellism in video applications. In *Proceedings of the 30th Annual International Symposium on Microarchitecture*, 1997, pp. 321–329.

S. Liao, S. Devadas, K. Keutzer, S. Tjiang, and A. Wang. Storage assignment to decrease code size. In *Proceedings of the ACM SIGPLAN Conference on Programming Language Design and Implementation*, 1995, pp. 186–195.

C. Liem, P. Paulin, and A. Jerraya. Address calculation for retargetable compilation and exploration of instruction-set architectures. In *Proceedings of the 33rd Design Automation Conference*, 1996, pp. 597–600.

A. Ligtenberg and M. Vetterli. A discrete fourier/cosine transform chip. *IEEE Journal on Selected Areas in Communications*, 4(1):49–61, Jan. 1986.

A. Linden and J. Fenn. Understanding Gartner's hype cycles. *Gartner Research Report*, May 2003.

F. Lohr, A. Fauth, and M. Freericks. *Sigh/sim: an Environment for Retargetable Instruction Set Simulation*. Technical report 1993/43, Department of Computer Science, TU Berlin, Berlin, 1993.

M. Lorenz, P. Marwedel, T. Dräger, G. Fettweis, and R. Leupers. Compiler based exploration of DSP energy savings by SIMD operations. In *Proceedings of the Asia and South Pacific Design Automation Conference*, 2004, pp. 838–841.

O. Lüthje. A methodology for automated test generation for LISA processor models. In *Proceedings of the 12th Workshop on Synthesis and System Integration of Mixed Information Technologies*, 2004, pp. 266–273.

D. MacKay and R. Neal. Near Shannon limit performance of low-density parity-check codes. *Electronic Letters*, 32(18):1645–1646, Mar.1996.

J. C. Madre and J. P. Billion. Proving circuit correctness using formal comparison between expected and extracted behaviour. In *Proceedings of the 25th Design Automation Conference*, 1988, pp. 205–210.

J. C. Madre, O. Coudert, and J. P. Billion. Automating the diagnosis and the rectification of design errors with PRIAM. In *Proceedings of the International Conference on Computer Aided Design*, 1989, pp. 30–33.

P. Marwedel and L. Nowak. Verification of hardware descriptions by retargetable code generation. In *Proceedings of the 26th Design Automation Conference*, 1989, pp. 441–447.

W. J. Masek and M. S. Peterson. A faster algorithm computing string edit distances. *Journal of Computer System Sciences*, 20:18–31, 1980.

T. Mason and D. Brown. *lex & yacc*. O'Reilly & Associates, 1991.

S. McFarling. *Combining Branch Predictors*. Technical report TN-36, Digital Equipment Corporation, Palo Alto, California, Jun. 1993.

S. O. Memik, E. Bozorgzadeh, R. Kastner, and M. Sarrafzadeh. A superscheduler for embedded reconfigurable systems. In *Proceedings of the International Conference on Computer-Aided Design*, 2001, pp. 391–394.

D. Meng, R. Gunturi, and M. Castelino. IXP2800 Intel network processor IP forwarding benchmark full disclosure report for OC192-POS. In *Intel Corp. Technical Report for the Network Processing Forum (NPF)*, 2003.

H. Meyr, M. Moeneclaey, and S. Fechtel. *Digital Communication Receivers—Synchronization, Channel Estimation and Signal Processing*. John Wiley & Sons, 1998.

H. Michel. *Implementation of Turbo-Decoders on Programmable Architectures*. PhD dissertation, Microelectronic System Design Reseach Group, Department of Electrical Engineering and Information Technology, University of Kaiserslautern, Kaiserslautern, Germany, 2002.

MIPS Technologies, Inc. http://www.mips.com.

P. Mishra, A. Kejariwal, and N. Dutt. Synthesis-driven exploration of pipelined embedded processors. In *Proceedings of the International Conference on VLSI Design*, 2004, p. 921.

P. Mishra, M. Mamidipaka, and N. Dutt. Processor-memory co-exploration using an architecture description language. *ACM Transactions on Embedded Computing Systems (TECS)*, 3(1): 140–162, Feb. 2004.

P. Mishra and N. Dutt. Automatic modeling and validation of pipeline specifications. *ACM Transactions on Embedded Computing Systems (TECS)*, 3(1):114–139, Feb. 2004.

——. Graph-based functional test program generation for pipelined processors. In *Proceedings of the Design, Automation and Test in Europe Conference and Exhibition, 2004*, pp. 182–187.

——. Functional coverage driven test generation for validation of pipelined processors. In *Proceedings of the Design Automation and Test in Europe Conference and Exhibition*, 2005, pp. 678–683.

——. *Functional Verification of Programmable Embedded Architectures: A Top-Down Approach*. Springer, 2005.

P. Mishra, N. Dutt, N. Krishnamurthy, and M. Abadir. A top-down methodology for validation of microprocessors. *IEEE Design & Test of Computers*, 21(2):122–131, Mar. 2004.

P. Mishra, N. Dutt, and A. Nicolau. Functional abstraction driven design space exploration of heterogenous programmable architectures. In *Proceedings of the 14th International Symposium on System Synthesis*, 2001, pp. 256–261.

R. Morelos-Zaragoza. Encoder/decoder for binary BCH codes in C (version 3.1), 1997. http://www.eccpage.com/bch3.c.

M. W. Moskewicz, C. F. Madigan, Y. Zhao, L. Zhang, and S. Malik. Chaff: engineering an efficient SAT solver. In *Proceedings of the 38th Design Automation Conference*, 2001, pp. 530–535.

S. S. Muchnik. *Advanced Compiler Design & Implementation*. Morgan Kaufmann, 1997.

T. Ngo and I. Verbauwhede. Turbo codes on the fixed point DSP TMS320C55x. In *Proceedings of the 2000 Workshop on Signal Processing Systems*, 2000, pp. 255–264.

A. Nohl, G. Braun, O. Schliebusch, R. Leupers, H. Meyr, and A. Hoffmann. A universal technique for fast and flexible instruction-set architecture simulation. In *Proceedings of the 39th Design Automation Conference*, 2002, pp. 22–27.

Open SystemC Initiative. *Functional Specification for SystemC 2.0.* http://www.systemc.org.

A. Orailoglu and A. Veidenbaum. Application specific microprocessors (guest editors introduction). *IEEE Design & Test Magazine*, 20(1):6–7, Jan. 2003.

J. Paakki. Attribute grammar paradigms–a high level target methodology in language implementation. *ACM Computing Surveys*, 27(2): 196–256, Jun. 1995.

I. Park, M. D. Powell, and T. N. Vijaykumar. Reducing register ports for higher speed and lower energy. In *Proceedings of the 35th Annual Symposium on Microarchitecture*, 2002, pp. 171–182.

P. Paulin, C. Pilkington, E. Bensoudane, M. Langevin, and D. Lyonnard. Application of a multi-processor SoC platform to high-speed packet forwarding. In *Proceedings of the Design Automation and Test in Europe Conference and Exhibition (Designer Forum)*, 2004, vol. 3, pp. 58–63.

D. Pearson and V. Vazirani. Efficient sequential and parallel algorithms for maximal bipartite sets. *Journal of Algorithms*, 14:171:179, 1993.

S. Pees, A. Hoffmann, and H. Meyr. Retargetable compiled simulation of embedded processors using a machine description language. *ACM Transactions on Design Automation of Electronic Systems (TODAES)*, 5(4):815–834, Oct. 2000.

A. Peymandoust, L. Pozzi, P. Ienne, and G. Micheli. Automatic instruction-set extension and utilization for embedded processors. In *Proceedings of the International Conference on Application Specific Array Processors*, 2003, pp. 103–114.

PowerEscape, Inc. http://www.powerescape.com.

L. Pozzi, K. Atasu, and P. Ienne. Optimal and approximate algorithms for the extension of embedded processor instruction sets. *IEEE Transactions on Computer-Aided Design of Integrated Circuits and Systems*, 25(7):1209–1229, Jul. 2006.

L. Pozzi and P. Ienne. Exploiting pipelining to relax register-file port constraints of instruction-set extensions. In *Proceedings of the International Conference on Compilers, Architectures, and Synthesis for Embedded Systems*, 2005, pp. 2–10.

W. Qin and S. Malik. Architecture description languages for retargetable compilation. In *The Compiler Design Handbook: Optimizations & Machine Code Generation*. CRC, 2002.

W. Qin, S. Rajagopalan, and S. Malik. A formal concurrency model based architecture description language for synthesis of software development tools. In *Proceedings of ACM Conference on Languages, Compilers, and Tools for Embedded Systems* 2004, pp. 47–56.

B. Rau and M. Schlansker. *Embedded Computing: New Directions in Architecture and Automation*. Technical report HPL-2000–115, HP Labs, Palo Alto, California, Sep. 2000.

M. Ravasi. *An Automatic C-code Instrumentation Framework for High Level Algorithmic Complexity Analysis and System Design*. PhD dissertation, Ecole Polytechnique Fédérale de Lausanne, Lausanne, Switzerland, 2003.

M. Ravasi and M. Mattavelli. High-level algorithmic complexity evaluation for system design. *Journal of Systems Architecture*, 48(13–15): 403–427, May 2003.

K. Ravindran, N. Satish, Y. Jin, and K. Keutzer. An FPGA-based soft multiprocessor for IPv4 packet forwarding. In *Proceedings of the 15th International Conference on Field Programmable Logic and Applications*, 2005, pp. 487–192.

R. Razdan and M. D. Smith. A high-performance microarchitecture with hardware-programmable functional units. In *Proceedings of the 27th International Symposium on Microarchitecture*, 1994, pp. 172–180.

M. Reshadi, P. Mishra, and N. Dutt. Instruction set compiled simulation: A technique for fast and flexible instruction set simulation. In *Proceedings of the 40th Design Automation Conference*, 2003, pp. 758–763.

T. Ristimäki and J. Nurmi. Reconfigurable IP blocks: a survey, In *Proceedings of the International Symposium on System-on-Chip*, 2004, pp. 117–122.

P. Robertson, P. Hoeher, and E. Villebrun. Optimal and sub-optimal maximum a posteriori algorithms suitable for turbo decoding. *European Transactions on Telecommunications*, 8(2):119–125, Mar.–Apr. 1997.

C. Rowen and S. Leibson. *Engineering the Complex SoC. Fast, Flexible Design with Configurable Processors*. Prentice Hall, 2004.

R. Rudell. Dynamic variable ordering for ordered binary decision diagrams. In *Proceedings of the International Conference on Computer Aided Design*,1993, pp. 42–47.

RWTH Aachen University. LANCE Retargetable C compiler. http://www.lancecompiler.com.

C. Sauer, M. Gries, and S. Sonntag. Modular domain-specific implementation and exploration framework for embedded software platforms. In *Proceedings of the Design Automation Conference*, 2005, pp. 254–259.

H. Scharwächter, D. Kammler, A. Wieferink, M. Hohenauer, K. Karuri, J. Ceng, R. Leupers, G. Ascheid, and H. Meyr. ASIP architecture exploration for efficient IPSec encryption: a case study. In *Proceedings of the International Workshop on Software and Compilers for Embedded Systems*, 2004.

O. Schliebusch, A. Chattopadhyay, D. Kammler, R. Leupers, G. Ascheid, H. Meyr, and T. Kogel. A framework for automated and optimized ASIP implementation supporting multiple hardware description languages. In *Proceedings of the Asia and South Pacific Design Automation Conference*, 2005, pp. 280–285.

O. Schliebusch, A. Chattopadhyay, E. M. Witte, D. Kammler, G. Ascheid, R. Leupers, and H. Meyr. Optimization techniques for ADL-driven

RTL processor synthesis. In *Proceedings of the 16th IEEE International Workshop on Rapid System Prototyping*, 2005, p. 165–171.

O. Schliebusch, A. Chattopadhyay, M. Steinert, G. Braun, A. Nohl, R. Leupers, G. Ascheid, and H. Meyr. RTL processor synthesis for architecture exploration and implementation. In *Proceedings of the Design Automation and Test in Europe Conference and Exhibition*, 2004, pp. 156–160.

B. Schneier. Description of a new variable-length key, 64-bit block cipher (Blowfish). In *Proceedings of the Cambridge Security Workshop on Fast Software Encryption*, LNCS-809, 1994, pp. 191–204.

R. Schreiber, S. Aditya, B. Rau, V. Kathail, S. Mahlke, S. Abraham, and G. Snider. *High-Level Synthesis of Nonprogrammable Hardware Accelerators*. Technical report HPL-2000–31, HP Labs, Palo Alto, California, May 2000.

N. Shah. *Understanding Network Processors*. Master's thesis, University of California, Berkeley, California, 2001.

N. Shah, W. Plishker, K. Ravindran, and K. Keutzer. NP-Click: A productive software development approach for network processors. *IEEE Micro*, 24(5):45–54, Sept.-Oct. 2004.

C. E. Shannon. A mathematical theory of communication. *Bell System Technical Journal*, 27:379–423 and 623–656, Jul. and Oct. 1948.

M. Sheeran, S. Singh, and G. Stalmarck. Checking safety properties using induction and a SAT-solver. In *Proceedings of the Third International Conference on Formal Methods in Computer-Aided Design*, 2000, pp. 108–125.

T. Sherwood, M. Oskin, and B. Calder. Balancing design options with Sherpa. In *Proceedings of the International Conference on Compilers, Architectures, and Synthesis for Embedded*, 2004, pp. 57–68.

D. Sima, T. Fountain, and P. Kacsuk. *Advanced Computer Architectures: A Design Space Approach*. Addison-Wesley, 1997.

I. Sourdis, D. Pnevmatikatos, and K. Vlachos. An efficient and low-cost input/output subsystem for network processors. *In Proceedings of the 1st Workshop on Application Specific Processors*, 2002, pp. 56–64.

ST Microelectronics N. V. http://www.st.com.

StarCore LLC. http://www.starcore-dsp.com.

S. Steinke, M. Knauer, L. Wehmeyer, and P. Marwedel. An accurate and fine grain instruction-level energy model supporting software optimizations. In *Proceedings of the 11th International Workshop on Power and Timing Modeling, Optimization and Simulation*, 2001.

S. Steinke, N. Grunwald, L. Wehmeyer, R. Banakar, M. Balakrishnan, and P. Marwedel. Reducing energy consumption by dynamic copying of instructions onto onchip memory. In *Proceedings of the 15th International Symposium on System Synthesis*, 2002. pp. 213–218.

Stretch, Inc. http://www.stretchinc.com.

L. Su. Digital media—The new frontier for supercomputing. Opening Keynote, *MPSoC*, 2005.

F. Sun, S. Ravi, A. Raghunathan, and N. Jha. Synthesis of custom processors based on extensible platforms. In *Proceedings of the International Conference on Computer Aided Design*, 2002, pp. 641–648.

Synopsys Inc. VCS documentation. *SystemVerilog Assertion Library*, 2004.

Taiwan Semiconductor Manufacturing Company, Inc. *TSMC 0.13-micron Technology Platform*.

Target Compiler Technologies N. V. http://www.retarget.com.

D. Taylor, J. Lockwood, T. Sproull, J. Turner, and D. Parlour. Scalable IP lookup for internet routers. *IEEE Journal on Selected Areas in Communications*, 21(4):522–534, May 2003.

Tensilica, Inc. http://www.tensilica.com.

——. Xtensa processor. http://www.tensilica.com.

——. Tensilica Xtensa LX processor tops EEMBC networking 2.0 benchmarks, May 2005. http://www.tensilica.com/html/pr_2005_05_16.html.

Third Generation Partnership Project (3GPP). www.3gpp.org.

——. 3GPP Technical Specification 25.213, Spreading and modulation (FDD). V7.0.0, Mar. 2006.

——, 3GPP Technical Specification 25.913, Requirements for Evolved UTRA (E-UTRA) and Evolved UTRAN (E-UTRAN). V7.0.0, Jun. 2005.

M. J. Thul, F. Gilbert, T. Vogt, G. Kreiselmaier, and N. Wehn. A scalable system architecture for high-throughput turbo-decoders. In *Proceedings of the 2002 Workshop on Signal Processing Systems*, 2002, pp. 152–158.

M. J. Thul, C. Neeb, and N. Wehn. Network-on-chip-centric approach to interleaving in high throughput channel decoders. In *Proceedings.of the 2005 IEEE International Symposium on Circuits and Systems*, 2005, pp. 1766–1769.

H. Tomiyama, A. Halambi, P. Grun, N. Dutt, and A. Nicolau. Architecture description languages for systems-on-chip design. In *Proceedings of the Asia Pacific Conference on Chip Design Language*, 1999, pp. 109–116.

Trimaran Consortium. *The MDES User Manual*, 1997. http://www.trimaran.org.

S. Trimberger, editor. *Field Programmable Gate Array Technology*. Kluwer, 1994.

P. C. Tseng and L. G. Chen. Perspectives of multimedia SoC. In *Proceedings of the IEEE Workshop on Signal Processing Systems*, 2003, p. 3.

Tuulos, K. *3GPP FDD Baseband Test Data Generator*. Master's thesis, University of Turku, Turku, Finland, February 2002, p. 74.

E. Ukkonen. Online construction of suffix trees. *Algorithmica*, 14:249–260, 1995.

J. Van Praet, D. Lanneer, G. Goossens, W. Geurts, and H. De Man. A graph based processor model for retargetable code generation. In *Proceedings of the European Design and Test Conference*, 1996, p. 102.

I. Verbauwhede, P. Schaumont, and H. Kuo. Design and performance testing of a 2.29-GB/s Rijndael processor. *IEEE Journal of Solid-State Circuits*, 38(3):569–572, Mar. 2003.

J. Vogt, K. Koora, A. Finger, and G. Fettweis. Comparison of different turbo decoder realizations for IMT-2000. In *Proceedings of the Global Telecommunications Conference*, 1999, vol. 5, pp. 2704–2708.

J. Vuillemin, P. Bertin, D. Roncin, M. Shand, H. Touati, and P. Boucard. Programmable active memories: reconfigurable systems come of age. *IEEE Transactions on Very Large Scale Integration (VLSI) Systems*, 4(1):56–69, Mar. 1996.

Xilinx, Inc. PicoBlaze soft processor, Sep. 2000. http://www.origin.xilinx.com/bvdocs/app-notes/xapp213.pdf.

——, Virtex4 and other FPGA families, 2005. http://www.xilinx.com.

J. Wagner and R. Leupers. C compiler design for a network processor. *IEEE Transactions on Computer-Aided Design of Integrated Circuits and Systems*, 20(11):1302–1308, Nov. 2001.

O. Wahlen, M. Hohenauer, G. Braun, R. Leupers, G. Ascheid, H. Meyr, and X. Nie. Extraction of efficient instruction schedulers from cycle-true processor models. In *Proceedings of the 7th International Workshop on Software and Compilers for Embedded Systems*, 2003, pp. 167–181.

Y. Wand and R. Weber. An ontological model of an information system. *IEEE Transactions on Software Engineering*, 16(11):1282–1292, Nov. 1990.

A. Wang, E. Killian, D. Maydan, and C. Rowen. Hardware/software instruction set configurability for system-on-chip processors. In *Proceedings of the 38th Design Automation Conference*, 2001, pp. 184–190.

S.-H. Wang, W.-H. Peng, Y. He, G.-Y. Lin, C.-Y. Lin, S.-C. Chang, C.-N. Wang, and T. Chiang. A software-hardware co-implementation of MPEG-4 advanced video coding (AVC) decoder with block level pipelining. *Journal of VLSI Signal Processing*, 41(1):93–110, Aug. 2005.

S. Weber and K. Keutzer. Using minimal minterms to represent programmability. In *Proceedings of the International Conference on Hardware/Software Codesign and System Synthesis*, 2005, p. 63–66.

S. Weber, M. Moskewicz, M. Gries, C. Sauer, and K. Keutzer. Fast cycle-accurate simulation and instruction set generation for constraint-based descriptions of programmable architectures. In *Proceedings of the International Conference on Hardware/Software Codesign and System Synthesis*, 2004, p. 18–23.

J. Wei and C. Rowen. Implementing low-power configurable processors—practical options and tradeoffs. In *Proceedings of the 42nd Design Automation Conference*. 2005, pp. 706–711.

R. P. Weicker. Dhrystone benchmark: rationale for version 2 and measurement rules. *ACM SIGPLAN Notices*, 23(8):49–62, Aug. 1988.

R. Wilhelm and D. Maurer. *Compiler Design*. Addison-Wesley, 1995.

E. Willink, J. Eker, and J. Janneck. Programming specifications in CAL. In *Proceedings of the OOPSLA Workshop on Generative Techniques in the Context of Model Driven Architecture*, 2002.

K. Winkelmann, H.-J. Trylus, D. Stoffel, and G. Fey. A cost-efficient block verification for a UMTS up-link chip-rate coprocessor. In In *Proceedings of the Design, Automation and Test in Europe Conference and Exhibition*, 2004, pp. 162–167.

T. Wilson, G. Grewal, B. Halley, and D. Banerji. An integrated approach to retargetable code generation. In *Proceedings of the 7th International Symposium on High-Level Synthesis*, 1994, pp. 70–75.

M. J. Wirthlin and B. L. Hutchings. A dynamic instruction set computer. In *Proceedings of the IEEE Workshop on FPGAs for Custom Computing Machines*, 1995, pp. 99–107.

E. M. Witte, A. Chattopadhyay, O. Schliebusch, D. Kammler, R. Leupers, G. Ascheid, H. Meyr, and T. Kogel. Applying resource-sharing algorithms to ADL-driven automatic ASIP implementation. In *Proceedings of the International Conference of Computer Design*, 2005, pp. 193–199.

A. Worm. *Implementation Issues of Turbo-Decoders*. PhD dissertation, Institute of Microelectronic Systems, Department of Electrical Engineering and Information Technology, University of Kaiserslautern, Kaiserslautern, Germany, 2001.

P. Yu and T. Mitra. Scalable custom instructions identification for instruction set extensible processors. In *Proceedings of the International Conference on Compilers, Architectures, and Synthesis for Embedded Systems*, 2004, pp. 69–78.

H. Zima and B. Chapman. *Supercompilers for parallel and vector computers*. ACM, 1990.

V. Zivojnovic, S. Pees, and H. Meyr. LISA—machine description language and generic machine model for HW/SW co-design. In *Proceedings of the IEEE Workshop on VLSI Signal Processing*, 1996, pp. 127–136.

INDEX